Terramechanics and Off-Road Vehicle Engineering

Terrain Behaviour, Off-Road Vehicle Performance and Design

To May

   Chak

   Anna

   Amanda

   the memory of my parents

   and the glory of the Almighty

Chapter 4. Characterization of the Response of Terrains to Normal and Repetitive Loadings 75
4.1 Response of Mineral Terrains 76
    4.1.1 Pressure–Sinkage Relationship 76
    4.1.2 Response to Repetitive Loading 85
4.2 Response of Muskegs 88
    4.2.1 Physical Features of the Muskegs Tested 88
    4.2.2 Pressure–Sinkage Relationship 90
    4.2.3 Response to Repetitive Loading 100
4.3 Response of Snow Covers 102
    4.3.1 Bearing Capacity of an Ice Layer in a Snow Cover 104
    4.3.2 Pressure–Sinkage Relationship of Snow Covers with a Significant Ice Layer 110
    4.3.3 Response to Repetitive Loading 112

Chapter 5. Characterization of the Shearing Behaviour of Terrains 115
5.1 Characterization of the Shear Stress–Displacement Relationships 115
5.2 Shearing Behaviour of Various Types of Terrain 123
5.3 Behaviour of Terrain under Repetitive Shear Loading 123

Chapter 6. Performance of Off-Road Vehicles 129
6.1 Factors Affecting Off-Road Vehicle Performance 129
    6.1.1 Aerodynamic Resistance 130
    6.1.2 Motion Resistance of Vehicle Running Gear 131
    6.1.3 Thrust (Tractive Effort or Propelling Force) 132
6.2 Drawbar Performance 133
    6.2.1 Drawbar Pull Coefficient and Drawbar Power 133
    6.2.2 Tractive (Drawbar) Efficiency 135
    6.2.3 Four-Wheel-Drive 139
6.3 Transport Productivity and Transport Efficiency 147
6.4 Mobility Map and Mobility Profile 149
Problems 151

Chapter 7. Methods for Evaluating Tracked Vehicle Performance 155
7.1 Empirical Methods 155
7.2 Methods for Parametric Analysis 163
7.3 Methods for Predicting Static Pressure Distributions beneath Tracks 171
Problems 175

Chapter 8. Computer-Aided Method NTVPM for Evaluating the Performance of Vehicles with Flexible Tracks 177
8.1 Basic Approach to the Prediction of Normal Pressure Distribution under a Flexible Track 178
8.2 Prediction of Shear Stress Distribution under a Flexible Track 188

Contents

Preface to the Second Edition .... *ix*
Preface to the First Edition .... *xi*
Conversion Factors .... *xiv*
Nomenclature .... *xv*

Chapter 1. Introduction 1
  1.1 Role of Terramechanics 2
  1.2 Some Basic Issues in Terramechanics 6
    1.2.1 Modelling of Terrain Behaviour 7
    1.2.2 Measurement and Characterization of Terrain Properties 8
    1.2.3 Modelling of Vehicle–Terrain Interaction 12
  1.3 Approaches to Terramechanics 15
    1.3.1 Empirical Methods 15
    1.3.2 Computational Methods 16
    1.3.3 Methods for Parametric Analysis 17

Chapter 2. Modelling of Terrain Behaviour 21
  2.1 Modelling Terrain as an Elastic Medium 22
  2.2 Modelling Terrain as a Plastic Medium 31
  2.3 Modelling Terrain Behaviour based on the Critical State Soil Mechanics 45
  2.4 Modelling Terrain using the Finite Element Method (FEM) 48
  2.5 Modelling Terrain Behaviour using the Discrete (Distinct) Element
      Method (DEM) 55
  Problems 62

Chapter 3. Measurement of Terrain Properties 65
  3.1 Cone Penetrometer Technique 65
  3.2 Bevameter Technique 68
    3.2.1 Basic Features of the Bevameter 69
    3.2.2 A Portable Automatic Data Acquisition and Processing System 72

Butterworth-Heinemann is an imprint of Elsevier
The Boulevard, Langford Lane, Kidlington, Oxford OX5 1GB, UK
Radarweg 29, PO Box 211, 1000 AE Amsterdam, The Netherlands

First edition 1989
Second edition 2010

Copyright © 2010 Elsevier Ltd. All rights reserved

No part of this publication may be reproduced, stored in a retrieval system or transmitted in any form or by any means electronic, mechanical, photocopying, recording or otherwise without the prior written permission of the publisher

Permissions may be sought directly from Elsevier's Science & Technology Rights Department in Oxford, UK: phone (+44) (0) 1865 843830; fax (+44) (0) 1865 853333; email: permissions@elsevier.com. Alternatively you can submit your request online by visiting the Elsevier web site at http://elsevier.com/locate/permissions, and selecting *Obtaining permission to use Elsevier material*

Notice
No responsibility is assumed by the publisher for any injury and/or damage to persons or property as a matter of products liability, negligence or otherwise, or from any use or operation of any methods, products, instructions or ideas contained in the material herein. Because of rapid advances in the medical sciences, in particular, independent verification of diagnoses and drug dosages should be made

British Library Cataloguing in Publication Data
A catalogue record for this book is available from the British Library

Library of Congress Cataloging-in-Publication Data
A catalog record for this book is available from the Library of Congress

ISBN: 978-0-7506-8561-0

For information on all Elsevier publications
visit our web site at books.elsevier.com

Printed and bound in the UK

10 11 12 13 14   10 9 8 7 6 5 4 3 2 1

Working together to grow
libraries in developing countries

www.elsevier.com | www.bookaid.org | www.sabre.org

ELSEVIER   BOOK AID International   Sabre Foundation

Terramechanics and Off-Road Vehicle Engineering

Terrain Behaviour, Off-Road Vehicle Performance and Design

Second Edition

J.Y. Wong, Ph.D., D.Sc.
Professor Emeritus and Distinguished Research Professor
Department of Mechanical and Aerospace Engineering
Carleton University, Ottawa
Ontario K1S 5B6
Canada

AMSTERDAM • BOSTON • HEIDELBERG • LONDON • NEW YORK • OXFORD
PARIS • SAN DIEGO • SAN FRANCISCO • SINGAPORE • SYDNEY • TOKYO

Butterworth-Heinemann is an imprint of Elsevier

Contents   vii

    8.3 Effects of Shear Stresses on the Normal Pressure Distribution............................... 191
    8.4 Prediction of Motion Resistance and Drawbar Pull 193
    8.5 Experimental Substantiation.. 195
    8.6 Comparisons of Predictions by NTVPM with Test Data 207

Chapter 9. Applications of the Computer-Aided Method NTVPM to Parametric Analysis of Vehicles with Flexible Tracks 211

    9.1 Effects of Initial Track Tension and Track System Configuration 214
    9.2 Effects of Suspension Setting 239
    9.3 Effects of Longitudinal Location of the Centre of Gravity 247
    9.4 Effects of Vehicle Total Weight 257
    9.5 Effects of Track Width 268
    9.6 Effects of Sprocket Location 281
    9.7 Concept of a High-Mobility Tracked Vehicle for Operation on Soft Ground 293
    9.8 Effects of Design Features on the Performance of Two-Unit Articulated Vehicles 295
    9.9 Analysis and Evaluation of Detracking Risks 305
        9.9.1 The Necessary Condition for Detracking 306
        9.9.2 Detracking Risk Indicators for Track Drive Lugs/Track Guides Overriding the Sprocket/Idler 308
        9.9.3 Risk Indicators for Disengagement of the Leading Track Drive Lug/Track Guide with the Sprocket/Idler 310
        9.9.4 Basic Features of the Module DETRACK for Evaluating Detracking Risks 312
        9.9.5 Applications of DETRACK to Evaluating Detracking Risks of Track Systems 313
    9.10 Applications of NTVPM to Product Development in the Off-Road Vehicle Industry ... 319
    9.11 Concluding Remarks 325

Chapter 10. Computer-Aided Method RTVPM for Evaluating the Performance of Vehicles with Long-pitch Link Tracks 333

    10.1 Basic Approach to the Development of the Computer-Aided Method RTVPM 333
        10.1.1 Analysis of the Upper Run of the Track 334
        10.1.2 Analysis of the Lower Run of the Track in Contact with the Terrain 339
        10.1.3 Analysis of the Links in Contact with the Idler 343
        10.1.4 Analysis of the Links in Contact with the Sprocket 347
        10.1.5 Analysis of the Complete Track System with Rigid Links 348
    10.2 Experimental Substantiation 349
    10.3 Applications of the Computer-Aided Method RTVPM to Parametric Analysis 353
        10.3.1 Effects of the Ratio of Roadwheel Spacing to Track Pitch 355
        10.3.2 Effects of Initial Track Tension 364
        10.3.3 Concept of a Vehicle with Enhanced Performance on Soft Ground 366
    10.4 Concluding Remarks 368

Chapter 11. Methods for Evaluating Wheeled Vehicle Performance 371
  11.1 Empirical Methods 372
  11.2 Methods for Parametric Analysis 378
    11.2.1 Rigid Wheel–Terrain Interaction 379
    11.2.2 Flexible Tyre–Terrain Interaction 391
  Problems 401

Chapter 12. Computer-Aided Method NWVPM for Evaluating the Performance of Tyres and Wheeled Vehicles 403
  12.1 Basic Features of NWVPM for Evaluating Tyre Performance 403
    12.1.1 Normal and Shear Stress Distributions on the Tyre–Terrain Interface 404
    12.1.2 Tyre Performance Prediction 409
    12.1.3 Effects of Tyre Lugs 412
  12.2 Basic Features of NWVPM for Evaluating Wheeled Vehicle Performance 414
  12.3 Experimental Substantiation 417
  12.4 Applications of NWVPM to Parametric Analysis of Wheeled Vehicle Performance 419
  12.5 Applications of NWVPM to the Evaluation of Lunar Vehicle Wheels 424
    12.5.1 Lunar Vehicle Wheels 425
    12.5.2 Procedures and Soil Conditions for Testing Lunar Vehicle Wheels 426
    12.5.3 Correlations between the Measured Performance of Lunar Vehicle Wheels and Predicted Performance by NWVPM 428
  12.6 Applications of the Finite Element Technique to Tyre Modelling in the Analysis of Tyre–Terrain Interaction 432
    12.6.1 Modelling Tyre Behaviour using the Finite Element Technique 434
    12.6.2 Analysis of Tyre–Terrain Interaction using the Tyre Model 436
  12.7 Wheeled Vehicles vs Tracked Vehicles from the Traction Perspective 439
    12.7.1 General Analysis of Wheeled Vehicles vs Tracked Vehicles 440
    12.7.2 Using Computer-Aided Methods for the Analysis of Wheeled Vehicles vs Tracked Vehicles 446
  Problems 450

References *451*
Index *461*

Preface to the Second Edition

Since the publication of the first edition of this book in 1989, notable progress has been made in terramechanics, which is the study of the dynamics of an off-road vehicle in relation to its environment – the terrain. Understanding of the mechanics of vehicle–terrain interaction has been improved. New techniques have been introduced into the modelling of terrain behaviour. A series of computer-aided methods for performance and design evaluation of off-road vehicles from the traction perspective, incorporating recent advancements in terramechanics, have been further developed. These methods have been gaining acceptance in industry in the development of new products. Continual interest in improving vehicle mobility over a wider range of environments and renewed enthusiasm for the exploration of the Moon, Mars and beyond shown by an increasing number of nations have given new impetus to the further development of terramechanics. To reflect these and other advancements in the field and to serve the changing needs of the professional and higher educational communities, time is ripe for this second edition.

While new topics are introduced and data are updated in this edition, the objective and format remain similar to those of the previous edition. The fundamentals of terramechanics underlying the rational development and design of off-road vehicles are emphasized. As the performance of off-road vehicles over unprepared terrain constitutes a basic issue in vehicle mobility, this book focuses on the study of vehicle–terrain interaction from the traction perspective.

To better serve the higher educational community in the fields of automotive engineering, off-road vehicle engineering, and agricultural and biological engineering, examples of the applications of the principles of terramechanics to solving engineering problems are given. Practical problems that may be assigned to senior undergraduate or postgraduate students as part of their study programme are also included in this new edition.

The number of chapters has been expanded to 12 in this edition from nine in the previous edition. Chapter 1 provides an introduction to the subject of terramechanics, outlines its roles, and presents outstanding examples of its practical applications. A brief review of the modelling of terrain behaviour is presented in Chapter 2. The fundamentals of the theories of elasticity, plastic equilibrium and critical state soil mechanics, as applied to the study of vehicle–terrain interaction, are outlined. The applications of the finite element method (FEM) and the discrete (distinct) element method (DEM) to the modelling of terrain are reviewed. While these theories or modelling techniques provide a foundation for an understanding of some aspects of the physical nature of vehicle–terrain interaction, there are limitations to their applications in practice, particularly in modelling behaviour of natural terrain. Chapter 3 describes the techniques and instrumentation currently used for measuring terrain behaviour in the field. The responses of various types of natural terrain to normal and repetitive loading observed in the field are discussed in Chapter 4. This provides the terrain information needed for predicting the sinkage of the vehicle running gear and the normal pressure distribution on the vehicle–terrain interface. Chapter 5 describes the shear strengths of various types of natural terrain measured in the field and their characterization. This provides the required terrain information for predicting the tractive capability of off-road vehicles in the field. Criteria commonly used for evaluating the performance of various types of off-road vehicle are reviewed in Chapter 6. Empirical and semi-empirical methods for predicting tracked vehicle performance are discussed in Chapter 7. Chapter 8 outlines the analytical basis for the computer-aided method NTVPM for performance and design evaluation of vehicles with flexible tracks, such as military and cross-country transport vehicles. The experimental validation of NTVPM is also described. Applications of NTVPM to parametric analyses of vehicle designs are discussed in Chapter 9. Examples of its applications to the development of new products in off-road vehicle industry are presented. Chapter 10 outlines the analytical basis for the computer-aided method RTVPM for performance and design evaluation of vehicles with long-pitch link tracks, such as industrial and agricultural tractors. The experimental validation of RTVPM and its applications to parametric analyses are presented. Chapter 11 presents empirical and semi-empirical methods for predicting wheel and wheeled vehicle performance. The analytical basis for the computer-aided method NWVPM for predicting the performances of wheels and wheeled vehicles is outlined in Chapter 12. As an example, the application of NWVPM to the evaluation of the performance of lunar vehicle wheels is presented.

Some of the material included in this new edition has been presented at professional development programmes and seminars in many countries. These included staff training programmes on the applications of terramechanics to the evaluation of planetary rover mobility, presented at the European Space Research and Technology Centre (ESTEC) of the European Space Agency (ESA) and at the Glenn Research Center, National Aeronautics and Space Administration (NASA), USA.

This new edition includes some of the results of recent research on off-road vehicle mobility carried out by the author together with his associates at Carleton University and at Vehicle Systems Development Corporation (VSDC), Ottawa, Canada. The author wishes to express his appreciation to his former research staff, postdoctoral fellows and graduate students at Carleton, and to his associates at VSDC for their contributions, particularly Jon Preston-Thomas, the late Michael Garber, Yuli Gao, Mike Galway and Wei Huang. Appreciation is due also to many organizations, in private and public sectors, for their generous support for our research over the years.

Jo Yung Wong
Ottawa, Canada

Preface to the First Edition

In the past few decades, the continual demand for greater mobility over a wider range of terrains and in all seasons by agricultural, construction and cross-country transport industries and by the military has stimulated a great deal of interest in the study of vehicle mobility over unprepared terrain. A large volume of research papers on this subject has been published in journals and conference proceedings of learned societies. A variety of methods for predicting and evaluating off-road vehicle performance, ranging from entirely empirical to highly theoretical, has been proposed or developed. However, methods that will enable the design engineer or the procurement manager to conduct a comprehensive and yet realistic evaluation of competing vehicle designs appear to be lacking. This prompted the author of this book to embark, more than a decade ago, on a series of research programmes aimed at filling this gap. The objective is to establish mathematical models for vehicle–terrain systems that will enable the engineering practitioner to evaluate, on a rational basis, a wide range of options and to select an appropriate vehicle configuration for a given mission and environment. To be useful to the design engineer or the procurement manager, the models should take into account all major vehicle design and operational parameters as well as pertinent terrain characteristics.

After more than a decade of intense effort, a series of computer-aided methods (computer simulation models) for predicting and evaluating the performance of tracked and wheeled vehicles, which meet the basic objective outlined above, have emerged. These methods have since been used to assist off-road vehicle manufacturers in developing new products and governmental agencies in evaluating vehicle candidates with most encouraging results. The encouragement that these developments have effected has convinced the author to put these pages together, with the hope that this book may enhance the interest of the professionals engaged in the field of off-road vehicle mobility.

This book summarizes some of the research and development work on the computer-aided methods for evaluating off-road vehicle performance carried out by the author and his associates at the Transport Technology Research Laboratory, Carleton University and Vehicle Systems Development Corporation, Ottawa, Canada. Chapter 1 provides an introduction to the subject of terramechanics, and outlines its roles and basic issues. Chapter 2 describes the techniques and instrumentation for measuring terrain behaviour. An understanding of the mechanical properties of the terrain is of importance in the prediction and evaluation of off-road vehicle performance, as the behaviour of the terrain quite often imposes severe limitations to vehicle mobility. Chapter 3 describes the responses of various types of natural terrain to normal and repetitive loading. This provides information for predicting the sinkage of the vehicle running gear and the normal pressure distribution on the vehicle–terrain interface. Chapter 4 describes the shear strength of various types of natural terrain. This provides information for predicting the tractive capability of off-road vehicles. Chapter 5 reviews some of the methods previously developed for predicting the performance of tracked vehicles. Chapter 6 outlines the analytical framework for the development of computer-aided methods for evaluating tracked vehicle performance, while Chapter 7 illustrates some of the applications of the computer-aided methods to the parametric analysis of tracked vehicle design and performance. Chapter 8 reviews some of the methods previously developed for predicting the performance of tyres, while Chapter 9 outlines the recently developed computer-aided methods for evaluating the performance of tyres and wheeled vehicles and illustrates their applications.

Some of the material included in this book has been presented at seminars and professional development programmes in Canada, China, Italy, Germany, Singapore, Spain, Sweden, the United Kingdom and the United States. Some of these seminars were jointly offered with the late Dr M.G. Bekker during the period from 1976 to 1985.

The computer-aided methods presented in this book represent recent advances in the methodology for predicting and evaluating off-road vehicle performance. This does not mean, however, that further development of the methods described is not required. If and when better mathematical models for vehicle–terrain interaction and for characterizing terrain behaviour are available, they could readily be fitted into the framework presented here to make an even more comprehensive and precise picture.

Many organizations have supported the research upon which this book is based. In particular, the author wishes to record the support provided by the Canadian Department of National Defence, National Research Council of Canada, Natural Sciences and Engineering Research Council of Canada, and Vehicle Systems Development Corporation. In writing this book, the author has drawn much on the experience acquired from working with many industrial and research organizations, including Hagglunds Vehicle AB of Sweden, US Naval Civil Engineering Laboratory, Institute for Earthmoving Machinery and Off-Road Vehicles (CEMOTER) of the Italian National Research Council, and Vehicle Mobility Section, Defence Research Establishment Suffield and other branches of the Canadian Department of National Defence. This acknowledgement does not imply, however, that the views expressed in this book necessarily represent those of these organizations.

The author acknowledges with gratitude the inspiration derived from collaboration and discussions with many colleagues in industry, research organizations and universities. He is indebted to Dr A.R. Reece, formerly with the University of Newcastle upon Tyne and now Managing Director, Soil Machine Dynamics Ltd, England, and the late Dr M.G. Bekker for their valued encouragement and stimulation. The author also wishes to express his appreciation to the staff members and graduate students at the Transport Technology Research Laboratory, Carleton University and to his associates at Vehicle Systems Development Corporation for their contributions to the research work presented in this book. He is especially indebted to Mr J. Preston-Thomas of Vehicle Systems Development Corporation for his contributions to the development of the computer-aided methods for evaluating off-road vehicle performance and for reviewing the manuscript.

J.Y. Wong
Ottawa, Canada
June 1989

Conversion Factors

Quantity	US customary unit	SI equivalent
Acceleration	ft/s^2	0.3048 m/s^2
Area	ft^2	0.0929 m^2
	in^2	645.2 mm^2
Energy	ft·lb	1.356 J
Force	lb	4.448 N
Length	ft	0.3048 m
	in	25.4 mm
	mile	1.609 km
Mass	slug	14.59 kg
	ton	907.2 kg
Moment of a force	lb·ft	1.356 N·m
Power	hp	745.7 W
Pressure or stress	lb/ft^2	47.88 Pa (N/m^2)
	lb/in^2 (psi)	6.895 kPa (kN/m^2)
Speed	ft/s	0.3048 m/s
	mph	1.609 km/h
Volume	ft^3	0.02832 m^3
	in^3	16.39 cm^3
	gal (liquids)	3.785 litre

Nomenclature

A	area
A_l	rigid area of a track link as a proportion of its nominal contact area
A_f	vehicle frontal area
A_u	parameter characterizing terrain response to repetitive loading
a	half width of loading area; distance defining the longitudinal location of the centre of gravity; acceleration
B	wheelbase
b	smaller dimension of a rectangular plate or the radius of a circular plate; width
b_b	belly width
b_{ti}	tyre width
b_{tr}	track width
C, CI	cone index
c	cohesion
D	diameter
D_h	hydraulic diameter
D_r	relative density
d	diameter
E	modulus of elasticity
e	void ratio; base for the natural logarithm
F	function; thrust, tractive effort
F_d	drawbar pull
F_v	tractive effort developed on the vertical shear surfaces on both sides of a track
f_o	yield strength of an ice layer in tension; coefficient of track internal resistance

f_t	radial deflection of the roadwheels of a track system
G	sand penetration resistance gradient
G_e	effective sand penetration resistance gradient
G_{ey}	revised effective sand penetration resistance gradient
H	horizontal component of a tension force
h	thickness; tyre section height
h_l	lug height
i	slip
i_s	skid
j	shear displacement
j_0	shear displacement where shear stress peaks
K	shear deformation parameter
K_1, K_2	parameters characterizing the shear stress–shear displacement relationship
K_r	ratio of residual shear stress to maximum shear stress
K_ω	shear displacement where shear stress peaks
k	stiffness; resultant pressure–sinkage parameter
k_c, k_ϕ	pressure–sinkage parameters in the Bekker equation
$k_c', k_\phi', k_c'', k_\phi''$	pressure–sinkage parameters in the Reece equation
k_e	tyre carcass flexing resistance coefficient
k_0	parameter characterizing terrain response to repetitive loading
$k_{p1}, k_{p2}, k_{z1}, k_{z2}$	pressure–sinkage parameters for snow cover
k_u	parameter characterizing terrain stiffness during the unloading–reloading cycle
L	perimeter; characteristic length for an ice layer
L_b	belly contact length
L_t	length of track in contact with terrain
l	length
M_o	limit bending moment per unit length of an ice layer
MI	mobility index
m, m_m	parameters characterizing the relation between the strength of the muskeg mat and that of the underlying peat
N	number
N_c	clay–tyre numeric
N_{cs}	cohesive-frictional soil–tyre numeric

N_s	sand-tyre numeric
N_{se}, N_{sey}	revised sand–tyre numerics
n	exponent of sinkage
n_{av}	average exponent of sinkage
n_r	number of wheel stations in a track system
P	load; power; spherical pressure
P_{co}	collapse load for an ice layer
P_d	drawbar power
P_e	engine power
P_{ro}	transport productivity
P_{us}	ultimate load due to local shear failure for an ice layer
P_{ut}	ultimate load due to circumferential tension failure for an ice layer
p	pressure
p''	reaction of sublayer
p_b	pressure on the belly–terrain interface
p_c	pressure due to carcass stiffness
p_{ca}	pressure exerted on the carcass by the terrain
$p_{c\ell}$	calculated pressure
p_{c0}, p_{c1}, p_{c2}	collapse pressures for an ice layer
p_{cr}	critical inflation pressure
p_g	ground pressure
p_{gcr}	critical ground pressure
p_i	tyre inflation pressure
p_m	measured pressure
p_p	punching pressure
p_u	pressure at the beginning of unloading in a loading–unloading–reloading cycle
p_w	pressure–sinkage parameter for a snow cover
q	surcharge; pressure exerted on the muskeg mat by the underlying peat
R	radius; deviatoric stress
R_a	aerodynamic drag
R_{bc}	belly drag
R_c	resistance due to terrain compaction
R_f	tyre carcass flexing resistance
R_g	grade resistance
R_{in}	internal resistance of gear running
R_{ob}	obstacle resistance

R_t	resistance due to vehicle running gear–terrain interaction
R_v	motion resistance of vehicle running gear
RCI	rating cone index
r	radius
S_v	shear force per unit track length developed on the vertical shear surfaces on both sides of a track
s	shear stress
s_b	shear stress on the belly–terrain interface
s_c	calculated shear stress
s_m	measured shear stress
s_{max}	maximum shear stress
s_r	residual shear stress
T	tension
t_o	thickness of muskeg mat
t_t	track pitch
V	vertical component of a tension force; actual forward speed of a vehicle; specific volume
V_a	absolute velocity
V_j	slip velocity
V_r	vehicle speed relative to wind
V_t	theoretical speed
VCI	vehicle cone index
W	load, weight
W_p	payload
w_r	weighting factor
z, z_o	sinkage
z_e	sinkage of a tyre in the elastic operating model
z_m	mean sinkage of a grouser
z_r	sinkage of a tyre in the rigid operating mode
z_u	sinkage at the beginning of unloading in a loading–unloading–reloading cycle
z_w	pressure–sinkage parameter for a snow cover
α	angle with the horizontal
α_b	vehicle belly inclination angle; rake angle
γ	density

δ	inclination angle; interface friction angles; tyre deflection
δ_t	tyre deflection
ε	goodness-of-fit; coefficient for tyre flexing resistance
η	efficiency
η_d	tractive (drawbar) efficiency
η_{do}	tractive (drawbar) efficiency overall
η_m	efficiency of motion
η_p	propulsive efficiency
η_s	slip efficiency
η_{st}	structural efficiency
η_t	transmission efficiency
η_{tr}	transport efficiency
θ	angle
λ	ratio of total lug tip area to total tyre tread area
ν	concentration factor
μ	Poisson's ratio; drawbar coefficient; coefficient of friction
σ	normal stress
τ	shear stress
τ_r	shear strength of muskeg mat
ϕ	angle of shearing resistance
φ	roadwheel contact angle
ω	angular speed

CHAPTER 1

Introduction

Man has a long history of involvement in off-road locomotion, perhaps since the invention of the wheel about 3500 BC. Powered off-road vehicles have come into wide use in many parts of the world in agriculture, construction, cross-country transportation and military operations since the turn of last century. In spite of rapid progress in technology, the development of cross-country vehicles has, for a long period of time, been guided by empiricism and the 'cut and try' methodology. Systematic studies of the principles underlying the rational development of off-road vehicles did not receive significant attention until the middle of the 20th century. The publication of Dr M.G. Bekker's classic treatises, Theory of Land Locomotion in 1956 and Off-the-Road Locomotion and Introduction to Terrain–Vehicle Systems in the 1960s, stimulated a great deal of interest in the systematic development of the principles of land locomotion mechanics (Bekker, 1956, 1960, 1969). His pioneering work and unique contributions laid the foundation for a distinct branch of applied mechanics, which has now become known as 'Terramechanics'.

In a broad sense, terramechanics is the study of the overall performance of a machine in relation to its operating environment – the terrain. It has two main branches: terrain–vehicle mechanics and terrain–implement mechanics. Terrain–vehicle mechanics is concerned with the tractive performance of a vehicle over unprepared terrain, ride quality over unprepared surfaces, handling, obstacle negotiation, water-crossing and other related topics. Terrain–implement mechanics, on the other hand, deals with the performance of terrain-working machinery, such as soil cultivating and earthmoving equipment.

The aim of terramechanics is to provide guiding principles for the rational development, design, and evaluation of off-road vehicles and terrain-working machinery. In recent years, the growing concern over energy conservation and environmental preservation has further stimulated the development of terramechanics. In addition to being a good engineering design in the traditional sense, an off-road machine is now expected to attain a high level of energy efficiency and not to cause undue damage to the operating environment, such as excessive soil compaction in agriculture. Increasing activity in the exploration and exploitation of natural resources in new frontiers, including remote areas and the seabed, and the growing demand for greater mobility over a wider range of terrains and in all seasons have also given much new impetus to the development of terramechanics.

Continuing interests of the USA, European Union and Russia, as well as programmes initiated by China, Japan, India and other nations, in the exploration of the Moon, Mars and beyond, have further stimulated advancements in terramechanics and its applications to the development of extraterrestrial vehicles, including manned and unmanned rovers (Wong and Asnani, 2008).

Terrain–vehicle mechanics is the prime subject of this book. It introduces the reader to the basic principles of terramechanics, which include the modelling of terrain behaviour, measurement and characterization of the mechanical properties of terrain pertinent to vehicle mobility, and the mechanics of vehicle–terrain interaction. As the performance of off-road vehicles over unprepared terrain constitutes a central issue in vehicle mobility, this book focuses on the study of vehicle–terrain interaction from the traction perspective. It provides the knowledge base for the prediction of off-road vehicle performance. Through examples, this book also demonstrates the applications of terramechanics to parametric analyses of terrain–vehicle systems and to the rational development and design of off-road vehicles from the traction perspective. The handling and ride of off-road vehicles are discussed in a separate book, *Theory of Ground Vehicles* (Wong, 2008).

1.1 Role of Terramechanics

The industries that manufacture and operate off-road equipment are multibillion dollar businesses. By considering the number of tractors and soil-cultivating implements used in agriculture, the number of earthmoving machines used in the construction industry, the number of off-highway trucks used in the off-road transport industry, and the number of combat and logistic vehicles used in the military, one can appreciate the scope for the applications of terramechanics.

Terramechanics, coupled with a systems analysis approach, can play a significant role in the development and evaluation of off-road equipment for a given mission and environment. Systems analysis is a methodology that provides a quantitative and systematic assessment of clearly defined issues and alternatives for decision makers. The knowledge of terramechanics can be applied, directly or indirectly, to the development, evaluation or selection of the following:

(a) vehicle concepts and configurations, defined in terms of form, size, weight and power;

(b) the running gear (or terrain-engaging elements) of a vehicle;

(c) the steering system of a vehicle;

(d) the suspension system of a vehicle;

(e) the power transmission and distribution system of a vehicle;

(f) the performance, handling and ride quality of a vehicle.

The role of terramechanics is illustrated in Figure 1.1.

There are many examples of the successful application of terramechanics and systems analysis methodology to the development and evaluation of off-road vehicles. One of the most striking examples is, perhaps, the development of the Lunar Roving Vehicle for the Apollo programmes under the guidance of Dr M.G. Bekker (1964, 1967, 1969, 1981). In a search for the optimum form of a vehicle for lunar surface exploration, walking machines, screw-driven vehicles, and a variety of tracked and wheeled vehicles were examined in detail (Asnani, Delap and Creager, 2009). Their performances were evaluated using the principles of terramechanics. The exhaustive studies led to the selection of a four-wheel vehicle with a unique type of tyre woven of steel wire and girded with titanium chevrons, as shown in Figure 1.2 (Cowart, 1971). It was found that this type of tyre produced optimum elasticity, traction, strength and durability with minimum weight, and was compatible with the vacuum and temperature extremes of the moon. This vehicle configuration was proved highly successful in operation on the lunar surface.

Another example of the application of terramechanics to the evaluation of terrain–vehicle systems was described by Sohne (1976) in connection with the studies of the optimum

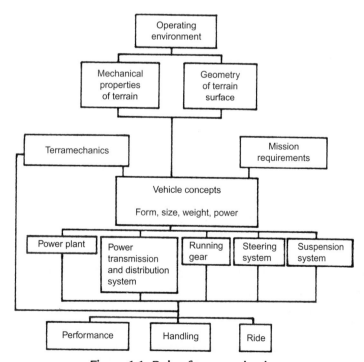

Figure 1.1: Role of terramechanics

(a) (b)

Figure 1.2: (a) The lunar roving vehicle for the Apollo missions, and (b) tyre woven of steel wire for the lunar roving vehicle (Reprinted by permission of the Council of the Institution of Mechanical Engineers from Cowart, 1971)

configuration for agricultural tractors. Based on the principles of terramechanics, he performed an analysis of the drawbar pull–slip characteristics and tractive efficiency of five configurations ranging from rear-wheel-drive with front-wheel-steering to six-wheel-drive with four-wheel-steering, as shown in Figure 1.3. Based on the results of the analysis, a comparison of the technical as well as economic performance of the various configurations was made. This type of analysis provides the designer with quantitative information upon which a rational decision may be made.

More recent examples of the application of terramechanics principles and systems analysis methodology to the parametric evaluation of tracked vehicles were reported by Wong (1992a, 1995, 2007, 2008). Using a computer-aided method, known as NTVPM, the validity of which has been substantiated by field test data, the effects of tracked vehicle design on performance can be quantitive by evaluated.

Figure 1.4 shows the effects of the initial track tension coefficient (i.e. the ratio of initial track tension to vehicle weight) on the drawbar pull coefficient (i.e. the ratio of drawbar pull to vehicle weight) of three vehicle configurations on deep snow, designated as Hope Valley snow (Wong, 2007). The three tracked vehicle configurations Vehicle A, Vehicle A (6 W) and Vehicle A (8 W) with five, six and eight overlapping roadwheels, respectively, are shown in Figure 1.5. It shows that the initial track tension has a significant effect on soft ground mobility of tracked vehicles with different design configurations. This finding has led to the development of an innovative device—a central initial track tension regulating system. This remotely control device enables the driver to increase the initial track tension for improving

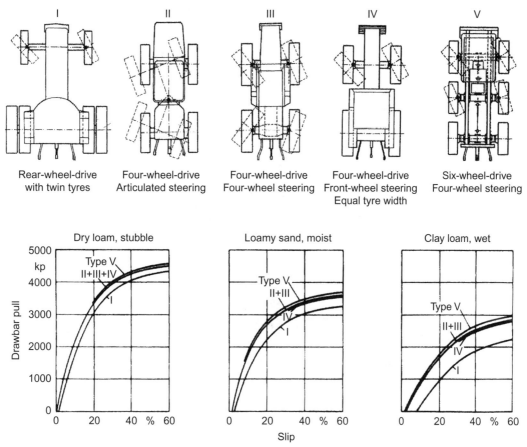

Figure 1.3: Comparison of various configurations for agricultural tractors (Reprinted by permission of *ISTVS* from Sohne, 1976)

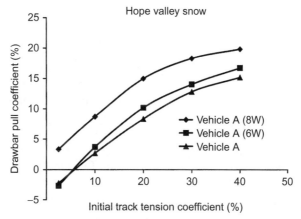

Figure 1.4: Variations of the drawbar pull coefficient with the initial track tension coefficient for the three vehicle configurations at 20% slip on Hope Valley snow, predicted by the computer-aided method NTVPM

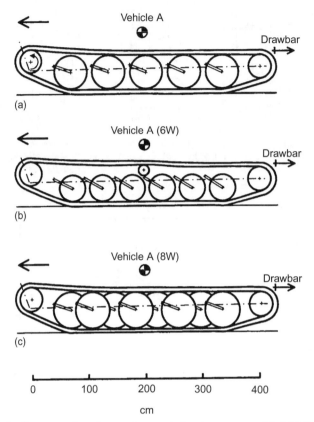

Figure 1.5: Schematic diagrams for the track–roadwheel systems of (a) Vehicle A, (b) Vehicle A (6W) and (c) Vehicle A (8W)

vehicle mobility on soft ground. This is analogous to the central tyre inflation system for improving wheeled vehicle mobility. The central initial track tension regulating system has been installed in a new generation of military vehicles (Wong, 1995). Figure 1.6(a) and (b) show the effects of the location of the centre of gravity (CG) and track width on the drawbar pull coefficient of Vehicle A, respectively, on Hope Valley snow (Wong, 2007). The analytical framework and the basic features of NTVPM are described in Chapter 7.

The computer-aided method NTVPM has been successfully used in the development of new products by off-road vehicle manufacturers and in the assessment of vehicle candidates from a procurement perspective for governmental agencies in Europe, North America, and Asia.

1.2 Some Basic Issues in Terramechanics

The study of the performance of a vehicle in relation to its operating environment — the terrain is a major focus in terramechanics. Accordingly, the modelling of terrain behaviour,

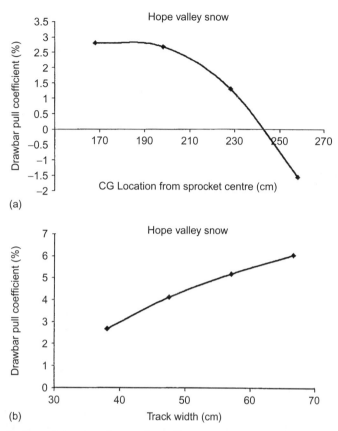

Figure 1.6: (a) Variation of the drawbar pull coefficient with the longitudinal location of the CG, and (b) variation of the drawbar pull coefficient with track width for Vehicle A at 20% slip on Hope Valley snow, predicted by the computer-aided method NTVPM

the measurement and characterization of terrain properties, the identification of pertinent parameters of the vehicle that affect its performance, and the elucidation of the interaction between the vehicle and the terrain are some of the basic issues in terramechanics. In the study of these issues, one should bear in mind that the prime objective is to provide the knowledge base upon which advancements in the design of off-road vehicles may be made.

1.2.1 Modelling of Terrain Behaviour

An understanding of terrain behaviour under vehicular load is of importance to the study of terramechanics. In the past, modelling the terrain as an elastic medium or as a rigid, perfectly plastic material has been widely used. Modelling the terrain as an elastic medium, together with the theory of elasticity, has provided a theoretical basis for the study of soil compaction. However, it is applicable only to dense terrain with vehicular load not exceeding a certain level. Modelling the terrain as a rigid, perfectly plastic material, together with the theory of plastic equilibrium, has found applications to the estimation of the maximum traction of an off-road vehicle, to the prediction of the forces acting on a bulldozer blade, or to the assessment of tractive force developed by a lug (or grouser) of a wheel. However, it can only be applied to estimating the maximum force acting on a soil-engaging element that the terrain can support, but cannot be employed to predict terrain deformation.

In recent decades, attempts have been made to apply the concept of critical state soil mechanics to modelling terrain behaviour. It is based on the assumption that the terrain is homogeneous and isotropic. It has the potential of modelling terrain behaviour over a wide range of conditions, from the loose to the dense state. However, in many types of natural terrain encountered in off-road operations, they are seldom homogeneous or isotropic. As a result, the critical state soil mechanics has so far found few practical applications to the study of vehicle-terrain interaction in the field.

With the rapid advance in computer technology and computational techniques in recent years, it has become feasible to model the terrain as an assemblage of finite elements. However, some basic issues remain to be resolved, such as the development of a robust method for determining the values of the parameters of the finite element model to properly represent terrain properties. Furthermore, the finite element method is developed on the premise that the terrain is a continuum. Consequently, it has inherent limitations in simulating large, discontinuous terrain deformation that usually occurs in vehicle-terrain interaction. To study the interaction between a vehicle and granular terrain, such as sand and the like, the discrete element method has been introduced. While the discrete element modelling technique has certain unique features, several key issues remain to be resolved before it can be generally accepted as a practical engineering tool. These include the development of a reliable method for determining the values of model parameters to realistically represent terrain properties in the field. In addition, improvements are needed in computing technique for full-scale simulations of vehicle-terrain interaction, which would require millions of discrete elements to represent the terrain and lengthy computation even on supercomputers. The modelling of terrain behaviour is discussed in detail in Chapter 2.

In view of the limitations of the techniques for modelling terrain behaviour described above, to study vehicle mobility in the field, practical techniques for measuring and characterizing terrain properties are required.

1.2.2 Measurement and Characterization of Terrain Properties

Currently, the cone penetrometer technique, the bevameter technique and the traditional technique of civil engineering soil mechanics are used for measuring the mechanical properties of the terrain for the study of vehicle mobility in the field. The selection of a particular type of technique is, to a great extent, influenced by the intended purpose of the method of approach to the study of vehicle mobility. For instance, if the method is intended to be used by the off-road vehicle engineer in the development and design of new products, then the technique selected for measuring and characterizing terrain properties would be quite different from that intended to be used by the military personnel for vehicle traffic planning on a go/no go basis. Currently, there are two major techniques used in measuring and characterizing terrain properties for evaluating off-road vehicle mobility in the field: the cone penetrometer technique and the bevameter technique. These techniques are briefly reviewed below and are discussed in detail later in Chapters 3, 4, and 5.

A. Cone penetrometer technique

The cone penetrometer technique was developed during the Second World War by the Waterways Experiment Station (WES) of the US Army Corps of Engineers. The original intention was to provide military intelligence and reconnaissance personnel with a simple field device for assessing vehicle mobility and terrain trafficability on a 'go/no go' basis.

The cone penetrometer developed by WES has a 30-degree right circular cone with a 3.23 cm^2 (0.5 in^2) base area (Figure 1.7). With the penetrometer, a parameter called the 'cone index' can be obtained. It represents the resistance to penetration into the terrain per unit cone base area. The index reflects the combined shear and compressive characteristics of the terrain and the adhesion and fiction on the cone-terrain interface. However, the contributions of these factors cannot be readily differentiated. The cone index and its gradient with respect to penetration depth have been used as a basis for predicting off-road vehicle performance in fine-grained soil (clay) and in coarse-grained soil (sand), respectively.

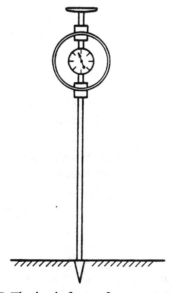

Figure 1.7: The basic form of a cone penetrometer

While the cone index of a terrain can readily be obtained, the issue of whether it can be used to adequately identify terrain characteristics from the vehicle mobility or terrain trafficability viewpoint remains controversial. For instance, work by Reece and Peca (1981) indicates that while the cone index may be useful in identifying the shear strength of remoulded frictionless clay, it is inadequate for characterizing the properties of sand. Prompted by these findings, Turnage (1984) of WES reanalysed a sizable body of experimental data previously obtained. Based on the results of his re-examination, Turnage concluded that to achieve better accuracy in predicting tyre performance in a given sand with a particular moisture content, additional laboratory testing is required to define the relationship between the before-tyre-pass cone penetration resistance gradient and the corresponding relative density, compactability and grain size distribution. This implies that the original concept of using a simple, single cone penetrometer measurement to define the properties of coarse-grained soil has to be replaced by a series of extensive geotechnical testing and analysis, including *in situ* measurements, sample acquisition and laboratory testing.

B. Bevameter technique

The bevameter technique pioneered by Bekker (1956, 1960, 1969) is based on the premise that terrain properties pertinent to terramechanics can best be measured under loading conditions similar to those exerted by an off-road vehicle. A vehicle exerts normal and shear loads to the terrain surface. To simulate these, the bevameter technique comprises two separate sets of tests. One is a plate penetration test and the other is a shear test. In the penetration test, the pressure–sinkage relationship of the terrain is measured using a plate of suitable size to simulate the contact area of the running gear of a vehicle. Based on the measurements, vehicle sinkage and motion resistance may be predicted. In the shear test, the shear stress–shear displacement relationship and the shear strength of the terrain are measured, upon which the tractive effort–slip characteristics and the maximum traction of a vehicle may be estimated. To provide data for predicting the multipass performance of vehicle running gear and the additional vehicle sinkage due to slip, the response of the terrain to repetitive loading and the slip–sinkage characteristics of the terrain are also measured (Wong et al., 1984).

Figure 1.8 shows a vehicle-mounted bevameter in field operation. To facilitate the processing of terrain data, a portable, computerized data acquisition and processing system and the associated software have been developed and successfully used in the field (Wong, 1980; Wong et al., 1981). Figure 1.9 shows the system installed in a vehicle for field operation. The use of a computerized data acquisition and processing system not only greatly reduces the effort required to obtain and process the data but also makes it possible to employ more rational procedures for deriving the values of terrain parameters (Wong, 1980; Wong et al., 1982; Wong and Preston-Thomas, 1983a,b).

To reduce the uncertainty in extrapolating terrain data obtained in the field for use in the prediction of the performance of full-scale machines, the size of the test piece used in the bevameter measurements should be comparable to that of the contact area of a tyre or that of a track link.

Figure 1.8: A vehicle-mounted bevameter for measuring terrain properties in the field

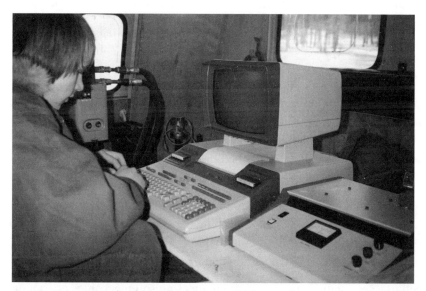

Figure 1.9: A computerized terrain data acquisition and processing system for the vehicle-mounted bevameter

For instance, measurements of the ground contact pressure under a track, shown in Figure 1.10, confirm that the idealization of a track as a strip footing is unrealistic, particularly for tracks with relatively short track pitch commonly used in high speed vehicles (Wong et al., 1984). The actual contact pressure under the track is not uniformly distributed, but rather is concentrated

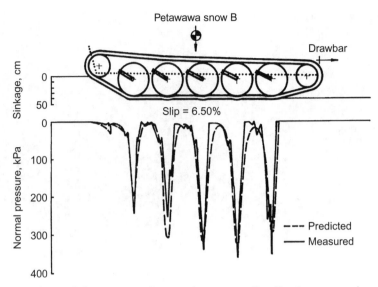

Figure 1.10: Comparison of the measured normal pressure distribution on track–terrain interface with that predicted using the computer-aided method NTVPM on snow

on the track links immediately under the roadwheels. This indicates that to better simulate track–terrain interaction, the size of the plate used in pressure–sinkage tests should be similar to that of a track link.

C. Techniques used in civil engineering soil mechanics

In civil engineering, the properties of soil are often described in terms of shear strength, shear modulus, density, void ratio, etc. To measure these parameters, soil samples are usually taken from the field and tested in a laboratory. The shear strength of the terrain is usually measured using a triaxial apparatus or a direct shear box.

The measurement of terrain properties in the field has certain advantages over that in the laboratory. The major advantage is that measurements are taken when the terrain is in its natural state, thus eliminating the possibility of disturbing the terrain samples during the sampling process for laboratory testing. Furthermore, field testing is generally less expensive and faster, particularly when a portable data acquisition and processing system is used.

As the procedures and facilities used in civil engineering soil mechanics are not particularly suited to the study of vehicle mobility in the field, they are only in limited use.

1.2.3 Modelling of Vehicle–Terrain Interaction

There are two prime objectives of the study of vehicle–terrain interaction. One is to establish the functional relationship between the performance of an off-road vehicle and its design parameters and terrain characteristics. These will enable the engineering practitioner to realistically predict vehicle performance under different operating conditions. An accurate method for predicting off-road vehicle performance is of prime interest to the designer, as well as to the user, of off-road vehicles. The other objective is to establish a procedure with which the changes in terrain conditions caused by the passage of an off-road vehicle or soil working machinery may be predicted. This is of great interest to the agricultural engineer in the evaluation of soil compaction caused by farm vehicles and to the construction equipment engineer in the assessment of the effectiveness of soil compactors and the like.

In this book, emphasis is placed on the discussion of the interrelationships between off-road vehicle performance and its design and operational parameters and terrain characteristics.

On a given terrain, the performance of an off-road machine is, to a great extent, dependent upon the manner in which the machine interacts with the terrain. Figures 1.11 and 1.12 show the flow patterns of soil under the action of a rigid wheel and of a wheel with grousers, respectively (Wong and Reece, 1966; Wong, 1967; Wu et al., 1984). As a result of the interaction between the machine element and the terrain, normal and shear stresses are developed on the machine–terrain interface. Figure 1.10 shows the variation of the normal pressure exerted by a flexible track on a snow-covered terrain. The normal and shear stress distributions on the contact area of a tyre in a sandy loam are shown in Figure 1.13 (Krick, 1969). It can be seen that in most cases the interaction between a machine element and terrain is very complex. For example, the flow patterns of soil under a wheel depend upon its kinematics as

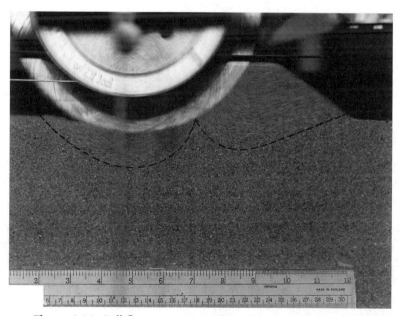

Figure 1.11: Soil flow patterns under a driven rigid wheel in sand

Figure 1.12: Soil flow under the action of grousers of a wheel in sand (Reprinted by permission of *ISTVS* from Wu et al., 1984)

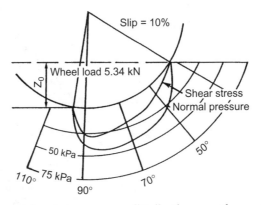

Figure 1.13: Measured normal and shear stress distributions on the contact patch of a tyre on a sandy loam (Reprinted by permission of *ISTVS* from Krick, 1969)

defined by its slip, while the stress conditions at the interface are determined by terrain conditions as well as the design and operational parameters of the wheel, including its dimensions and the vertical load and torque applied to it.

Since the performance of an off-road vehicle, defined in terms of its motion resistance, tractive effort, drawbar pull, tractive efficiency, etc., is determined by the normal and shear stresses on the vehicle–terrain interface, a central issue in terramechanics is the formulation of a mathematical model to predict the interacting forces between the vehicle and the terrain, based on the knowledge of terrain behaviour and pertinent design and operational parameters of the vehicle.

Identifying the design parameters of the vehicle that significantly influence vehicle–terrain interaction is of importance in formulating a mathematical model for off-road vehicle performance. For instance, tyre diameter, section width, section height, and lug angle and spacing are considered to have varying degrees of influence on the tyre–terrain interaction. For tracked vehicles, design parameters, such as the number of roadwheels, roadwheel dimensions and spacing, track geometry and dimensions, initial track tension, suspension characteristics, and arrangements for the sprocket, idler and supporting rollers, are shown to have an effect on the tractive performance.

1.3 Approaches to Terramechanics

A variety of methods of approach to the study of off-road vehicle mobility has been developed over the years. They range from entirely empirical to highly analytical. The selection of the method of approach is greatly influenced by the following factors.

A. Intended purposes

Depending upon whether the method is to be used by the development and design engineer in the optimization of vehicle design, by the procurement manager in the evaluation of vehicle candidates for a given mission, or by the vehicle operator in assessing vehicle mobility on a 'go/no go' basis, the method of approach varies greatly. For instance, a method intended for use in the design and development of off-road machinery requires a level of sophistication, accuracy, and detail that differs substantially from that intended for use by military intelligence and reconnaissance personnel in evaluating vehicle mobility in the field.

B. Environmental, economic and operational constraints

As in any other branches of engineering, the method of approach to the study of vehicle-terrain interaction is subject to constraints. For instance, in the selection of the techniques for identifying the mechanical properties of terrain in remote areas, on the seabed, or on the surface of other planets, environmental, economic, operational and other constraints may be the most important factors to be considered.

Comparison and evaluation of different methods of approach to terramechanics should, therefore, be made in the context of their intended purposes and constraining factors.

1.3.1 Empirical Methods

It is generally recognized that the interaction between an off-road machine and the terrain is complex and difficult to model accurately. To circumvent this difficulty, empirical methods for the study of vehicle mobility have been developed.

Following the empirical approach, vehicles are tested in a range of terrains considered to be representative and at the same time the terrain is identified by field observations and simple measurements. The results of vehicle performance testing and terrain measurements are then empirically correlated. This can lead to the development of a scale for evaluating terrain trafficability on the one hand and vehicle mobility on the other. This approach is best exemplified by the work of the US Army Waterways Experiment Station (WES) based on the cone index. The method was first developed during the Second World War and was originally intended to provide military intelligence and reconnaissance personnel with a simple means to assess terrain trafficability and vehicle mobility on a 'go/no go' basis. Recently, this approach has been extended, for instance, to empirically correlate certain dimensionless performance parameters of tyres with mobility numbers (numerics) based on the cone index or the cone index gradient. Some success has been reported in applying this empirical method to the prediction of tyre performance on remoulded frictionless soils. However, difficulties have been encountered in the application of the method to the evaluation of tyre performance in certain types of sand, as reported by Reece and Peca (1981). It has also been reported by Gee-Clough (1978) that this empirical approach does not give sufficiently accurate predictions of certain performance parameters of tyres.

Within the context of their intended purposes, well-developed empirical methods are useful in estimating the performance of vehicles with design features similar to those that have been tested under similar operating conditions. It is by no means certain, however, that empirical relations can be extrapolated beyond the conditions upon which they were derived. Consequently, it is uncertain that an entirely empirical approach could play a useful role in the evaluation of new design concepts or in the prediction of vehicle performance in new operating environments. Furthermore, an entirely empirical approach is only feasible where the number of variables involved in the problem is relatively small. If a significant number of parameters are required to define the problem, then an empirical approach may not be cost effective.

To provide a more general approach to terramechanics, particularly for parametric analyses of terrain–vehicle systems, other methods of approach have been developed.

1.3.2 Computational Methods

With rapid progress in computer technology and the availability of commercial computer codes, the finite element method (FEM), and the discrete (distinct) element method (DEM) have been introduced into the analysis of vehicle-terrain interaction. These methods generally involve intensive computation and may be referred to as computational methods. They have the potential of providing a tool with which certain aspects of the mechanics of vehicle-terrain interaction may be examined in detail.

Based on a review of the state-of-art in the applications of these computational methods to the study of vehicle-terrain interaction, it appears that the finite element method or the discrete element method may be applied to evaluating, on a relative basis, the design and performance of tyres or soil engaging implements of simple form. Predictions of tyre performance based on these methods have been shown to be in qualitative agreement with experimental data on certain types of terrain (Liu and Wong, 1996; Seta et al., 2003). For a track system which is a complex mechanical system, its interaction with the terrain involves not only the part of the track system in contact with the terrain, but also other factors, such as roadwheel system configuration, suspension characteristics, locations of the sprocket and idler, initial track tension, arrangements for the supporting rollers on the top run of track, etc. To make the analysis amenable to the finite element method, however, the track usually has to be simplified as a rigid footing with either uniform or trapezoidal form of normal pressure distribution (Karafiath, 1984). In many cases, the ratio of the shear stress to normal pressure has also to be specified at the outset of the prediction process. It will be shown later that these are unrealistic. Consequently, it cannot provide the off-road vehicle engineer with a realistic tool for design and performance evaluation of track systems. Furthermore, as pointed out previously, the finite element method is based on the premise that the terrain is a continuum. As a result, it has inherent limitations in simulating large, discontinuous terrain deformation which often occurs in off-road operations.

Several other issues must also be resolved before these computational methods can be considered as a practical engineering tool. These include the development of a robust method for determining the values of the parameters in the finite element or discrete element method that realistically represent the mechanical properties of the terrain in the field. This poses one the greatest challenges, in view of the variability and complexity of terrain behaviour in the natural environment. Furthermore, it is estimated that to conduct a realistic three-dimensional simulation of full-scale machine-terrain interaction problem by the discrete element method, the number of elements required would be in the order of 10^6 to 10^8. This would require high-power computing resources, usually supercomputers. For instance, it has been shown that in a simulation of the interaction between a mine plow and soil, ten million elements are used and the simulation takes just over 16,000 CPU hours on a 256-processor Cray T3E supercomputer (Horner, Peters and Carrillo, 2001).

In summary, the applications of the finite element method or the discrete element method to the study of vehicle-terrain interaction are still in the nascent stage. Prior to being considered acceptable as a useful tool for design and performance evaluation of off-road vehicles, several challenging issues facing these computation methods have to be resolved, as outlined above.

Detailed discussions of the finite element method and the discrete (distinct) element method and their applications to vehicle mobility study are presented in Chapter 2.

1.3.3 Methods for Parametric Analysis

In view of the limitations of the empirical and computational methods noted above, mathematical models for parametric analysis of the performance of off-road vehicles have been developed. A pioneering effort in this area was made by Bekker (1956, 1960, 1969). Lately, a series of computer-aided methods for parametric analysis of the performance and design of both tracked vehicles and off-road wheeled vehicles, from a traction perspective, have emerged (Wong, et al., 1984; Wong and Preston-Thomas, 1986 and 1988; Wong, 1992; Gao and Wong, 1994; Wong, 1995, 1998, 2007, 2008; Wong and Huang, 2005, 2006a and b, 2008). These include methods for performance and design evaluation of vehicles with flexible tracks (NTVPM), vehicles with long-pitch link tracks (RTVPM), and off-road wheeled vehicles (NWVPM). The basic features of NTVPM, RTVPM and NWVPM are presented in Chapters 8, 10 and 12, respectively.

These methods are based on an understanding of the physical nature of vehicle-terrain interaction and on the principles of terramechanics. They take into account all major design features of the vehicle that affect its performance. All pertinent terrain characteristics, such as pressure-sinkage and shearing characteristics and response to repetitive loading, measured by the bevameter technique (Bekker, 1960, 1969; Wong, 2008) are taken into account. The predictive capabilities of these computer-aided methods have been verified by field test data obtained on various types of terrain.

These computer-aided methods are particularly suited for the evaluation of competing designs, optimization of design parameters, and selection of vehicle candidates for a given mission and environment They have been successfully used to assist off-road vehicle manufacturers in the development of new products and governmental agencies in the evaluation of vehicle candidates in Europe, North America, Asia and Africa. Examples of the applications of NTVPM, RTVPM and NWVPM to performance and design evaluations are presented in Chapters 9, 10 and 12, respectively.

In summary, the introduction given above indicates that since the founding of terramechanics in the 1960s, a great deal of progress has been accomplished. A number of outstanding examples have demonstrated its relevance to engineering practice and its growing acceptance in industry. With the continual enthusiasm shown by an increasing number of countries in exploration on the surfaces of the Moon, Mars and other planets, terramechanics is playing an increasingly significant role in the development of the mobility systems of extraterrestrial rovers (Wong and Asnani, 2008). Figure 1.14 shows the image of the NASA Mars Exploration Rovers (Spirit and Opportunity) launched from earth in the middle of 2003 and landed on Mars in early 2004.

Looking ahead, it appears that terramechanics has to provide an improved knowledge base for the further development of engineering tools for performance and design evaluation of off-road vehicles operating in a wider spectrum of environment, and with greater accuracy and reliability. In addition, theses tools have to be user-friendly that will appeal to a wide range

Figure 1.14: NASA Mars Exploration Rover

of engineering practitioners, including vehicle designers, researchers and procurement managers. In the further development of terramechanics, one should bear in mind that the prime objective is to provide the engineering practitioner with useful and reliable tools that will lead to innovations in the design and development of off-road vehicles to meet society's changing needs for environmental protection, energy conservation and sustainable development.

CHAPTER 2
Modelling of Terrain Behaviour

An understanding of terrain behaviour under vehicular load is of importance to the study of vehicle–terrain interaction. In this chapter, various approaches to modelling terrain behaviour are reviewed.

In the past, modelling the terrain as an elastic medium or as a rigid, perfectly plastic material has been widely adopted. Modelling the terrain as an elastic medium, together with the theory of elasticity, has found applications in the study of soil compaction and terrain damage due to vehicular traffic. Modelling the terrain as a rigid, perfectly plastic material, together with the theory of plastic equilibrium, has found applications in the prediction of the maximum traction developed by off-road vehicles and of the thrust developed by lugs (grousers) of a vehicle running gear. It has also been employed in the prediction of the resistance of a bulldozer blade. While the idealization of the terrain as an elastic medium or as a rigid, perfectly plastic material may provide a basis for elucidating certain aspects of the physical nature of vehicle–terrain interaction, there are limitations. For instance, the theory of elasticity may only be applied to dense terrain with vehicular load not exceeding a certain level, so that the terrain may be considered elastic. On the other hand, the theory of plastic equilibrium can only be employed in estimating the maximum vehicle load that the terrain can support without causing its failure, but cannot be used to predict the sinkage of the vehicle due to its normal load or the slip of the vehicle due to the shearing action of its running gear.

To overcome the limitations of modelling the terrain as an elastic medium or as a rigid, perfectly plastic material, attempts have been made to model the terrain based on the concept of the critical state soil mechanics, as it has the potential capability to predict both the stress and strain in the terrain under vehicular load. However, due to the complexity and the variability of terrain behaviour in the field, so far its applications to the study of vehicle–terrain interaction are limited.

With advancements in computer technology and computational techniques in recent years, modelling the terrain using the finite element method (FEM) or using the discrete (distinct) element method (DEM) has emerged. These methods have the potential capability to examine certain aspects of the physical nature of vehicle–terrain interaction in great detail. Their basic concepts and applications to the study of vehicle–terrain interaction are outlined in this chapter.

2.1 Modelling Terrain as an Elastic Medium

For dense terrain, such as compact sand and the like, its behaviour under certain circumstances may be compared with that of an ideal elastoplastic medium with a stress–strain relationship shown in Figure 2.1. When the vehicular load applied to the terrain does not exceed a certain level and the corresponding stress in the medium is lower than that denoted by 'A' in the figure, the terrain may exhibit elastic behaviour. The idealization of the terrain as an elastic medium, together with the classical theory of elasticity, has been employed in the prediction of stress distribution in the terrain and in the assessment of the effect of vehicular traffic on soil compaction or on terrain damage.

The prediction of stress distribution in a semi-infinite elastic medium subject to any form of loading on its surface may be based on that subject to a point load. For a homogeneous, isotropic, elastic medium subject to a point load on the surface, the stress distribution in the medium may be predicted using the Boussinesq equation. The expression for the vertical stress σ_z at a point in the medium below the surface defined by the coordinates shown in Figure 2.2 is given below:

$$\sigma_z = \frac{3W}{2\pi\left[1 + (r/z)^2\right]^{5/2} z^2}$$

$$= \frac{3W}{2\pi R^2}\left(\frac{z}{R}\right)^3 = \frac{3W}{2\pi R^2}\cos^3\theta \qquad (2.1)$$

where $r = \sqrt{x^2 + y^2}$ and $R = \sqrt{z^2 + r^2}$.

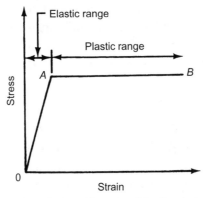

Figure 2.1: Stress–strain relationship of an idealized elastoplastic material

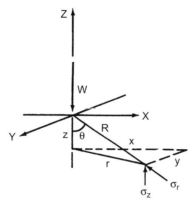

Figure 2.2: Stresses in a semi-infinite elastic medium subject to a point load on the surface

When polar coordinates are used, the radial stress σ_r (Figure 2.2) is expressed by

$$\sigma_r = \frac{3W}{2\pi R^2} \cos\theta \tag{2.2}$$

It should be noted that the stresses in the medium are independent of the modulus of elasticity of the material. They are only functions of the load applied and the distance from the point of application of the load. It should be noted that Eqns (2.1) and (2.2) can only be used for predicting stresses at points not too close to the location of the point load. The material in the vicinity of the point load does not exhibit elastic behaviour.

From the analysis of stress distribution under a point load, the stress distribution in an elastic medium under a variety of loading forms may be predicted following the principle of superposition. For instance, for a circular loading area with radius r_0 and with uniform contact pressure p_0 (Figure 2.3), the vertical stress at a depth z below the centre of the circular loading area may be calculated following the procedure given below (Bekker, 1956; Wong, 2008).

The load applied on the contact area may be considered to be an assemblage of a number of discrete point loads, $dW = p_0\, dA = p_0 r\, dr\, d\theta$. Thus, in accordance with Eqn (2.1),

$$d\sigma_z = \frac{3 p_0 r\, dr\, d\theta}{2\pi \left[1 + (r/z)^2\right]^{5/2} z^2} \tag{2.3}$$

The resultant vertical stress σ_z at a depth z below the centre of the circular loading area is equal to the sum of the stresses produced by a series of point loads represented by $p_0 r\, dr\, d\theta$ and may be calculated by a double integration (Bekker, 1956):

$$\sigma_z = \frac{3 p_0}{2\pi} \int_0^{r_0} \int_0^{2\pi} \frac{r\, dr\, d\theta}{\left[1 + (r/z)^2\right]^{5/2} z^2} = 3 p_0 \int_0^{r_0} \frac{r\, dr}{\left[1 + (r/z)^2\right]^{5/2} z^2}$$

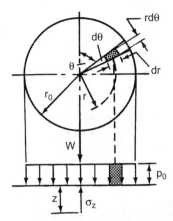

Figure 2.3: Vertical stresses in a semi-infinite elastic medium below the centre of a circular loading area (Reprinted by permission from M.G. Bekker, *Theory of Land Locomotion*, University of Michigan Press, 1956, copyright © by the University of Michigan)

By substituting $(r/z)^2 = u^2$, the above equation may be rewritten by

$$\sigma_z = 3p_0 \int_0^{r_0/z} \frac{u\,du}{(1+u^2)^{5/2}} = p_0 \left[1 - \frac{z^3}{(z^2 + r_0^2)^{3/2}} \right] \qquad (2.4)$$

The computation of stresses at points other than those directly below the centre of the circular loading area cannot be expressed by a simple equation like Eqn (2.4). The stress distribution in an elastic medium under distributed loads over an area of elliptic or super-elliptic shape (Hallonborg, 1996), similar to that of the contact patch of a pneumatic tyre, may be determined by numerical methods following a similar approach.

Another case of interest from the viewpoint of terramechanics is the distribution of stresses in a semi-infinite elastic medium under the action of a strip load on the surface (Figure 2.4). Such a strip load may be considered as an idealization of that applied by a tracked vehicle, with the track assumed to be equivalent to a rigid footing with uniform normal pressure. It can be shown that the stresses in the elastic medium due to uniform pressure p_0 exerted over a strip of infinite length and of constant width b (Figure 2.4) may be expressed by the following equations (Bekker, 1956):

$$\sigma_x = \frac{p_0}{\pi}(\theta_2 - \theta_1 + \sin\theta_1 \cos\theta_1 - \sin\theta_2 \cos\theta_2) \qquad (2.5)$$

$$\sigma_z = \frac{p_0}{\pi}(\theta_2 - \theta_1 - \sin\theta_1 \cos\theta_1 + \sin\theta_2 \cos\theta_2) \qquad (2.6)$$

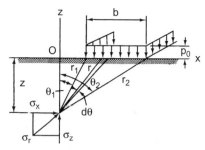

Figure 2.4: Stresses at a point in a semi-infinite elastic medium under a uniform strip load (Reprinted by permission from M.G. Bekker, *Theory of Land Locomotion*, University of Michigan Press, 1956, copyright © by the University of Michigan)

$$\tau_{xz} = \frac{p_0}{\pi}(\sin^2 \theta_2 - \sin^2 \theta_1) \qquad (2.7)$$

The points in the medium that experience the same level of stress may be described in the form of a family of isostress lines (or surfaces), commonly referred to as pressure bulbs. The general characteristics of the vertical stress bulbs under a vehicle with two tracks, idealized as two rigid footings, are illustrated in Figure 2.5. It is interesting to point out that at a depth equal to the width of the track, the vertical stress under the centre of the loading area is approximately 50% of the pressure p_0 exerted on the surface by the track. It diminishes at a depth equal to twice the width of the track. The boundaries of the vertical pressure bulbs may be considered as being sloped at an angle of 45° with the horizontal as shown in Figure 2.5 (Bekker, 1956).

It should be pointed out that by modelling the terrain as an elastic medium and by applying the theory of elasticity to predict the stress distribution in the terrain produces only approximate results. Measurements have shown that the stress distribution in the terrain deviates from that predicted using the Boussinesq equation, dependent on terrain conditions (Sohne, 1958). There is a tendency for the stress in the terrain to concentrate around the central axis of the loading area. It becomes greater as the moisture content of the terrain increases. Based on these observations, various semi-empirical factors (or parameters) have been introduced to the Boussinesq equation, to account for the behaviour of different types of terrain. For instance, Frohlich introduced a concentration factor v to the Boussinesq equation. By introducing the concentration factor v, the expressions for the vertical and radial stresses in the terrain due to a point load applied on the surface take the following forms:

$$\sigma_z = \frac{vW}{2\pi R^2}\cos^v \theta = \frac{vW}{2\pi z^2}\cos^{v+2}\theta \qquad (2.8)$$

$$\sigma_r = \frac{vW}{2\pi R^2}\cos^{v-2}\theta = \frac{vW}{2\pi z^2}\cos^v \theta \qquad (2.9)$$

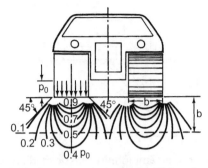

Figure 2.5: Distribution of vertical stresses in a semi-infinite elastic medium under a tracked vehicle (Reprinted by permission from M.G. Bekker, *Theory of Land Locomotion*, University of Michigan Press, 1956, copyright © by the University of Michigan)

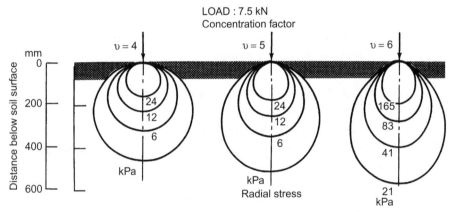

Figure 2.6: Distribution of radial stresses under a point load in soils with different concentration factors (Reprinted by permission of *ASABE* from Sohne, 1958)

The value of υ depends on the type of terrain and on its moisture content. For instance, for hard, dry soil, the value of υ is 4; for farm soil with normal density and moisture content, the value of υ is 5; and for wet soil, the value of υ may be 6 (Sohne, 1958). If the soil is perfectly elastic, the value of υ is 3. In this case, Eqns (2.8) and (2.9) are identical to Eqns (2.1) and (2.2), respectively. Figure 2.6 shows the distribution of radial stress σ_r in the terrain with different values of concentration factor υ under a point load applied on the surface. It can be seen that when the value of υ is 4, the lines of equal radial stress (or pressure bulbs) are approximately circular. As the value of υ increases, the shape of the pressure bulbs becomes narrower and the pressure bulbs penetrate deeper into the terrain.

In practice, a tyre applies load to the terrain surface through a finite contact area rather than a point. To determine the stress distribution in the terrain due to tyre loading, the shape and size of the contact area and the pressure distribution over the contact patch must be known. Figure 2.7 shows the measured contact areas of a tyre under different terrain conditions (Sohne, 1958). The shape of the contact areas shown in the figure may be described as the combination of two super ellipses of different shapes (Hallonborg, 1996). For instance, in the first quadrant shown in Figure 2.8, where $0 < x < a_1$, the shape of the curve may be described by

$$y = b\left(1 - \frac{x^n}{a_1^n}\right)^{1/n} \qquad (2.10)$$

In the second quadrant, where $-a_2 < x < 0$, the shape of the curve may be expressed by

$$y = b\left(1 - \frac{x^m}{a_2^m}\right)^{1/m} \qquad (2.11)$$

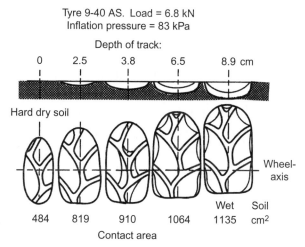

Figure 2.7: Contact areas of a tyre for different soil conditions (Reprinted by permission of *ASABE* from Sohne, 1958)

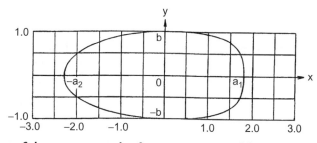

Figure 2.8: Shape of the contact patch of a tyre represented by parts of two super ellipses (Reprinted by permission of *ISTVS* from Hallonborg, 1996)

where a_1 and a_2 are axes of the super ellipses along the x direction in the first and second quadrant, respectively; the axes of the two super ellipses in the y direction must have the same value b, in order to ensure the continuity of the two curves at $x = 0$. As indicated in Eqns (2.10) and (2.11), the exponents of the two curves n and m may be different. The corresponding curves below the x-axis, with y being negative, are the mirror images of those above the x-axis.

The tyre contact patches with areas ranging from 819 to 1135 cm² shown in Figure 2.7 may be approximated by the combination of two super ellipses with the values of exponents m and n shown in Figure 2.9 (Hallonborg, 1996). The corresponding total areas of the combination of two super ellipses are also given in the figure. It can be seen that the super ellipses provide a reasonably good representation of the shape of tyre contact areas shown in Figure 2.7.

With respect to the pressure on the contact patch of a tyre without lugs, an approximately uniform pressure over the entire contact area may be assumed on hard, dry soil. On soft soils, the pressure on the contact patch varies with the depth of the rut. It generally decreases towards the boundary of the contact area and is more concentrated at the centre of the contact area. Representative pressure distributions over the contact patch of a tyre without lugs on hard, dry soil, on fairly moist, relatively dense soil, and on wet soil are illustrated in Figure 2.10 (Sohne, 1958).

When the shape and size of the tyre contact patch and the pressure distribution over it are known, it is possible to predict the pressure distribution in the terrain with a given concentration factor, by following the procedure outlined above. Figure 2.11 shows the distributions of the major principal stress under a tyre on various types of soil with different values of concentration factor v.

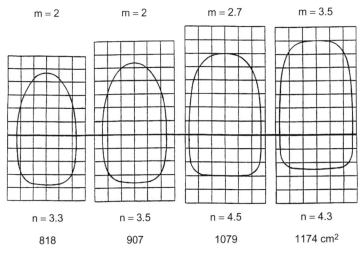

Figure 2.9: Contact areas of a tyre for different soil conditions represented by the parts of two super ellipses (Reprinted by permission of *ISTVS* from Hallonborg, 1996)

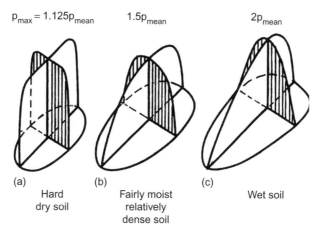

Figure 2.10: Pressure distributions on the tyre contact patch for different soil conditions (Reprinted by permission of *ASABE* from Sohne, 1958)

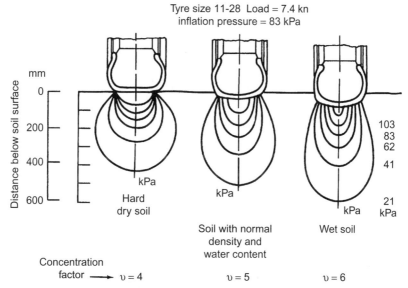

Figure 2.11: Distributions of major principal stress under a tyre for different soil conditions (Reprinted by permission of *ASABE* from Sohne, 1958)

■ Example 2.1

The shape of the contact patch of a tyre without lugs on a hard, dry soil is assumed to be circular with radius of 15 cm. The contact pressure is assumed to be a uniform 103 kPa.

(a) Calculate the vertical stress at a depth of 15 cm in a hard, dry soil directly below the centre of the circular contact area. For the hard, dry soil, the value of concentration factor υ is assumed to be 4.

(b) Determine the depth below the centre of the contact area at which the vertical stress becomes 5% of the pressure applied on the tyre contact patch.

Solution

(a) When the concentration factor υ is 4, the vertical stress σ_z at a point in the soil due to a point load W applied on the contact surface is given by

$$\sigma_z = \frac{4W}{2\pi R^2}\cos^4\theta = \frac{4W z^4}{2\pi(z^2+r^2)^3} = \frac{4W}{2\pi z^2 \left(1+(r/z)^2\right)^3}$$

The vertical stress σ_z at a depth z directly below the centre of a circular contact area of radius r_0 and with a contact pressure p_0 is expressed by

$$\sigma_z = \frac{4p_0}{2\pi}\int_0^{r_0}\int_0^{2\pi}\frac{r\,dr\,d\theta}{z^2\left[1+(r/z)^2\right]^3} = 4p_0\int_0^{r_0/z}\frac{u\,du}{(1+u^2)^3}$$

$$= p_0\left[1-\frac{1}{\left[1+(r_0/z)^2\right]^2}\right]$$

where $u^2 = r^2/z^2$. For $p_0 = 103$ kPa, $r_0 = 15$ cm and $z = 15$ cm,

$$\sigma_z = 103\left[1-\frac{1}{\left[1+(15/15)^2\right]^2}\right] = 77\text{ kPa}$$

(b) The depth z at which the vertical stress σ_z is 5% of the pressure p_0 on the contact surface can be obtained by solving the following equation:

$$\frac{\sigma_z}{p_0} = 0.05 = 1 - \frac{1}{\left[1+(r_0/z)^2\right]^2}$$

Solving the above equation, it is found that under the centre of the contact area at a depth $z = 93$ cm approximately, the vertical stress σ_z is 5% of the pressure p_0. ∎

In summary, modelling the terrain as an elastic medium, together with the theory of elasticity, has found applications in predicting stress distributions in the terrain under different forms of pressure on the contact patch. However, the applications are limited to dense terrain with vehicular load not exceeding a certain level, so that the terrain may be considered to exhibit elastic behaviour.

2.2 Modelling Terrain as a Plastic Medium

When the load applied to the terrain surface exceeds a certain limit, the stress level within a certain boundary of the terrain reaches that denoted by 'A' on the idealized stress–strain relationship shown in Figure 2.1. An infinitely small increase in stress beyond point 'A' produces a rapid increase in strain, which constitutes plastic flow. The state preceding plastic flow is usually referred to as plastic equilibrium. The transition from the state of plastic equilibrium to that of plastic flow represents the failure of the terrain material (Wong, 2008).

A number of criteria have been proposed or developed for defining the failure of soils or other types of terrain material. Among them, the Mohr-Coulomb failure criterion is one of the most widely used. It postulates that the material at a point will fail if the shear stress at that point satisfies the following condition:

$$\tau = c + \sigma \tan \phi \tag{2.12}$$

where τ is shear stress, c is the cohesion, σ is the normal stress on the shearing surface, and ϕ is the angle of internal shearing resistance of the material.

Cohesion is the bond that cements particles of the material together, regardless of the normal pressure between the particles. On the other hand, particles of frictional material can be held together only when a normal pressure is present between them. Thus, the shear strength of saturated clay and the like does not depend on the normal pressure, whereas the shear strength of dry sand or similar material increases with an increase in the normal pressure.

For saturated clay, its shear strength is given by

$$\tau = c \tag{2.13}$$

and for dry sand, its shear strength is expressed by

$$\tau = \sigma \tan \phi \tag{2.14}$$

Terrains that cover most of the trafficable earth surface generally have both cohesive and frictional properties and their shear strength is described by Eqn (2.12).

The meaning of the Mohr-Coulomb failure criterion may be further illustrated with the aid of the Mohr circle of stress. If specimens of a terrain material are subject to different states of stress, for each mode of failure a Mohr circle can be constructed, as shown in Figure 2.12. If a straight line is drawn to envelope the set of Mohr circles so obtained, it will be of the form of Eqn (2.12), with cohesion of the terrain defined by the intercept of the straight line with the shear stress axis and the angle of internal shearing resistance being represented by the slope of the straight line. The Mohr-Coulomb failure criterion simply implies that if a Mohr circle representing the state of stress at a point in the terrain touches the enveloping line, failure will take place at that point.

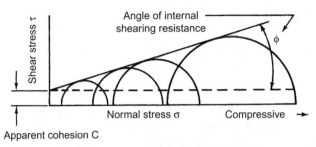

Figure 2.12: Mohr-Coulomb failure criterion (Reprinted by permission from J.Y. Wong, *Theory of Ground Vehicles*, 4th Ed., Wiley, 2008, copyright © 2008 by John Wiley)

The shear strength parameters c and ϕ in Eqn (2.12) may be measured by a variety of devices. The triaxial apparatus and the translational shear box are commonly used in soil mechanics studies related to civil engineering. For terramechanics studies, however, rectangular or annular shear plates are generally employed. The measurement and characterization of the shearing behaviour of various types of terrain are discussed Chapter 5.

The thrust (tractive effort) of a vehicle running gear, such as a tyre or a track, is developed through its shearing action on the terrain surface, as shown in Figure 2.13. Equation (2.12) can be applied to estimating the maximum thrust that can be developed by a vehicle running gear (Micklethwaite, 1944). For instance, if the contact area of a tyre or a track is known and the pressure on the contact patch is uniform, then its maximum thrust can simply be predicted by the following equation:

$$F_{max} = \tau A = (c + \sigma \tan\phi)A = cA + W\tan\phi \qquad (2.15)$$

where A is the contact area of a tyre or a track; the product of contact pressure and contact area is equal to the normal load on the tyre or the track W.

Equation (2.15) indicates that on a frictional terrain, such as dry desert sand and the like, the maximum thrust of a vehicle running gear is dependent on its normal load and the angle of internal shearing resistance of the terrain. It is independent of its contact area. Thus, on dry desert sand and the like, the maximum thrust developed by a vehicle running gear is expressed by

$$F_{max} = W\tan\phi \qquad (2.16)$$

Equation (2.16) indicates that on a frictional terrain, the higher the normal load, the higher the maximum thrust of a vehicle running gear can be developed.

On a cohesive terrain, such as saturated clay and the like, the maximum thrust of a vehicle running gear is dependent on its contact area and the cohesion of the terrain. It is independent of its normal load. Thus, on saturated clay and the like, the maximum thrust of a vehicle running gear is given by

$$F_{max} = cA \tag{2.17}$$

Equation (2.17) implies that on a cohesive terrain, the larger the contact area, the higher the thrust of a vehicle running gear can be developed.

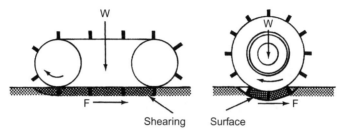

Figure 2.13: Shearing action of a track and a wheel (Reprinted by permission from J.Y. Wong, *Theory of Ground Vehicles*, 4th Ed., Wiley, 2008, copyright © 2008 by John Wiley)

■ Example 2.2

A tracked vehicle with two tracks weighs 110 kN. Each of the two tracks is 0.38 m wide and 2.65 m long. Estimate the maximum thrust of the vehicle and the ratio of the maximum thrust to vehicle weight on a sandy loam with cohesion of 1.72 kPa and angle of internal shear resistance of 29°.

Solution

The maximum thrust of the tracked vehicle can be predicted using Eqn (2.15):

$$F_{max} = cA + W\tan\phi$$
$$= 1.72 \times 2 \times (0.38 \times 2.65) + 110 \times \tan 29° = 3.464 + 60.974$$
$$= 64.438 \, kN$$

The ratio of the maximum thrust to vehicle weight may be expressed by

$$F_{max}/W = (cA + W\tan\phi)/W = \frac{c}{W/A} + \tan\phi = \frac{c}{p_{ave}} + \tan\phi$$
$$= \frac{1.72}{54.618} + 0.5543 = 58.58\%$$

where p_{ave} is usually referred to as the average (or nominal) ground pressure.

■

Another example of the application of the Mohr-Coulomb failure criterion may be illustrated by examining the plastic equilibrium of a prism in a semi-infinite soil mass shown in Figure 2.14.

The prism of soil with weight density γ_s having depth z and width equal to unity is in a state of incipient plastic failure due to lateral pressure. There are no shear stresses on the vertical sides of the prism; the normal stress on the base of the prism and that on the vertical sides are therefore the principal stresses. The prism may be set into a state of plastic equilibrium by two different operations: one is to stretch it and the other is to compress it in the horizontal direction. If the prism is being stretched, the normal stress on both vertical sides decreases until the conditions for plastic equilibrium are satisfied, while the normal stress at the bottom remains unchanged. Any further expansion merely causes a plastic flow. In this case, the weight of the soil mass assists in the expansion and this type of failure is called the active failure. On the other hand, if the prism is compressed, the normal stress on both vertical sides increases, while that at the bottom remains unchanged. In this case, lateral compression of the soil is resisted by its own weight and the resulting failure is called the passive failure. The two states of stress prior to plastic flow caused by expansion and compression of the soil are generally referred to as the Rankine active and passive state, respectively. In most problems in terramechanics, passive failure is involved. The application of the Mohr-Coulomb failure criterion to the analysis of passive failure is presented below.

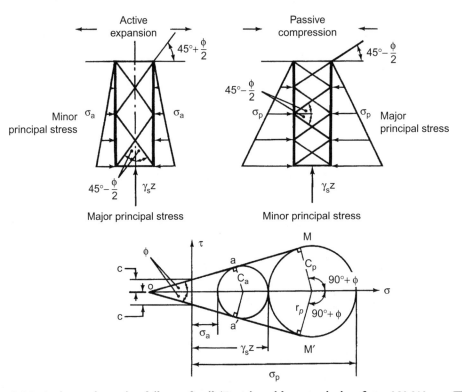

Figure 2.14: Active and passive failure of soil (Reprinted by permission from J.Y. Wong, *Theory of Ground Vehicles*, 4th Ed., Wiley, 2008, copyright © 2008 by John Wiley)

As passive failure is caused by lateral compression, the normal stress acting at the bottom of an element of soil is equal to $\gamma_s z$ and is the minor principal stress, as shown in Figure 2.14. Circle C_p in the figure represents the state of stress of an element at incipient passive failure. The point of intersection between circle C_p and the horizontal axis of the Mohr diagram determines the major principal stress, which is the lateral compressive stress on both vertical sides required to set the element into passive failure. This normal stress is referred to as the passive earth pressure σ_p. From the geometric relationships shown in Figure 2.14, the expression for the passive earth pressure σ_p is given by

$$\sigma_p = \gamma_s z + 2r_p \tag{2.18}$$

where r_p is the radius of circle C_p and is expressed by

$$r_p = \frac{c\cos\phi + \gamma_s z \sin\phi}{1 - \sin\phi} \tag{2.19}$$

Therefore,

$$\begin{aligned}\sigma_p &= \gamma_s z \frac{1+\sin\phi}{1-\sin\phi} + 2c\frac{\cos\phi}{1-\sin\phi} = \gamma_s z \tan^2(45° + \phi/2) + 2c\tan(45° + \phi/2) \\ &= \gamma_s z N_\phi + 2c\sqrt{N_\phi}\end{aligned} \tag{2.20}$$

where $N_\phi = \tan^2(45° + \phi/2)$ and is called the flow value.

For passive failure, since the major principal stress σ_p is horizontal, the slip lines (or slip surfaces) on which failure takes place (or on which the shear stress τ satisfies the Mohr-Coulomb failure criterion) are sloped at an angle of $(45° - \phi/2)$ with the horizontal, as shown in Figure 2.14.

If a pressure q is applied to the terrain surface, which is usually referred to as the surcharge, then the normal stress at the base of the element at depth z is

$$\sigma = \gamma_s z + q \tag{2.21}$$

Accordingly, the passive earth pressure σ_p is given by

$$\sigma_p = \gamma_s z N_\phi + q N_\phi + 2c\sqrt{N_\phi} \tag{2.22}$$

Modelling the terrain as a plastic medium, together with the theory of plastic equilibrium, has found applications, for instance, in the prediction of the forces acting on a bulldozer blade and of the thrust developed by a lug (or grouser) of a cage wheel used in a paddy field, as shown in Figure 2.15.

Consider a vertical bulldozer blade or a lug of a wheel in the vertical position being pushed against the soil, due to the force applied to the blade from a tractor or due to a driving torque

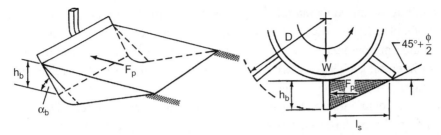

Figure 2.15: Interaction of a soil cutting blade and a grouser of a wheel with soil (Reprinted by permission from J.Y. Wong, *Theory of Ground Vehicles*, 4th Ed., Wiley, 2008, copyright © 2008 by John Wiley)

applied to the wheel, respectively. The soil in front of the blade or behind the lug will be brought into a state of passive failure. If the blade or the lug is relatively wide in comparison with the cutting depth of the blade or the penetration depth of the lug, the case may be treated as two-dimensional. Furthermore if the surface of the blade or of the lug is relatively smooth, then the normal pressure exerted on the soil by the blade or by the lug will be the major principal stress and equal to the passive earth pressure σ_p. If on the soil surface there is no surcharge q, the resultant horizontal force acting on the blade or on the lug F_p may be determined by integrating the passive earth pressure σ_p over the cutting depth of the blade or the penetration of the lug h_b and multiplying by the width of the blade or the lug b. From Figure 2.15,

$$F_p = b \int_0^{h_b} \sigma_p \, dz = b \int_0^{h_b} \left(\gamma_s z N_\phi + 2c \sqrt{N_\phi} \right) dz \qquad (2.23)$$

$$= b \left(\frac{1}{2} \gamma_s h_b^2 N_\phi + 2 c h_b \sqrt{N_\phi} \right)$$

If there is a surcharge q acting on the terrain surface in front of the blade or behind the lug, the resultant force F_p may be expressed by

$$F_p = b \int_0^{h_b} (\gamma_s z N_\phi + q N_\phi + 2c \sqrt{N_\phi}) \, dz \qquad (2.24)$$

$$= b \left(\frac{1}{2} \gamma_s h_b^2 N_\phi + q h_b N_\phi + 2 c h_b N_\phi \right)$$

If the ratio of width to cutting depth of a blade or that of width to penetration depth of a lug is relatively small, than the end effects may be significant and should be taken into account in predicting the resultant force acting on the blade or lug.

It should be pointed out that in general the lug on a wheel may behave in one of two ways. If the spacing between two adjacent lugs is sufficiently large that it enables the soil to fail in the manner shown in Figure 2.15, then the thrust developed by the lug in the vertical position may be estimated using either Eqn (2.23) or (2.24). On the other hand, if the spacing between two adjacent lugs is too small to allow the soil between them to fail in a manner shown in Figure 2.15, then shearing action will occur across the lug tips. Under these circumstances, the major function of the lugs would be to increase the effective diameter of the wheel.

If the soil behind the lug fails in the manner shown in Figure 2.15 and the width of the lug is the same as that of the rim, there will be a surcharge acting on the soil surface behind the lug due to the load applied by the wheel rim. In this case, Eqn (2.24) is applicable. For a cage wheel or a traction-aid device attached to a tyre, the rim is usually narrower than the lug. In this case, the benefit of the surcharge would not be obtained. It should be noted that the shearing forces developed on the vertical surfaces on both sides of the lug would increase the total thrust and that they should be taken into account when the penetration depth of the lug is relatively large, particularly on cohesive soil, such as clay (Wong, 2008).

■ Example 2.3

A traction-aid device with 18 lugs on a narrow rim is to be attached to the tyre of an off-road wheeled vehicle to increase its traction on wet soil. The outside diameter of the device measured from the lug tips is 1.5 m. The lugs are 0.25 m wide and penetrate 0.10 m into the soil at the vertical position. Estimate the thrust that a lug can develop at the vertical position in a clayey soil with $c = 25$ kPa, $\phi = 5°$ and $\gamma_s = 16$ kN/m³. The surface of the lug is relatively smooth and the friction and adhesion between the lug and the soil may be neglected.

Solution

The spacing between two adjacent lugs at the tip is 0.262 m. The rupture distance l_s shown in Figure 2.15 with a penetration depth $h_b = 0.10$ m is

$$l_s = \frac{h_b}{\tan(45° - \phi/2)} = 0.109 \text{ m}$$

This indicates that the spacing between two adjacent lugs is large enough to allow the soil between them to fail in a manner shown in Figure 2.15. Since the rim is narrow, the effect of surcharge may be neglected. Also the surface of the lug is relatively smooth, so the friction and adhesion between the lug and the soil may be neglected. The tractive force that can be developed by the lug in the vertical position is therefore given by

$$F_b = b \left(\frac{1}{2} \gamma_s h_b^2 N_\phi + 2 c h_b \sqrt{N_\phi} \right) = 1.388 \text{ kN}$$

As the wheel rotates, the inclination and the penetration depth of the lug changes. Thus, the thrust developed by a lug varies with its angular position. As more than one lug are in contact with the terrain at a given instance, the total thrust that the traction-aid device can develop is the sum of the horizontal forces acting on all the lugs in contact with the terrain; and it varies as the wheel rotates.

∎

There are limitations to the application of the simple passive earth pressure theory presented above to the solution of many practical engineering problems. For instance, the surface of a bulldozer blade or a wheel lug is usually not smooth, as assumed in the theory. As a result, there is friction and/or adhesion on the blade (or lug)–soil interface. It has been found that the angle of soil–metal friction δ may vary from 11° for a highly polished, chromium-plated steel with dry sand to almost equal to the angle of internal shearing resistance of the soil for a very rough steel surface (Osman, 1964). Because of the existence of friction and/or adhesion on the blade (or lug) surface, there are shear stresses acting on the interface. Consequently, the normal pressure on the interface is no longer a principal stress even for a vertical blade (or lug), and the failure pattern of the soil mass will not be the same as that shown in either Figure 2.14 or Figure 2.15. Furthermore, the bulldozer blade usually is not vertical and may have an angle α_b with the horizontal (usually referred to as the rake angle), as shown in Figure 2.16. The slip line fields in the failure zones in front of an inclined bulldozer blade with a rake angle α_b are shown in Figure 2.16. The soil in ABC is in the Rankine passive failure zone, which is characterized by straight slip lines inclined to the horizontal at an angle of $45°-\phi/2$. Zone ABD adjacent to the blade (or lug) is characterized by curved and radial slip lines and is usually referred to as the radial shear zone. The shape of the curved slip lines, such as BD in the figure may be considered as being either a logarithmic spiral (for frictional material) or an arc of a circle (for cohesive medium). On the blade (or lug)–soil interface, usually there are three force components: normal force F_{pn}, frictional force $F_{pn} \tan \delta$ (along the interface), and adhesive force F_{ca}, as shown in the figure. The resultant of the normal force and the frictional force acting on the interface is denoted by F_p in Figure 2.16. In the presence of friction and/or adhesion on the interface and/or with a rake angle, Eqn (2.23) or (2.24) can no longer be used to predict the forces acting on a bulldozer blade or on a wheel lug.

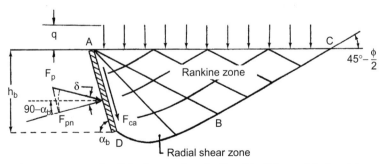

Figure 2.16: Failure patterns of soil in front of an inclined bulldozer blade with a rough surface

To address these issues, an analytical procedure, based on a rigorous mathematical solution applying the theory of plastic equilibrium, has been developed for predicting the force F_p acting at an angle δ to the normal of the interface shown in Figure 2.16 (Hettiaratchi and Reece, 1974). The analysis is based on the assumptions that the terrain is a rigid, perfectly plastic material and that the Mohr-Coulomb failure criterion applies. For a two-dimensional case where the contact width is much greater than the contact depth (for a bulldozer blade, this means that its width is much greater than its cutting depth), the force F_p is expressed by

$$F_p = b[\gamma_s h_b^2 K_\gamma + c h_b K_{ca} + q h_b K_q - \gamma_s h_b^2 K_s e^{-S}] \tag{2.25}$$

where b is the contact width of the blade (or lug), γ_s is the weight density of soil, h_b is the cutting depth of the blade (or penetration of the lug), c is cohesion, q is surcharge, S is the soil scale index equal to $(c + q)/(\gamma_s h_b)$, and K_γ, K_{ca}, K_q, and K_s are coefficients that are functions of the angle of internal shearing resistance of soil ϕ, interface friction angle δ, and rake angle α_b. The values of these coefficients may be obtained from the charts given in the reference (Hettiaratchi and Reece, 1974). The charts provide the values for the coefficients in Eqn (2.25) over the following ranges: $5° < \alpha_b < 170°$; $0 < \phi < 45°$; $0 < \delta < \phi$; and $0 < S <$ infinity.

Based on the charts, the values of K_γ, K_{ca}, K_q, and K_s for specific values of the angle of internal shearing resistance of the soil ϕ, the interface friction angle δ and the rake angle α_b are given in Tables 2.1, 2.2, 2.3 and 2.4, respectively.

Table 2.1: Coefficient K_γ

α_b		30°	50°	70°	90°	110°	130°	150°
$\phi = 30°$	$\delta = 0$	1.2	0.95	1.1	1.5	2.5	6.5	25
	$\delta = \phi$	1.7	1.7	2.3	3.5	7.0	20	90
$\phi = 15°$	$\delta = 0$	1.1	0.78	0.75	0.85	1.1	2.0	5.5
	$\delta = \phi$	1.22	0.95	0.95	1.15	1.7	3.1	9.0

Source: Hettiaratchi and Reece, 1974.

Table 2.2: Coefficient K_{ca}

α_b		30°	50°	70°	90°	110°	130°	150°
$\phi = 30°$	$\delta = 0$	0.5	0.95	1.9	3.4	6.3	13	31
	$\delta = \phi$	2	3.2	5.0	8.0	14	27	65
$\phi = 15°$	$\delta = 0$	0.5	0.95	1.7	2.6	4.2	7.3	14
	$\delta = \phi$	1.4	2.1	2.9	4.1	6.1	10	20

Source: Hettiaratchi and Reece, 1974.

Table 2.3: Coefficient K_q

α_b		30°	50°	70°	90°	110°	130°	150°
$\phi = 30°$	$\delta = 0$	2.3	1.9	2.2	3.0	4.8	8.8	20
	$\delta = \phi$	3.5	3.4	4.1	5.7	9.1	17	38
$\phi = 15°$	$\delta = 0$	2.2	1.6	1.5	1.7	2.2	3.2	6.0
	$\delta = \phi$	2.4	1.9	1.9	2.2	2.8	4.0	7.5

Source: Hettiaratchi and Reece, 1974.

Table 2.4: Coefficient K_s

α_b		30°	50°	70°	90°	110°	130°	150°
$\phi = 30°$	$\delta = 0$	0.53	0.18	0.07	–	0.25	2.5	15
	$\delta = \phi$	0.86	0.41	0.26	0.42	1.8	9.0	–
$\phi = 15°$	$\delta = 0$	0.51	0.17	0.04	–	0.09	0.65	3.5
	$\delta = \phi$	0.64	0.23	0.0.8	0.06	0.29	1.1	3.8

Source: Hettiaratchi and Reece, 1974.

The adhesive force on the interface F_{ca} is expressed by

$$F_{ca} = bc_a h_b / \sin \alpha_b \tag{2.26}$$

where c_a is adhesion on the interface between the blade (or lug) and soil.

Equations (2.25) and (2.26), together with the charts given in the reference or with Tables 2.1–2.4, provide an analytical framework for predicting the forces on the contact surface of a bulldozer blade (or lug) over a wide range of soil interface properties and rake angles.

■ Example 2.4

A bulldozer blade with width of 2.5 m and rake angle of 70° is used to remove a layer of dry, sandy soil of 0.3 m deep. The weight density of the soil γ_s is 16 kN/m³; its angle of internal shearing resistance ϕ is 30°; and its cohesion c is negligible. The blade surface is rough and the interface friction angle δ may be assumed equal to ϕ. As the soil is dry, the interface adhesion c_a may be neglected. Estimate the horizontal force required to push the bulldozer blade at the beginning of the bulldozing process.

Solution

At the beginning of the bulldozing process, there is no soil piled up on the horizontal soil surface in front of the blade, so the surcharge q is zero. With both q and c equal to zero, the soil scale index $S = (c + q)/(\gamma_s h_b) = 0$. As a result, for this case, Eqn (2.25) may be rewritten as

$$Fp = b[\gamma_s h_b^2 K_\gamma - \gamma_s h_b^2 K_s]$$

From Tables 2.1 and 2.4, for $\phi = \delta = 30°$ and $\alpha_b = 70°$, the value of K_γ is 2.3 and that of K_s is 0.26, therefore,

$$F_p = 2.5 \times 16 \times 0.3^2 (2.3 - 0.26) = 7.344 \text{ kN}$$

The horizontal force F_{ph} required to push the bulldozer blade is given by (referring to Figure 2.16)

$$F_{ph} = F_p \cos[\delta - (90° - \alpha_b)] = F_p \cos 10° = 7.232 \text{ kN}$$

∎

In many problems in vehicle–terrain interaction, the failure (flow) patterns of soil under the vehicle running gear are much more complex than those shown in Figure 2.14, 2.15, or 2.16. Figures 2.17–2.20 show the flow patterns of dry sand beneath a wheel under various operating conditions: driven, towed, spinning (100% slip), and locked (100% skid) (Wong and Reece, 1966; Wong, 1967). It can be seen that the flow patterns in the longitudinal plane depend on wheel slip or skid, among other factors. There are normally two flow zones under the wheel. In one zone, soil flows forward and in the other soil flows backward. These two zones degenerate into a single backward zone at 100% slip and a single forward zone for a locked wheel (100% skid). It can be seen that a wedge-shaped soil body is formed in front of a locked wheel; and it appears to be attached to the wheel and behaves like a vertical and rough bulldozer blade (with the friction angle δ of the interface equal to the angle of internal shearing resistance of the soil ϕ). Figure 2.21 shows the trajectories of clay particles beneath a rigid

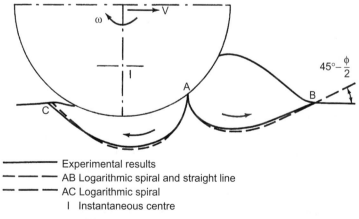

Figure 2.17: Flow patterns and bow wave under the action of a driven rigid wheel in sand

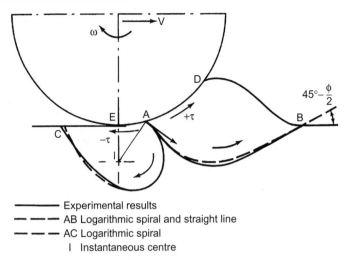

Experimental results
- - - - AB Logarithmic spiral and straight line
- - - AC Logarithmic spiral
I Instantaneous centre

Figure 2.18: Flow patterns and bow wave under the action of a towed rigid wheel in sand

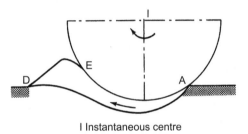

I Instantaneous centre

Figure 2.19: Flow patterns beneath a driven rigid wheel at 100% slip in sand

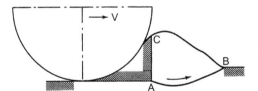

Figure 2.20: Flow patterns and soil wedge formed in front of a locked rigid wheel at 100% skid in sand

wheel under towed and driven conditions (Wong, 1967). The characteristics of the trajectories indicate that the soil at first is in the Rankine passive state, when it is in front of the oncoming wheel. As the wheel advances, the soil is driven backward. Under a free-rolling or towed wheel, the final position of a soil particle is in front of its initial position. On the other hand, under a driven wheel with slip, its final position is behind its initial position. The characteristics of the trajectories of soil particles shown in Figure 2.21 further confirm the existence of the two flow zones under a driven and a towed wheel, shown in Figures 2.17 and 2.18, respectively.

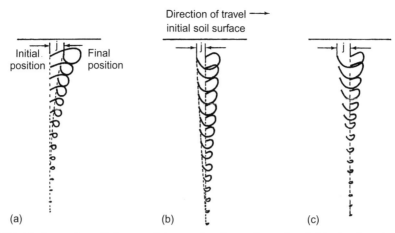

Figure 2.21: Trajectories of clay particles under the action of a rigid wheel under different operating conditions, (a) towed, (b) driven at 37% slip, and (c) driven at 63% slip

The publication of the research findings on soil flow patterns beneath a wheel described above has stimulated a great deal of interest in applying the theory of plastic equilibrium to the investigation of wheel–terrain interaction (Dagen and Tulin, 1969; Karafiath, 1971; Karafiath and Nowatzki, 1978). A set of equations that combine the differential equations of equilibrium of the soil mass, together with an appropriate failure criterion (such as the Mohr-Coulomb failure criterion), is first formulated. The boundary conditions, such as the friction angle (or more generally the direction of the major principal stress) on the wheel–soil interface, as well as the overall contact angle of the wheel and the separation angle of the front and rear failure zones shown in Figure 2.17 or 2.18, are then specified. The solutions to the set of differential equations with specified boundary conditions yield the geometry of the slip lines and associated stresses within the soil mass. As an example, Figure 2.22(a) and (b) show the slip line fields in the front and rear failure zones of the soil beneath a rigid wheel under driven and towed conditions, respectively (Karafiath, 1971).

It should be pointed out that the boundary conditions on the wheel–soil interface vary with design parameters and operating conditions of the wheel, as well as terrain characteristics. This makes it very difficult, if not impossible, to specify appropriate boundary conditions at the outset, as required by the solution process noted above. So far, the approach used to specify the boundary conditions is primarily based on empirical data and simplifying assumptions (Karafiath and Nowatzki, 1978). For instance, in the prediction of the performance of driven rigid wheels (or driven tyres), the value of the interface friction angle is assumed constant throughout the contact patch. On the other hand, in the prediction of the performance of towed rigid wheels, the interface friction angle is assumed to decrease linearly from the entry or exit point of wheel contact towards the centre, and to reduce to zero at the point of separation between the front and rear flow zones shown in Figure 2.18. The value of the interface

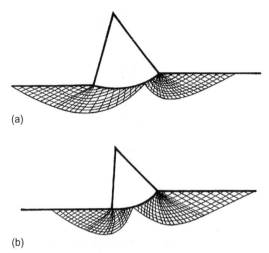

Figure 2.22: Slip line field in soil predicted using the theory of plastic equilibrium under (a) a driven rigid wheel and (b) a towed rigid wheel (Reprinted by permission of *ISTVS* from Karafiath, 1971)

friction angle at the entry point of wheel contact is assumed to be in the range of 1/3 to 1/2 of the angle of internal shearing resistance of the soil, while the same value but with opposite sign is assumed at the exit point of wheel contact. In the analysis of the mechanics of towed tyres, on the other hand, the value of the interface friction angle at the entry and exit points of tyre contact is assumed to be 1/4 of the angle of internal shearing resistance of the soil. Furthermore, the selection of some of the input parameters required for initiating the solution process, such as the rear contact angle for towed rigid wheels, the angle defining the deflected part of a tyre, or the angle defining the location where the tyre returns to its undeflected shape, is somewhat arbitrary (Karafiath and Nowatzki, 1978). This indicates that the elaborate solution procedure, based on the theory of plastic equilibrium, for predicting the performance of rigid wheels or tyres rely heavily on either empirical data or assumed boundary conditions on the wheel (tyre)–soil interface, some of which do not necessarily have any justifiable theoretical basis (Wong, 1972b, 1979).

In summary, the modelling of the terrain as a plastic medium, together with the theory of plastic equilibrium, has found applications in estimating the maximum traction of an off-road vehicle, the forces acting on a bulldozer blade, or the thrust developed by a lug on a wheel, etc. It is not capable, however, of predicting the deformation of the terrain caused by vehicular load. Furthermore, the theory of plastic equilibrium is based on the assumption that the terrain behaves like a rigid, perfectly plastic material. This means that the terrain does not deform significantly until the stresses within certain boundaries reach a certain level at which failure occurs. While dense terrain, such as compact sand and the like, may exhibit behaviour similar to that of a rigid, perfectly plastic medium, a wide range of natural terrains encountered in off-road operations, such as snow or organic terrain, have a high degree of compressibility and their behaviour does not conform to that of a rigid, perfectly plastic medium. Consequently, there are severe limitations to the applications of the theory of plastic equilibrium to the evaluation of vehicle–terrain interaction and to the prediction of vehicle performance in the field.

2.3 Modelling Terrain Behaviour based on the Critical State Soil Mechanics

As noted previously, modelling the soil as an elastic medium or as a rigid, perfectly plastic material has limitations in its applications to many practical engineering problems. In an attempt to overcome these limitations, the concept of critical state soil mechanics was developed by Roscoe and his associates at Cambridge University (Roscoe et al., 1958; Schofield and Wroth, 1968; Kurtay and Reece, 1970).

In classical soil mechanics the soil is usually described as dense or loose. However, these two states are not generally described quantitatively. The classical soil mechanics theories usually only apply to soil in the dense state, while soil in the loose state is dealt with more or less empirically. The critical state soil mechanics attempts to cover the soil behaviour in the whole range of states. In essence, the critical state soil mechanics establishes the relationship between the specific volume V, spherical pressure P, and the deviatoric stress R of the soil. The specific volume V is equal to $1 + e$, where e is the void ratio of the soil, which is the ratio of the volume of the voids to the volume of the solids. The spherical pressure P and the deviatoric stress R are defined as follows:

$$P = \frac{\sigma_1 + \sigma_2 + \sigma_3}{\sqrt{3}} \tag{2.27}$$

and

$$R = \frac{1}{\sqrt{3}} \left[(\sigma_1 - \sigma_2)^2 + (\sigma_2 - \sigma_3)^2 + (\sigma_3 - \sigma_1)^2 \right]^{1/2} \tag{2.28}$$

where σ_1, σ_2 and σ_3 are principal stresses acting on a cubic element of the soil.

There are four basic assumptions in the critical state soil mechanics:

(a) The soil is homogeneous and isotropic.

(b) The mechanical behaviour of soil depends only on effective stress, which is defined as the difference between the total stress and the pore water pressure (usually referred to as pore pressure). The presence or absence of pore pressure or moisture tension has no effect except in so far as they alter the effective stress.

(c) The mechanical behaviour of soil can be described by a macroscopic model. It is not necessary to relate soil behaviour to the properties of, and interactions between, individual particles.

(d) The mechanical behaviour of soil is not time dependent, and the soil is not viscous.

Based on experimental observations, it is found that for a given spherical pressure P and at a particular deviatoric stress R, the soil can reach a state known as the critical state at which any further change in specific volume will not occur with the further increase of deviatoric stress R. Distortion may, however, continue. The combination of the spherical pressure P, deviatoric stress R and specific volume V at which the soil reaches the critical state can be identified by a point, known as the critical state point, in the three-dimensional space of P-R-V. By connecting a family of critical state points for different combinations of P, R and V, a critical state line can be traced in the P-R-V space, as shown in Figure 2.23. Two regions, separated by the vertical wall beneath the critical state line and perpendicular to the P-V plane, can be identified, as shown in Figure 2.23. In these two regions, the soil material exhibits two different types of behaviour. On one side of the wall remote from the origin of the P-R-V space,

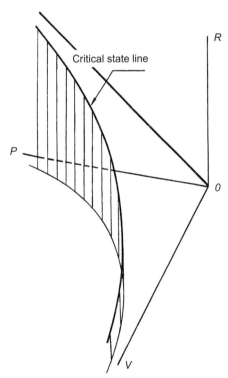

Figure 2.23: Critical state line in the *P-R-V* space (Reprinted by permission of *ISTVS* from Kurtay and Reece, 1970)

the material is initially loose. After being compressed under load, it becomes stronger. In this case, the soil is treated as a work-hardening material and its deformation under load may be predicted. On the other side of the wall, the material is relatively dense. When subject to load, it fails in a brittle manner and becomes weaker. Dilation of the material will occur on the thin failure plane. The rest of the material, however, tends to move as a solid block. This makes it difficult to prediction of the overall deformation of the soil under these circumstances. These two types of behaviour, in terms of stress–strain relationship, may be illustrated in Figure 2.24. The line identified by '*a*' represents the stress–strain relationship of the material in the dense state, while line '*b*' represents the behaviour of the material initially in a loose condition.

The critical state soil mechanics represents an advancement in the understanding of soil behaviour over a wide range of conditions. It provides an analytical framework within which the relationship between the load applied to soil and the resulting deformation may be studied. However, applying critical state soil mechanics to the prediction of off-road vehicle performance in the field, such as predicting vehicle sinkage due to normal load or slip of the running gear due to sprocket torque, faces significant challenges. One of them is that the critical state soil mechanics is based on the assumption that the soil is homogeneous and isotropic. In practice, however, an off-road vehicle may encounter a variety of natural terrains. These include snow or organic terrain, such as tundra or muskeg, which under most circumstances cannot possibly be idealized as homogeneous and isotropic. As a result, the critical

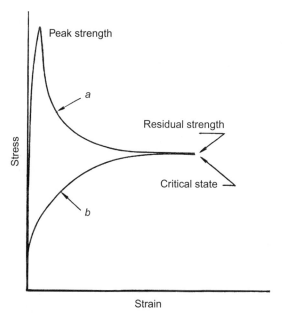

Figure 2.24: Stress–strain relationships of soil in (a) dense state and (b) loose state (Reprinted by permission of *ISTVS* from Kurtay and Reece, 1970)

state soil mechanics has so far found few practical applications to the study of vehicle–terrain interaction in the field.

2.4 Modelling Terrain using the Finite Element Method (FEM)

The finite element method (FEM) was initially developed for structural analysis. Over the years, its applications have been expanded to cover many fields in engineering. The application of the finite element method to the study of vehicle–terrain interaction was initially stimulated by the need for a numerical method to examine in detail the stress distribution and soil deformation under the loading of a tractor (Perumpral et al., 1971). Since then the finite element method has been applied to the study of the interactions between tyre and soil, tyre and snow, track and soil, etc.

The basic concept of the finite element method is the idealization of a continuum (such as soil mass under certain circumstances) as an assemblage of a finite number of elements. For a two-dimensional continuum, the finite elements may take the form of a triangle or quadrilateral. For three-dimensional analysis, the finite elements may be in the form of a tetrahedral, rectangular prism, or hexahedra. Triangular or quadrilateral shell or membrane elements are usually used in modelling tyre carcasses. These elements are interconnected at joints which are called nodes or nodal points. Simple mathematical functions, such as polynomials, are usually used to approximate the distribution of the displacements within each element. To describe the behaviour of the element under load, the stiffness matrix is formulated for each element. The stiffness relates the displacement to the applied force at the nodal point. The stiffness matrix consists of the coefficients of the equilibrium equations derived from the material and geometric properties of an element and usually obtained using the variational principle of mechanics (such as the principle of minimum potential energy). The assembly of the overall stiffness matrix for the entire body at issue is carried out from the individual element stiffness matrices. The assembly of the overall force or load vector for the whole body is made from the element nodal force vectors. For a given load applied to the continuum, the equilibrium equations noted above are solved for the unknown displacements of the nodal points. Based on the predicted displacements of the nodal points, the deformation of the elements can be computed. With the properties of the material known or given, the stresses and strains of the elements can then be determined. The finite element method offers the flexibility of assigning different mechanical properties to different elements, thus providing a means to examine engineering problems involving non-homogeneous or anisotropic media.

Early work on applying the finite element method to the study of wheel (tyre)–soil interaction assumes that the soil is either a linear or non-linear elastic continuum. The normal and shear stress distributions on the wheel–soil interface are required as input for initiating the solution process (Perumpral et al., 1971; Yong and Fattah, 1976). It should be pointed out that when the normal and shear stress distributions on the wheel–soil interface are specified, the performance of the wheel, expressed in terms of motion resistance, thrust, drawbar pull and tractive efficiency, is completely defined (Wong, 1977). Thus, in the early work the role of the finite element method is restricted to predicting the stress distribution and deformation in the soil mass for given normal and shear stress distributions on the wheel–soil interface.

In recent years, research on the applications of the finite element method to the study of wheel–terrain interaction has advanced considerably. To accommodate various types of terrain behaviour, a number of constitutive models have been introduced. Pressure-dependent elastoplastic models are considered to be appropriate representations of the behaviour of terrain materials, because of the existence of inelastic deformation of the terrain when subject to normal pressure and/or shear stress on the wheel–soil interface. Among the constitutive models, the Drucker-Prager cap model and the Cam-Clay critical state soil model are two that are widely used, although the Mohr-Coulomb yield model is also employed in some cases (Liu and Wong, 1996; Liu et al., 1999; Seta et al., 2003; Fervers, 2004; Zhang et al., 2005). A number of comprehensive finite element codes considered to be applicable to the study of machine–terrain interaction have become widely available in recent years. All of these have facilitated the advancements in the applications of the finite element method to the investigation of vehicle–terrain interaction.

The finite element technique has now advanced to the point that it can be applied to predicting the geometry of the contact patch and the normal and shear stress distributions on the wheel–terrain interface for a given load and wheel slip, hence the tractive performance of the wheel. Its application is, therefore, no longer restricted to only predicting the stress distribution and deformation in the soil mass for a given stress boundary condition, as in its early stage of development. For instance, in a two-dimensional study of wheel–sand interaction, a modified Cam-Clay critical state soil model, in conjunction with a new non-linear elastic law, has been implemented in the finite element program MARC (Liu and Wong, 1996; Liu, Wong and Mang, 2000). Since large soil deformation occurs, due to plastic deformation and localized failure, the soil model is implemented within the framework of large strains. Material parameters for the soil have been calibrated with experimental data from hydrostatic and triaxial tests. Simulations of the performance of a rigid wheel on loose sand and a 12.5/75R20 tyre on Ottawa sand have been conducted (Liu and Wong, 1996; Liu, Wong and Mang, 2000). In both cases, a vertical load is applied and then the wheel (tyre) rotates and moves horizontally at a low speed and at a given slip, until a steady-state condition is reached.

Figure 2.25 shows the deformed patterns of the finite element mesh and the vertical stress on loose sand under a rigid wheel with diameter of 1.245 m, width of 0.305 m and normal load of 9.28 kN. Figure 2.26 shows the variations of the predicted drawbar pull with horizontal movement from standstill for the rigid wheel at five different slips. Figure 2.27 shows a comparison of the measured relationship between the steady-state drawbar pull coefficient (i.e. the ratio of the drawbar pull to normal load) and slip of the rigid wheel with the predicted one, obtained

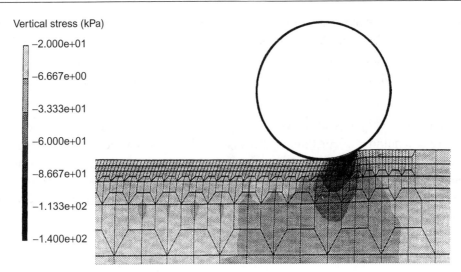

Figure 2.25: Finite element mesh and its deformed patterns and the distribution of vertical stress on loose sand beneath a driven rigid wheel at 3.1% slip (Reprinted by permission of *ISTVS* from Liu and Wong, 1996)

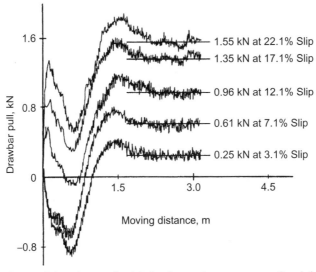

Figure 2.26: Variations of drawbar pull with horizontal movement of a rigid wheel at various slips on loose sand predicted using the finite element method (Reprinted by permission of *ISTVS* from Liu and Wong, 1996)

using the finite element method, for the rigid wheel on loose sand. It is shown that there is a reasonably close agreement between them. It should be mentioned, however, that there is only a qualitative agreement between the measured and predicted normal and shear stress distributions on the interface, particularly with respect to the locations of the maximum normal and shear stresses (Liu and Wong, 1996; Liu, Wong and Mang, 2000).

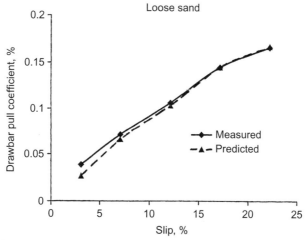

Figure 2.27: Comparison of the measured drawbar pull coefficient of a rigid wheel on loose sand with the predicted one obtained using the finite element method (Reprinted by permission of *ISTVS* from Liu and Wong, 1996)

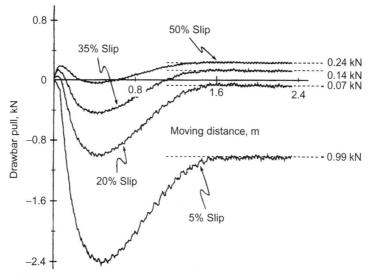

Figure 2.28: Variations of drawbar pull with horizontal movement of a tyre at various slips on Ottawa sand predicted using the finite element method (Reprinted by permission of *ISTVS* from Liu and Wong, 1996)

Figure 2.28 shows the variations of the predicted drawbar pull with horizontal movement from standstill for a 12.5/75R20 tyre at an inflation pressure of 345 kPa and a normal load of 9.23 kN at four different slips on the Ottawa sand. Figure 2.29 shows a comparison of the measured relationship between the steady-state drawbar pull coefficient and slip of the tyre with the predicted one, obtained using the finite element method. It is shown that there is a reasonably close agreement between them (Liu and Wong, 1996; Liu, Wong and Mang, 2000).

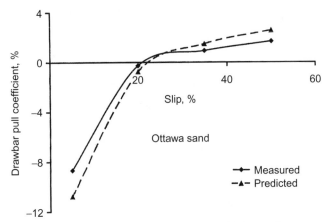

Figure 2.29: Comparison of the measured drawbar pull coefficient of a tyre on Ottawa sand with the predicted one obtained using the finite element method (Reprinted by permission of *ISTVS* from Liu and Wong, 1996)

Another example is the application of the finite element technique to the study of tyre–snow interaction. To provide guidance for improving tread pattern designs of winter tyres for passenger cars, a three-dimensional analysis of the mechanics of tyre–snow interaction has been performed (Seta et al., 2003). In the study, the tyre is modelled using the finite element method, while the snow is modelled using the finite volume method (FVM). Coupling between the tyre and snow is automatically computed by the coupling element. The snow is assumed to be a homogeneous elastoplastic material. The Mohr-Coulomb yield model is adopted. The shear strength parameters of the snow are: cohesion of 0.016 mPa and angle of internal shearing resistance of 31°. The tractive performance of a 195/65R15 tyre on snow, with a normal load of 4.0 kN, at an inflation pressure of 200 kPa, and operating at a speed of 60 km with 30% slip, has been simulated using the computer program MSC.Dytran. The number of elements used to represent the tyre carcass in three dimensions is nearly 60 000. The number of three-dimensional elements used for the snow domain is approximately 70 000. A supercomputer has been used to perform the simulations. Tables 2.5 and 2.6 show the measured and predicted traction forces of the 195/65R15 tyre with different groove widths and groove angles, respectively. Despite a noticeable difference between the predicted and measured traction forces in terms of absolute value, predictions are considered to be in qualitative agreement with experimental data, when comparing the predicted and measured traction forces in terms of index, as shown in the tables. It appears that the technique could be employed to evaluate the effects of design parameters on the traction of winter tyres on a relative basis.

In addition to applying the finite element method to the study of wheel (tyre)–terrain interaction, efforts have also been made in applying the technique to the study of track–terrain interaction (Karafiath, 1984). A track is a complex mechanical system. For instance, a segmented metal track, commonly used in agricultural and industrial tractors and in the current generation

Table 2.5: Comparison of the predicted and measured tyre traction forces for different groove widths on snow

Groove width	Predicted traction force			Measured traction force		
	kN	lb	Index	kN	lb	Index
Narrow	0.60	135	100	0.80	180	100
Wide	0.65	146	108	0.89	200	111

Source: Seta et al. (2003).

Table 2.6: Comparison of the predicted and measured tyre traction forces for different groove angles on snow

Groove angle, degrees	Predicted traction force			Measured traction force		
	kN	lb	Index	kN	lb	Index
0*	0.69	155	100	0.94	211	100
30**	0.65	146	94	0.92	207	97
−30***	0.68	153	98	0.93	209	99

Source: Seta et al. (2003).
*Straight grooves parallel to the tyre rotating axis.
**V-shape grooves in the direction of tyre rotation, with both sides at 30° with the tyre rotating axis.
***Reverse V-shape grooves in the direction of tyre rotation, with both sides at 30° with the tyre rotating axis.

of military fighting or logistic vehicles, consists of a number of metal links connected by pins (or with rubber bushings for military vehicles). It interacts with the roadwheels and through them with the suspension system, in addition to the terrain. Its operation is influenced by a number of design and operational parameters, such as roadwheel arrangements, suspension characteristics, initial track tension, locations of the sprocket and idler, configuration of supporting rollers for the top run of the track, etc. To make the analysis amenable to the finite element method, however, the track system has to be greatly simplified. For instance, in some studies, the track is idealized as a rigid footing. Furthermore, to predict the stress distribution and deformation of the soil mass under the track, the normal pressure and shear stress distributions on the track–terrain interface are required as input to the solution process at the outset, similar to that in the early stage of development of the application of the finite element technique to the wheel–soil interaction study discussed previously. This overlooks the crucial issue of relating the stress distributions on the interface to vehicle design and operational parameters and terrain characteristics.

In the study by Karafiath (1984), simplified normal pressure distributions on the interface, such as uniform and trapezoidal forms of pressure distributions, are used. Ratios of shear stress to normal pressure are specified. As will be shown later, idealizing the track as a rigid footing or with uniform or trapezoidal forms of normal pressure distribution, with fixed ratios of shear stress to normal pressure, are oversimplifications and unrealistic in most cases. To represent the behaviour of cohesive clay used in the study, the Ramberg-Osgood model, originally for modelling the strain hardening behaviour of metal, is used. A computer code known as the Dynamic Crash Analysis of Structures (DYCAST), originally developed for the evaluation of the crashworthiness of vehicles (or aircraft), is employed. Figure 2.30 shows the finite element mesh for the clay and its deformed patterns under a track 0.38 m wide and 3.05 m long, with simplified normal pressure distribution of trapezoidal form (with high and low pressures of 1.5 and 0.2 times of the average pressure of 51.5 kPa, respectively), at a ratio of horizontal load (thrust) to vertical load of 0.3, and moving at a speed of 9.14 m/s (Karafiath, 1984). It is claimed that using the finite element technique, the effects of normal pressure distribution and track forward speed on soil deformation, motion resistance, and thrust of the track can be predicted. It should be pointed out, however, that the methodology used in the study does not address the basic needs of the vehicle engineer for predicting vehicle performance using vehicle design parameters and terrain characteristics as direct inputs, rather than based on assumed normal and shear stress distributions on the interface.

From the examples presented above, as well as other research results reported in the literature, it appears that the capability of the finite element method to investigate certain aspects of the physical nature of wheel (tyre)–terrain interaction in detail have been demonstrated. It is shown that the technique could be a useful tool for comparing wheel (tyre) performances or designs on a relative basis on certain types of terrain.

For a track system, its interaction with the terrain is greatly influenced by a large number of design and operational factors. The assumption that a track is equivalent to a rigid footing or has uniform or trapezoidal form of normal pressure distribution with fixed ratios of shear stress to normal pressure is inadequate and unrealistic in most cases. Consequently, the results of simulations using the finite element technique based on oversimplifying assumptions for the track system would not be useful to the vehicle designer in realistically evaluating the performance or design of tracked vehicles.

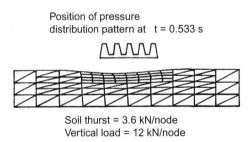

Figure 2.30: Finite element mesh and its deformed patterns beneath an idealized track with normal pressure distribution of trapezoidal form on clay (Reprinted by permission of *ISTVS* from Karafiath, 1984)

Further improvements in the applications of the finite element method to the study of vehicle–terrain interaction are generally required. These include improvements in modelling terrain behaviour, particularly in the natural environment, in the acquisition of reliable data on terrain model parameters, etc. It should be noted that conducting a realistic three-dimensional simulation of wheel (or track)–terrain interaction using the finite element technique requires significant computing resources.

It should be pointed out that the finite element method is developed on the premise that the terrain is a continuum and that the general principles of continuum mechanics are applicable. As a consequence, it has inherent limitations in simulating significant soil flow or large, discontinuous soil deformation. Significant soil flow beneath a vehicle running gear, such as a wheel, has been observed, as shown in Figures 2.17–2.20. To examine the interaction between a wheel (or track) and terrain, particularly granular terrain, such as sand and the like, the discrete (distinct) element method has been introduced in recent years. The basic principles of the discrete element method are discussed in the next section.

2.5 Modelling Terrain Behaviour using the Discrete (Distinct) Element Method (DEM)

The discrete (distinct) element method (DEM) was initially developed for the study of rock mechanics, as an alternative to modelling a granular (particulate) material as a continuum (Cundall and Strack, 1979). Since then its applications have been extended to other fields of engineering, including the study of vehicle–terrain interaction.

The basic concept of the discrete element method is the representation of soil (terrain) as an assemblage of a number of discrete elements. For two-dimensional analysis, the discrete elements are usually assumed to be of circular shape, although elements of other shapes, such as clumps of two circular elements joined together, elliptical shape elements, etc., have been used. For three-dimensional analysis, the discrete elements are often assumed to be of spherical form, although elements of other forms, such as ellipsoids, have been employed. By such an idealization, it becomes possible to analyse the characteristics of wheel (track)–soil interaction by examining the mechanical interactions between the wheel (track) and adjacent elements and those between the contacting elements. Elements in contact with the wheel (track) surface receive contact forces from the wheel (track). Elements not in contact with the wheel (track) surface receive contact forces from other contacting elements. The magnitudes of the contact forces are assumed to be related to the relative displacement and relative velocity of the contacting elements, dependent upon the model for the mechanical property of the elements used.

The discrete element method in its basic form assumes that each element has stiffness (in each of the normal and tangential directions) characterized by a spring constant k and possesses damping (in each of the normal and tangential directions) characterized by a viscous

56 Chapter 2

damping coefficient η. Between contacting elements or between the wheel (track) surface (or container wall) and adjacent elements, it is assumed that friction exists in the tangential direction, which is characterized by a coefficient of friction μ. The maximum tangential force is limited by the product of the coefficient of friction and the normal force.

The interactions between two contacting elements or between the wheel (track) surface (or container wall) and an adjacent element are schematically shown in Figure 2.31 (Tanaka et al., 2000). It should be noted that the spring elements in both normal and tangential directions shown in the figure cannot sustain tensile forces. In other words, between contacting elements or between the wheel (track) surface (or container wall) and adjacent elements, there are no-tension joints. This indicates that the cohesion between contacting elements or the adhesion between the wheel (track) surface (or container wall) and adjacent elements are not included in the models shown in Figure 2.31.

For the models shown in Figure 2.31, the interacting forces between two contacting elements are expressed by

In the normal direction:

$$F_{np} = k_{np} u_{np} \pm \eta_{np} v_{np} \tag{2.29}$$

where F_{np} is the normal (compressive) force between two contacting elements, k_{np} is the stiffness in the normal direction, u_{np} is the deformation (relative displacement) in the normal direction, η_{np} is the viscous damping coefficient in the normal direction, and v_{np} is the relative velocity in the normal direction. If the spring force $k_{np} u_{np}$ (assumed to be compressive) and the viscous damping force $\eta_{np} v_{np}$ are in the same direction, then the positive sign in Eqn (2.29) should be used; otherwise the negative sign should be applied.

In the tangential direction:

$$F_{sp} = k_{sp} u_{sp} \pm \eta_{sp} v_{sp} \quad \text{for } k_{sp} u_{sp} \pm \eta_{sp} v_{sp} < \mu F_{np} \tag{2.30}$$

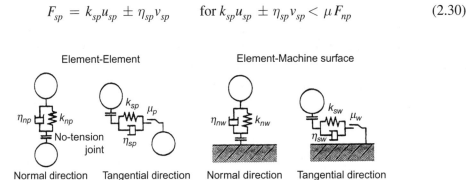

Figure 2.31: Representations of the interactions between two contacting discrete elements and between the wheel (track) surface (or container wall) and a discrete element (Reprinted by permission of *ISTVS* from Tanaka et al., 2000)

$$\text{or } F_{sp} = \mu F_{np}, \quad \text{for } k_{sp}u_{sp} \pm \eta_{sp}v_{sp} \geq \mu F_{np} \qquad (2.31)$$

where F_{sp} is the tangential force between two contacting elements, k_{sp} is the stiffness in the tangential direction, u_{sp} is the deformation (relative displacement) in the tangential direction, η_{sp} is the viscous damping coefficient in the tangential direction, v_{sp} is the relative velocity in the tangential direction, and μ is the coefficient of friction in the tangential direction. If the spring force $k_{sp}u_{sp}$ and the viscous damping force $\eta_{sp}v_{sp}$ are in the same direction, then the positive sign in Eqn (2.30) should be used; otherwise the negative sign should be applied.

The expressions for the interacting forces between the wheel (track) surface (or container wall) and an adjacent element in both the normal and tangential directions can be derived in a similar way.

In some models, such as those for dry granular materials, the viscous damping in both normal and tangential directions, shown in Figure 2.31, is replaced by the Coulomb damping. The Coulomb damping is a type of mechanical damping in which energy is dissipated through friction. The internal hysteresis loss due to deformation of the element can be represented by the Coulomb damping.

As noted previously, in the models shown in Figure 2.31, the interacting forces in both normal and tangential directions can only be compressive. As a result, the effects of cohesion between contacting elements or adhesion between the wheel (track) surface (or container wall) and adjacent elements cannot be simulated. The micro-mechanism for the formation of cohesion between soil particles has been discussed by Zhang and Li (2006).

To represent the effects of cohesion or adhesion, tensile springs are used. Figure 2.32(a) and (b) show the models used to represent the cohesion between contacting elements in the normal and tangential direction, respectively.

The cohesive forces between two contacting elements in the model shown in Figure 2.32 are given by (Asaf et al., 2006)

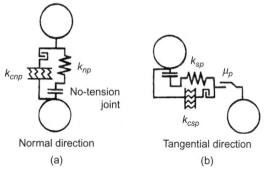

Figure 2.32: Representation of cohesion (adhesion) between two contacting discrete elements

In the normal direction:

$$F_{cnp} = k_{cnp} u_{np} A \quad \text{for } C_{np} \geq k_{cnp} u_{np} \quad (2.32)$$

$$\text{or } F_{csp} = 0 \quad \text{for } C_{np} < k_{cnp} u_{np} \quad (2.33)$$

where F_{cnp} is the cohesive force between two contacting elements in the normal direction, k_{cnp} is the stiffness of the tensile spring per unit contact area in the normal direction, u_{np} is the deformation (relative displacement) in the normal direction, A is contact area, and C_{np} is the maximum contact stress representing the cohesion in the normal direction. When the contact stress reaches the value of C_{np}, the bond due to cohesion between two contacting elements in the normal direction is broken.

In the tangential direction:

$$F_{csp} = k_{csp} u_{sp} A \quad \text{for } C_{sp} \geq k_{csp} u_{sp} \quad (2.34)$$

$$\text{or } F_{csp} = 0 \quad \text{for } C_{sp} < k_{csp} u_{sp} \quad (2.35)$$

where F_{csp} is the cohesive force between two contacting elements in the tangential direction, k_{csp} is the stiffness of the tensile spring per unit contact area in the tangential direction, u_{sp} is the deformation (relative displacement) in the tangential direction, A is contact area, and C_{sp} is the maximum contact stress representing the cohesion in the tangential direction. When the contact stress reaches the value of C_{sp}, the bond due to cohesion between two contacting elements in the tangential direction is broken.

The expressions for the adhesive forces between the wheel (track) surface (or container wall) and an adjacent element in both the normal and tangential directions can be derived in a similar manner.

To determine whether contact occurs between two elements or between the wheel (track) surface (or container wall) and an adjacent element, their geometrical relationships are examined. If the distance between two elements is less than the sum of the radii of these two elements, contact between them is considered to have occurred. On the other hand, if the distance between the wheel (track) surface (or container wall) and the centre of an adjacent element is less than the radius of the element, contact between them is considered to have been established.

The positions of each element are determined step by step at selected time intervals. The contact forces on each element determine the motion of the element. As an example, for two-dimensional analysis, Figure 2.33 shows the positions of two contacting elements i and j at time t (Tanaka et al., 2000). The direction of motion of each element during time Δt is indicated

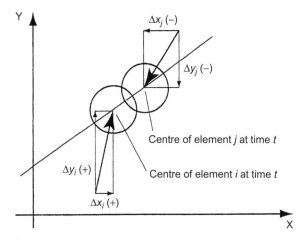

Figure 2.33: Directions of motion and displacement of two contacting discrete elements (Reprinted by permission of *ISTVS* from Tanaka et al., 2000)

by an arrow. Displacements of elements i and j in the X and Y directions during time Δt are expressed by Δx_i, Δx_j Δy_i, and Δy_j, respectively.

As noted previously, contact forces acting on an element are assumed to be related to the relative displacement and relative velocity between two contacting elements (if viscous damping is assumed) or between the wheel (track) surface (or container wall) and an adjacent element. For instance, the normal component of the relative displacement between two contacting elements determines the normal force due to elastic deformation. The normal component of the relative velocity determines the normal component of the viscous damping force. The tangential component of the relative displacement induces a tangential force due to elastic deformation. The tangential component of the relative velocity determines the tangential component of the viscous damping force. The resultant tangential force, however, cannot exceed the maximum frictional force between two contacting elements. For two-dimensional analysis, based on the resultant force acting on the element from all other contacting elements, three equations of motion for each element can be established: two for the linear motions of the mass centre in the X and Y directions and one for the rotation about the mass centre of the element. From these equations of motion, the linear acceleration components of the mass centre in the X and Y directions and the angular acceleration about the mass centre of the element can be determined. Integrating the accelerations over the time interval and adding the resulting values to the velocities at the previous step, the velocity of the element in each direction at the current step can be obtained. To enhance the stability of solution, the incremental displacements during the time interval are taken as the average of the incremental displacements at the previous step and the integrations of the velocities over the time interval.

Repeating these procedures for all elements involved and over the duration specified, the interacting forces on the wheel (track)–soil interface can be determined and the movements of soil particles (represented by the discrete elements) under the action of the wheel (track) can be identified. Usually, the solution process is initiated by specifying the vertical load, forward speed and slip of the vehicle running gear.

The discrete element method has been applied to the investigations into the interactions between vehicle running gear and soil. For instance, a two-dimensional simulation of the interaction between a lugged, rigid wheel for a lunar microrover and the lunar soil (regolith) stimulant has been performed. Figure 2.34 shows the simulation of the lugged, rigid wheel operating on the lunar soil stimulant represented by discrete circular elements (Nakashima et al., 2007). In the simulation, the mechanical behaviour of the elements is represented by the models shown in Figure 2.31. Parameters of the lugged, rigid wheel and of the elements used are given in Table 2.7. It is found that there is a qualitative agreement between the test results and predictions obtained using the discrete element method, with respect to the effects of lug height, lug thickness, number of lugs, and wheel diameter on traction (Nakashima et al., 2007). Applications of the discrete element method to the study of track link–soil interaction and of mine plough–soil interaction, etc. have also been made (Asaf et al., 2006; Horner et al., 2001).

It appears that at present, the discrete element modelling technique as applied to the study of vehicle–terrain interaction is still in its developmental stage. Several issues must be resolved before the technique can be considered as a practical tool. Some of these issues are outlined below:

1. Currently, the model with spring-damper (with viscous or Coulomb damping) is widely used to represent the mechanical behaviour of soil particles, primarily because of its simplicity. Elements of simple shape are employed and are assumed to be homogeneous. However, soil particles in the natural environment are seldom of simple shape and homogeneous. As a result, the behaviour of soil particles, such as the interlocking mechanism between particles, may not be properly represented. The size of the elements

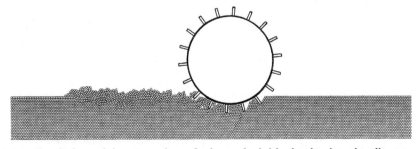

Figure 2.34: Simulation of the operation of a lugged, rigid wheel using the discrete element method (Reprinted by permission of *ISTVS* from Nakashima et al., 2007)

used in many simulations is often of an order of magnitude larger than the actual size of soil particles (see Table 2.7), primarily due to the constraint of computing resources available. It has been shown that this produces an error of scale, unless a one-to-one correspondence in size is maintained between the elements and actual soil particles (Horner et al., 2001). To improve the discrete element modelling technique, better understanding of the micromechanics of soil particle interaction is of importance.

2. The values of model parameters representing the behaviour of soil particles are often assumed or derived from the Hertz contact theory or from the elastic properties (i.e. the Young's modulus and Poisson's ratio) of the bulk soil material. At the present time there is a lack of generally accepted and robust methods for determining the values of model parameters, particularly those representing terrain in the natural environment. This poses one of the greatest challenges to practical applications of the discrete element modelling technique, particularly in view of the variability and complexity of terrain behaviour in the natural environment.

3. In many studies of vehicle–terrain interaction using the discrete element method performed so far, the number of elements used in the simulation is rather small

Table 2.7: Parameters of the wheel and of the element used in the simulation shown in Figure 2.34

Parameters	Values
Number of elements used in simulation	6986
Diameter of element, mm	4.0
Mass density of element, g/cm^3	1.6
Diameter of wheel, mm	200
Mass of wheel, g	500
Diameter of lug element, mm	2.5
Vertical load on wheel, N	9.8, 19.6
Traction load on wheel, N	Varied
Angular velocity of wheel, rad/s	0.138
Duration of soil consolidation, s	1.0
Duration of vertical sinkage, s	1.0
Simulation time for wheel travel, s	45.0
Time increment, s	0.0001
Normal spring constants*, N/m	10,000
Tangential spring constants*, N/m	500
Normal damping coefficient*, N s/m	8.97
Tangential damping coefficient*, N s/m	2.01
Friction coefficient between soil elements	0.9
Friction coefficient for wheel, lug and wall contact**	0.5

Source: Nakashima et al. (2007).
*Constants/coefficients are assumed to be the same for contact between soil elements, between soil elements and wheel or lug surface, and between soil elements and soil bin wall.
**Friction coefficients are assumed to be the same for contact between soil elements and wheel or lug surface, and between soil elements and soil bin wall.

(for instance, only about 7000 elements are used in the case shown in Figure 2.34 and Table 2.7), because of the limitation of computing resources available. This is one of the reasons why the elements used in many simulations are much larger in size than the actual soil particles noted earlier. It is estimated that to conduct a realistic three-dimensional simulation of full-scale vehicle–terrain interaction problems, the number of elements required may be in the order of 10^6–10^8. This would require high-power computing resources, usually supercomputers. For instance, in a three-dimensional simulation of the interaction between a mine plough and soil, ten million elements were used and the simulation took just over 16 000 CPU hours on a 256-processor Cray T3E supercomputer (Horner et al., 2001). This indicates that improvements in computing technology remain one of the key issues in making the discrete element modelling technique a practical tool for the engineer in performance and design evaluation of off-road vehicles.

4. In a number of vehicle–terrain interaction problems, the mechanical system involved is rather complex. As noted in Section 2.4, for instance, the study of track–terrain interaction involves not only that part of the track in contact with terrain, but also other factors, such as track–roadwheel interaction, suspension characteristics, supporting roller arrangements for the top run of the track, sprocket and idler configuration, variation of tension along the track due to sprocket torque, etc. This would make the issue even more complex and would require even more high-power computing resources.

In view of the limitations to the study of vehicle mobility in the field by modelling the terrain as an elastic or rigid, perfectly plastic medium, or based on the critical state soil mechanics, the finite element method, or the discrete element method, other practical techniques and approaches have been developed. In the following chapters, some of the practical techniques and approaches to the study of vehicle–terrain interaction are presented.

PROBLEMS

2.1 The contact area of a tyre on a hard, dry soil may be approximated by a circular area having an effective radius of 0.25 m. The contact pressure is assumed to be uniform of 200 kPa. For this type of soil, the concentration factor may be assumed to be 4. Calculate the vertical stress in the soil at a depth of 0.25 m below the centre of the contact area. Estimate the depth at which the vertical stress is 10% of the contact pressure on the soil surface.

2.2 A traction-aid device with 20 grousers (lugs) and a narrow rim is to be attached to the tyres of a wheeled vehicle to improve its traction on a muddy field. The tip of the grouser is 0.75 m from the wheel centre and the width of the grouser is 0.35 m. When the grouser is at the vertical position, it penetrates 0.15 m into the soil. The shear strength parameters of the soil are: cohesion of 10 kPa and angle of internal shearing resistance of 12°. The weight density of the soil is 16 kN/m³. The grouser surface is relatively smooth and the friction and adhesion between the grouser surface and the soil may be neglected. Estimate the thrust that a grouser in the vertical position develops and the wheel torque required to generate the thrust developed by the grouser.

2.3 An industrial tractor has a total weight of 340 kN and each of its two tracks has a ground contact length of 3.21 m and width of 0.56 m. Compare the maximum thrust that the tractor can develop on two terrains: (a) sandy terrain with an angle of internal shearing resistance of 35° and negligible cohesion; and (b) clayey soil with an angle of internal shearing resistance of 14° and cohesion of 7.58 kPa.

2.4 (a) If the tractor described in Problem 2.3 is required to generate a maximum thrust of 275 kN on the sandy terrain and 125 kN on the clayey soil, what changes to the vehicle and/or track parameters would you recommend for these two cases? State the reasons for your recommendations for each case.

(b) What measures would you suggest to increase the ratio of the maximum vehicle thrust to weight on the sandy terrain and on the clayey soil described in Problem 2.3?

2.5 A bulldozer blade mounted in front of a tractor has a width of 3 m and a rake angle of 70°. It is used to remove a layer of sandy soil of 0.35 m deep. The soil has a weight density of 16 kN/m³, an angle of internal shearing resistance of 30°, and cohesion of 5 kPa.

(a) If the blade is rough and the friction angle between the blade surface and the soil is the same as the angle of the internal shearing resistance of the soil, and the adhesion between the blade surface and the soil is the same as the cohesion of the soil, estimate the horizontal force that the tractor has to apply to the blade to initiate the bulldozing process.

(b) The blade becomes polished after being used for a period of time. The friction and adhesion between the blade surface and the soil may be assumed to be insignificant. Estimate the extent to which the horizontal force required to push the blade is reduced.

CHAPTER 3
Measurement of Terrain Properties

Currently, the cone penetrometer technique and the bevameter technique are widely used in the measurement of the mechanical properties of terrain for the study of vehicle mobility in the field. The selection of the technique for measuring terrain properties is closely related to the method of approach chosen for the study of vehicle–terrain interaction. For instance, the cone penetrometer technique is usually employed to identify terrain conditions for the study of vehicle mobility using empirical relations. On the other hand, the bevameter technique is primarily used in methods intended for detailed parametric analyses of vehicle performance and design.

While the cone penetrometer technique and the bevameter technique are widely used in practice at the present time, new techniques for measuring the mechanical properties of terrain in the field may emerge in the future, as our understanding of the physical nature of vehicle–terrain interaction improves and the method of approach to the study of vehicle mobility evolves.

3.1 Cone Penetrometer Technique

As mentioned previously, the cone penetrometer technique was originally developed to provide military intelligence and reconnaissance personnel with a simple method for evaluating vehicle mobility and terrain trafficability on a 'go/no go' basis. The cone penetrometer originally developed by the Waterways Experiment Station (WES) is a hand-held mechanical device, consisting of a 1.59 cm (5/8 in.) diameter rod, a 30-degree right circular cone having a base area of 3.23 cm² (0.5 in²) and a proving ring and a dial for indicating the force required to push the cone into the terrain (Figure 1.7). The recommended rate of penetration is approximately 3 cm/s (6 ft/min). The force per unit cone base area is called the cone index (CI). Although the CI is used as an undimensioned parameter, it is in fact the force in pounds exerted on the penetrometer divided by the area of the cone base in square inches. In operation, the first CI reading is taken when the base of the cone is flush with the terrain surface. Subsequent readings are taken at increments of 7.6 cm (3 in.) to a depth of 30.5 cm (12 in.) and then at increments of 15.2 cm (6 in.) to a depth of 76.2 cm (30 in.), or to a CI value of 300, which is the capacity of the cone penetrometer developed by WES.

With the rapid progress in electronics and computer technology, a variety of cone penetrometers using electrical (or electronic) sensors for monitoring the force and penetration depth and computer technology for storing and processing data have been developed (Anderson et al., 1980; Olsen, 1987). Some of these penetrometers are hand-held, while the others are mounted on a vehicle or on a carriage that can be moved manually in the field. Some of the recently developed penetrometers are driven electrically or hydraulically.

In addition to the CI, other indices can be obtained using the cone penetrometer. For instance, to evaluate the change in terrain strength that may occur under repeated vehicular traffic, a remoulding index (RI) is introduced. It is the ratio of the cone index of a sample of soil after remoulding to that before remoulding. For fine-grained soils, the remoulding is carried out with 100 blows of a 1.135 kg (2.5 lb) hammer falling from a height of 30.5 cm (12 in.) onto a sample in a remoulding cylinder (SAE, 1967). For coarse-grained soils with fines, the sample is remoulded by dropping the remoulding cylinder and base with the soil sample inside 25 times from a height of 15.2 cm (6 in.). Also in this case, a cone with a base area of 1.29 cm^2 (0.2 in^2) is used, instead of the 3.23 cm^2 (0.5 in^2) cone, to measure the cone index before and after remoulding. It is not necessary to measure the remoulding index for clean sands, such as those normally found in deserts or on beaches.

Remoulding may cause an increase or decrease in the strength of the terrain, depending upon the type and condition of the terrain. The rating cone index (RCI), which is the product of the remoulding index (RI) and the cone index (CI) measured before remoulding, is used to represent the strength of the terrain under repeated vehicular traffic.

To indicate terrain trafficability, a vehicle cone index (VCI) is used. It is the minimum cone index of a soil in the critical layer that permits a given vehicle to make a specific number of passes without immobilization. The depth of the critical layer varies with vehicle type and weight. For instance, it has been suggested that the depth of the critical layer for wheeled vehicles with weights up to 222.4 kN (50,000 lb) and tracked vehicles with weights up to 444.8 kN (100,000 lb) is 15.2 to 30.5 cm (6 to 12 in.), while for vehicles over these limits the depth of the critical layer is 22.9 to 38.1 cm (9 to 15 in).

The cone penetrometer has been widely used in vehicle mobility studies, primarily due to its simplicity. In spite of its wide use and acknowledged role in providing a measure of terrain strength, it is generally recognized that the cone index alone is not sufficient for adequately defining the mechanical properties of the terrain that are pertinent to vehicle mobility. The studies performed by Mulqueen et al. (1977) indicated that the cone index is a compound parameter reflecting the shear, compressive and tensile strengths of the terrain and soil–metal friction and adhesion. In homogeneous remoulded soils, it was observed that the relative proportions of the shear, compressive and tensile strengths reflected by the cone index vary with moisture content. As soil moisture content increases, the cone index becomes increasingly insensitive to changes in shear or compressive strength. Furthermore, the formation of soil bodies and compaction zones ahead of the cone effectively changes its geometry and so the relationship between penetration resistance and soil properties changes. In the field, where the terrain moisture content, bulk density, shear strength and structural state may vary significantly with depth, the interpretation of cone penetrometer data becomes even more difficult.

Traditionally, the cone index (or its derivatives) has been directly used as a descriptor of soil strength in formulating empirical relations for predicting vehicle mobility. While these empirical relations are useful in predicting the performance of vehicles similar to those that have been evaluated, they normally should not be extrapolated beyond the conditions upon which they were derived, such as for predicting the performance of vehicles with different design features or in new operating environments. Because of the limitations of empirical methods that use the cone index as a terrain descriptor, a number of other methods based on the detailed analysis of vehicle–terrain interaction have been developed in recent years. These methods require that the terrain be described in terms of its basic mechanical properties, such as the angle of internal shearing resistance, cohesion, shear modulus, etc. Attempts have been made to correlate the cone index with the basic mechanical properties of the terrain (Rohani and Baladi, 1981). This has been prompted by the desire to utilize the existing vast cone index database for the analytical studies of vehicle–terrain interaction and for performance prediction. It has also been motivated by the perception that if the basic mechanical properties of the terrain can be deduced from the cone index, then the process of evaluating terrain properties may be greatly expedited, as the cone index can easily be measured (Hettiaratchi and Liang, 1987).

Based on the assumption that the terrain may be idealized as either an elastoplastic or a rigid, perfectly plastic medium, attempts have been made to formulate the functional relationships between the cone index and the basic mechanical properties of the terrain. For instance, by assuming that the cone penetration process is equivalent to the expansion of a series of spherical cavities in an elastoplastic medium, Rohani and Baladi (1981) derived an expression for the cone index in terms of the angle of internal shearing resistance, cohesion, density, and apparent shear modulus of the terrain, and of the cone geometry and penetration depth. Idealizing the terrain as a rigid, perfectly plastic material and applying the classical theory of plastic equilibrium, Hettiaratchi and Liang (1987) formulated an expression for the load on an indenter (conical or wedge-shaped) in terms of its geometry and penetration depth, and the angle of internal shearing resistance and cohesion of the terrain.

It should be mentioned that while the calculation of the cone index values from the basic mechanical properties of the terrain is straightforward, the reverse is generally a very involved process. For instance, if the angle of internal shearing resistance, cohesion, density and shear modulus are to be derived from the cone index values, then measurements must be made using four cones with different geometry in order to establish four independent equations. Furthermore, the relationships between the cone index and the basic mechanical properties of the terrain are non-linear. In general, the cone index (or the load on an indenter) is a cumbersome transcendental function of the basic mechanical properties of the terrain. To derive the values of the terrain parameters, a set of non-linear simultaneous equations must be solved. Closed-form solutions for these equations have not been found and thus sophisticated numerical techniques must be employed. In practice, perhaps the most significant point to be made regarding this approach is that the established functional relationships between the mechanical properties of the terrain and the cone index are based on greatly simplified assumptions. The terrain is idealized as either an elastoplastic or a rigid, perfectly plastic medium. However, very little experimental evidence is available to show that the proposed functional relationships are valid for marginal terrains (such as soft muskeg and snow) that are of major concern to vehicle mobility.

In view of the uncertainties in, as well as the complexity of, the indirect methods for obtaining the values of the basic mechanical properties of the terrain from the cone index described above, it would seem more useful to concentrate on improving the techniques for direct measurements of the basic mechanical properties of the terrain in the field. Applications of modern electronic and computer technologies to the acquisition and processing of field data can greatly facilitate the *in situ* measurements of these properties.

To extend the capabilities of the hand-held penetrometer, a vane–cone device as shown in Figure 3.1 has been proposed (Yong et al., 1975). In comparison with the cone penetrometer developed at WES, the vane–cone device can measure the shear strength of the terrain. In penetration tests, the behaviour of the vane–cone is essentially the same as that of the cone penetrometer and is, therefore, subject to the same limitations. For shear tests, the vane–cone is rotated and both the applied load and torque are measured. From these, the shear strength of the terrain may be deduced.

The reported experience in applying the vane–cone device to vehicle mobility study is rather limited. Yong and Muro (1981) showed that the deformation energies obtained from the plate and vane–cone tests do not correspond with each other very well. They, therefore, concluded that since the mechanism of terrain deformation under plate loading is more akin to that under track loading, the prediction of the deformation energy loss under a tracked vehicle is more aptly obtained from the results of a plate loading test rather than a vane–cone penetration test. Regarding the use of the vane–cone device in determining the shear strength of the terrain, it should be pointed out that the shear strength parameters derived from the test data depend, to a great extent, on the assumed yield criterion for the terrain and the failure surface under the vane–cone. Since the loading conditions of the vane–cone are complex, uncertainty in the interpretation of test data can arise, particularly on non-homogeneous or stratified terrain.

3.2 Bevameter Technique

This technique, originally conceived and developed by Bekker, is based on the premise that terrain properties pertinent to vehicle mobility can best be measured under loading conditions similar to those exerted by an off-road vehicle. A vehicle exerts normal and shear loads on the terrain surface. To simulate these, the original bevameter technique comprises two separate

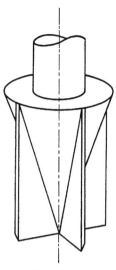

Figure 3.1: Vane–cone device (Reprinted by permission of *ISTVS* from Yong et al., 1975)

sets of tests. One is a set of plate penetration tests and the other is a set of shear tests. In the penetration tests, a plate of suitable size is used to simulate the contact area of a vehicle running gear, and the pressure–sinkage relationship of the terrain is measured. It is used for predicting the normal pressure distribution on the vehicle–terrain interface. To minimize the uncertainty in applying the measured data to the prediction of vehicle performance, it is preferable that the size of the plate used in the tests should be comparable to that of the contact patch of a tyre or a track link. In the shear tests, the shear stress–shear displacement relationship at various normal pressures is measured. This provides the required input for predicting the shear stress distribution on the vehicle running gear–terrain interface. In addition, to provide data for predicting the multipass performance of vehicle running gear and the additional vehicle sinkage due to slip, the behaviour of the terrain under repetitive normal and shear loadings and the slip–sinkage characteristics of the terrain are also measured (Wong et al., 1984).

The bevameter technique provides the closest simulation of vehicle loading conditions among the various measuring techniques presently in use. Accordingly, the terrain data obtained by the bevameter are used as input for a detailed examination of vehicle–terrain interaction presented later in this book.

3.2.1 Basic Features of the Bevameter

The basic features of a bevameter are illustrated in Figure 3.2.

A bevameter originally built at the University of Newcastle upon Tyne and extensively modified at Carleton University is shown in Figure 1.8.

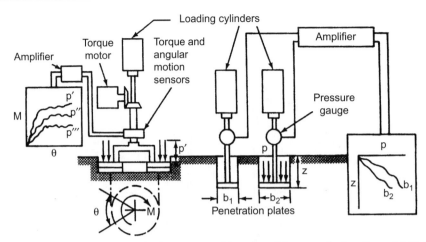

Figure 3.2: Schematic diagram of a bevameter (Reprinted by permission from M.G. Bekker, *Introduction to Terrain–Vehicle Systems*, University of Michigan Press, 1969, copyright © by the University of Michigan)

In the bevameter shown, a hydraulic ram is mounted vertically at one end of the frame and is used to apply normal load to the sinkage plate in pressure–sinkage tests. A shear head is mounted at the other end of the bevameter frame and is rotated by the hydraulic ram through a chain drive. The hydraulic ram has a 6.35 cm diameter bore with a 2.5 cm diameter piston rod. It has a maximum stroke of 76.2 cm and a loading capacity of 8.9 kN. A flow control valve allows the penetration rate to be adjusted within the range of 0 to 10 cm/s.

The bevameter shown in Figure 1.8 makes use of a series of circular plates with diameters ranging from 5 to 20 cm and a series of rectangular plates with widths ranging from 3.75 to 7.5 cm and having an aspect ratio (length to width ratio) of 6. The load applied to the sinkage plates is measured using a Lebow 3132 strain-gauge type load cell with a rated capacity of 13.34 kN. The sinkage of the plate is measured with a calibrated potentiometer using the terrain surface as a reference. Figure 3.3 shows a pressure–sinkage test being performed on muskeg.

A shear ring with an outside diameter of 34 cm and an inside diameter of 27 cm is used for shear tests with the bevameter described above. Shear plates with grouser heights of 1 cm and 2.5 cm and a grouser spacing of 30° are used for testing different types of terrain. A shear plate covered with rubber is also available for determining the rubber–terrain shearing characteristics. The normal load on the shear head is applied using dead weights. The torque applied to the shear ring is measured using a Lebow 2110-2 K strain-gauge type torque sensor with a rated capacity of 226 Nm. The angular displacement of the shear ring is measured using a potentiometer. When a torque is applied to the shear ring, additional sinkage of the shear ring, usually referred to as 'slip–sinkage', results. This 'slip–sinkage' is also measured using a calibrated potentiometer. Figure 3.4 shows a shear test being performed in snow.

Figure 3.3: Pressure–sinkage tests using a vehicle-mounted bevameter

In the bevameter described above, the pressure–sinkage and shear tests are carried out using two separate devices, and for shear tests the normal load on the shear head is applied using dead weights. In a bevameter developed by the Petawawa National Forest Institute (PNFI) of Canada, the devices for pressure–sinkage and shear tests have been integrated into a single mechanism capable of performing both test functions (Golob, 1981). This is achieved by using a specially developed hydraulic cylinder in combination with a hydraulic motor, as shown in Figure 3.5. In the pressure–sinkage tests, a sinkage plate is attached to the bottom of the cylinder rod, and force is applied through hydraulic pressure acting on the piston in the usual manner. In the shear tests, a shear head replaces the plate and a normal load is applied hydraulically. The top of the cylinder rod is connected to the

72   Chapter 3

Figure 3.4: Shear tests using a vehicle-mounted bevameter

hydraulic motor through a splined shaft. This allows the shear head to sink freely while being rotated by the motor.

During shear tests, the shear head may sink further as it rotates. This is the 'slip–sinkage' phenomenon mentioned previously. If the normal load is applied hydraulically, the sinking of the shear head causes fluctuations in the hydraulic pressure. To maintain a constant normal load on the shear head during these tests, either a load-sensing feedback control system or a large hydraulic accumulator must be employed.

3.2.2 A Portable Automatic Data Acquisition and Processing System

To streamline the processing of terrain data, to reduce data processing time, and to employ more rational data processing procedures, an automatic data acquisition system was developed and successfully used in field measurements (Wong, 1980; Wong et al., 1981). This system incorporates recent developments in computer technology and microelectronics. It can sample terrain data measured by the bevameter, the cone penetrometer or other similar devices and can

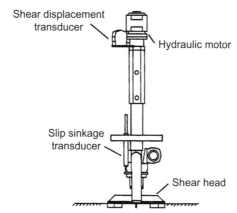

Figure 3.5: Schematic diagram of a bevameter developed by the Petawawa National Forest Institute, Canada (Reprinted by permission of *ISTVS* from Golob, 1981)

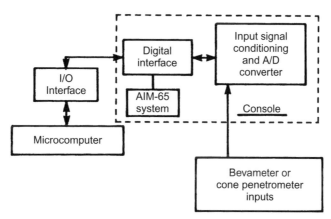

Figure 3.6: Block diagram for a computerized terrain data acquisition and processing system

process them using preprogrammed procedures. The test results, such as the pressure–sinkage and shear stress–displacement relationships obtained using the bevameter, and the cone index versus penetration depth relationships obtained using the cone penetrometer, are plotted on a computer monitor and then transferred to an integral printer for a permanent record. The parameters characterizing the behaviour of the terrain, such as the pressure–sinkage constants and shear strength parameters obtained using the bevameter technique, are printed on paper immediately after each test. On completion of testing in a particular area, a complete statistical description of terrain parameters for the area is provided.

The automatic data acquisition and processing system is self-contained and can readily be installed in a vehicle, as shown in Figure 1.9. Figure 3.6 shows the general features of the system. Together with various terrain measuring devices (the bevameter, the cone penetrometer, or other similar devices), the system can be used in the survey of terrain properties for the development of trafficability maps, and in defining terrain conditions in vehicle proving grounds.

The automatic data acquisition system comprises two major subsystems. One uses a microcomputer programmed to perform an extensive analysis of terrain data and to derive the values of the constants used to characterize the behaviour of the terrain from test data. The other is a microprocessor-based data acquisition console, the function of which is to sample, at a predetermined rate, the signals generated by the transducers of terrain measuring devices, store them in a temporary memory and at the conclusion of an experiment, transfer them to the microcomputer for analysis. A typical experiment may take six to eight seconds to complete and the subsequent data transfer, plotting, analysis, storage on tapes (or discs), and printout usually take less than a minute.

CHAPTER 4
Characterization of the Response of Terrains to Normal and Repetitive Loadings

A vehicle applies normal load to the terrain through the running gear, which results in sinkage. This in turn causes motion resistance. In addition, an element of the terrain is usually subject to the repetitive loading of the consecutive wheels of a multi-axle wheeled vehicle or of the roadwheels in a track system. To predict the normal pressure distribution on the vehicle–terrain interface and the tractive performance of a vehicle, the response of the terrain to normal load and to repetitive loading must be measured. This section describes the measurement and characterization of the response of a variety of terrains, including mineral terrain, muskeg, and snow-covered terrain. The measurements were made using the bevameter with the associated automatic data acquisition and processing system described in Chapter 3.

One of the fundamental tasks in the characterization of terrain behaviour is to establish functional relationships that can realistically describe the responses of the terrain to various types of loading, such as the stress–strain relationship, pressure–sinkage relationship, and shear stress–shear displacement relationship. Because the structure, as well as the behaviour, of natural terrains varies greatly, it seems unlikely that a unified theory or method can be developed in the foreseeable future to adequately describe the characteristics of different kinds of natural terrain. A practical approach to the characterization of terrain behaviour has, therefore, to be developed. It appears that a realistic approach would be to measure terrain responses under loading conditions similar to those exerted by an off-road machine and then to establish the pertinent functional relationships based on the measured data using established curve-fitting and statistical techniques. To obtain useful data, measurements should be made *in situ* using measuring devices that simulate the action of the vehicle running gear. The size of the measuring device should preferably be comparable to that of the contact patch of a tyre or that of a track link, so as to minimize the uncertainty in applying the derived functional relationships to the prediction of the performance of full-scale vehicles in the field.

76   Chapter 4

Since one of the prime purposes for describing terrain behaviour using a mathematical function is to enable the engineering practitioner to predict or evaluate the performance of off-road machines in a systematic and convenient manner, the selection of a particular function or method to characterize terrain behaviour should be based not only on how accurately it represents the terrain response, but also on how conveniently it can be incorporated into the selected framework for predicting machine performance.

In the following, a brief description of the methods for characterizing the responses of different types of terrain to various kinds of loading is presented.

4.1 Response of Mineral Terrains

The responses of a variety of mineral terrains in Eastern Ontario, Canada, including sand (LETE sand), sandy loam (Upland sandy loam and Rubicon sandy loam), clayey loam (North Gower clayey loam) and loam (Grenville loam), to normal and repetitive loadings were measured and characterized. The grain size distributions for the mineral terrains examined are shown in Figures 4.1 to 4.5.

These are used as examples to illustrate the characteristics of mineral terrains and the procedures for characterizing their behaviour.

4.1.1 Pressure–Sinkage Relationship

Representative sets of the pressure–sinkage curves for the various types of mineral terrain described above are shown in Figures 4.6 to 4.10.

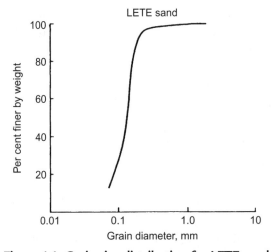

Figure 4.1: Grain size distribution for LETE sand

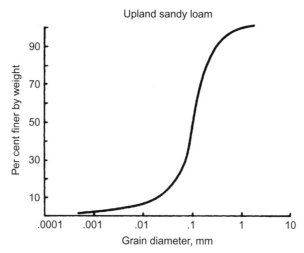

Figure 4.2: Grain size distribution for Upland sandy loam

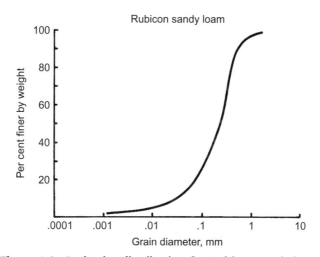

Figure 4.3: Grain size distribution for Rubicon sandy loam

The pressure–sinkage relationship for the mineral terrain may be characterized by the following equation proposed by Bekker (1960):

$$p = (k_c/b + k_\phi)z^n = k_{eq}z^n \tag{4.1}$$

where p is pressure; b is the radius of a circular plate or the smaller dimension of a rectangular plate; n, k_c and k_ϕ are pressure–sinkage parameters for the Bekker equation; $k_{eq} = k_c/b + k_\phi$, and z is sinkage.

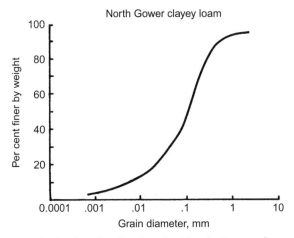

Figure 4.4: Grain size distribution for North Gower clayey loam

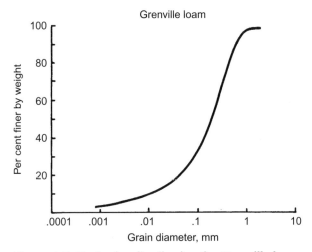

Figure 4.5: Grain size distribution for Grenville loam

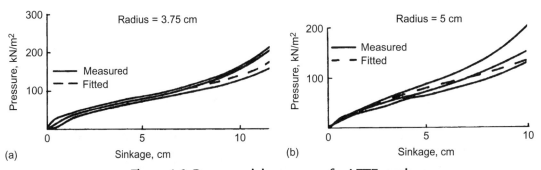

Figure 4.6: Pressure–sinkage curves for LETE sand

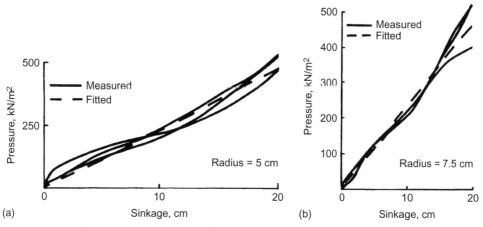

Figure 4.7: Pressure–sinkage curves for Upland sandy loam

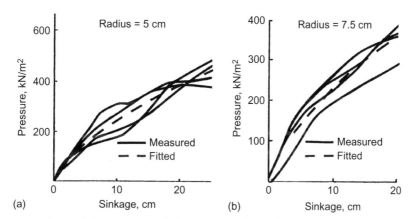

Figure 4.8: Pressure–sinkage curves for Rubicon sandy loam

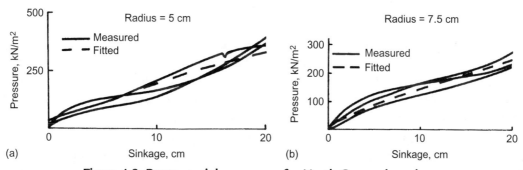

Figure 4.9: Pressure–sinkage curves for North Gower clayey loam

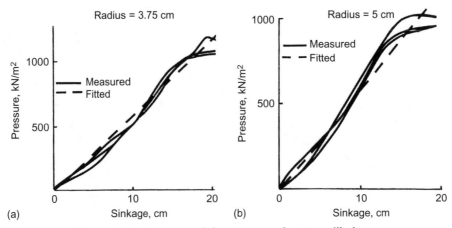

Figure 4.10: Pressure–sinkage curves for Grenville loam

It has been shown by Bekker that the pressure–sinkage parameters are insensitive to the width of rectangular plates with large aspect ratios (i.e. the length to width ratio of the plate is between 5 and 7). A number of tests have been performed to determine the degree of dependence of the values of the pressure–sinkage parameters on the shape of the test plate. Test results obtained so far seem to indicate that there is little difference between the values of the pressure–sinkage parameters obtained using a set of rectangular plates of large aspect ratios and those obtained with a set of circular plates having radii equal to the widths of the rectangular plates. Because of this, circular plates are commonly used, since they require lower load than the corresponding rectangular plates for the same average contact pressure.

It should be emphasized that Eqn (4.1) is essentially an empirical equation. Furthermore, the parameters k_c and k_ϕ have variable dimensions, depending on the value of the exponent n. Influenced by the work of a more fundamental nature in soil mechanics and by experimental evidence, Reece proposed the following equation for the pressure–sinkage relationship for homogeneous soil (Reece, 1965):

$$p = (ck_c' + \gamma_s b k_\phi')(z/b)^n \tag{4.2}$$

where n, k_c' and k_ϕ' are the pressure–sinkage parameters for the Reece equation; γ_s is the weight density of the terrain; and c is the cohesion of the terrain. A series of penetration tests were carried out to verify the basic features of Eqn (4.2). Plates with various widths and with an aspect ratio of at least 4.5 were used. For frictionless clay, the term k_ϕ' should be negligible and the relationship between p and (z/b) is not affected by the plate width b. For dry, cohesionless sand, the term k_c' should be negligible and the pressure p increases linearly with the increase in width of the plate. These predictions were well borne out by experimental data (Reece, 1965).

It should be noted that while Eqn (4.2) only differs from Eqn (4.1) in using the ratio of sinkage to plate width (z/b) to replace sinkage z, it is sufficient to mark an improvement (Reece, 1965). The parameters k'_c and k'_ϕ in Eqn (4.2) are dimensionless, whereas parameters k_c and k_ϕ in Eqn (4.1) have dimensions dependent upon the value of n. Furthermore, Eqn (4.2) seems to allow itself to fit in with classical soil mechanics theories. For instance, Eqn (4.2) and the Terzaghi bearing capacity equation have a similar form (Wong, 2008). In frictionless clay, both equations show that the increase of width has no effect on pressure. On the other hand, in dry, cohesionless sand, both equations show that the increase in width can cause a linear increase in pressure. It should be noted, however, that Eqn (4.2) applies only to homogeneous terrain. For non-homogeneous (layer) terrain, the pressure–sinkage relationship obtained using a smaller-size plate may not be extrapolated to that of a larger-size contact area. This is because for the same ratio of z/b, a plate with a wider width b has to penetrate to a deeper layer of the terrain, which may have different mechanical properties from those of a shallower layer (Wong, 2008).

When using Eqn (4.1) or (4.2) to characterize the pressure–sinkage relationship of the terrain, one of the basic tasks is to derive the proper values of the terrain parameters from experimental data. Traditionally, obtaining these values involves considerable manual manipulation of the test data. The experimental data are first plotted on a log–log scale and a straight line is fitted to the data by eye. There are no generally accepted guidelines for obtaining the best fitting line. Thus, this procedure heavily relies on the skill of the investigator and is liable to produce errors for inexperienced personnel.

To provide a more rational approach for deriving the values of the parameters characterizing the pressure–sinkage relation from experimental data, a weighted least squares method has been developed. This data processing procedure has been computerized and incorporated into the automatic data acquisition and processing system described in Chapter 3 (Wong, 1980; Wong et al., 1981).

Consider the pressure–sinkage equation proposed by Bekker as an example. Taking the logarithms of both sides of Eqn (4.1), one obtains

$$\ln p = \ln(k_c/b + k_\phi) + n \ln z \qquad (4.3)$$

In the above equation, p and z are measured values and b is the smaller dimension (or radius) of the plate. k_c, k_ϕ and n are constants to be determined from experimental data. To obtain the best fitting line based on Eqn (4.3), the conventional least squares technique cannot be directly employed, as p and z are the measured quantities, not $\ln p$ and $\ln z$. Applying the conventional least squares principle to the logarithms of the deviations will give excessive weight to deviations at low pressures which are usually the least significant. However, one may approach the problem by introducing a weighting factor and by minimizing the value of

$$w_r \left[\ln p - \ln(k_c/b + k_\phi) - n \ln z\right]^2 \tag{4.4}$$

where w_r is an appropriate weighting factor.

If an equal liability to error for all observations is assumed, the weighting factor w_r may be taken as (Wong, 1980)

$$w_r = p^2 \tag{4.5}$$

Thus to obtain the best values of k_c, k_ϕ and n for a particular set of test data, the following function must be minimized:

$$F = \sum p^2 \left[\ln p - \ln(k_c/b + k_\phi) - n \ln z\right]^2$$

or

$$F = \sum p^2 \left[\ln p - \ln k_{eq} - n \ln z\right]^2 \tag{4.6}$$

where $k_{eq} = k_c/b + k_\phi$.

To minimize the value of the function F, the first partial derivatives of F with respect to n and to k_{eq} are taken and set to zero. This leads to the following two equations:

$$\ln k_{eq} \sum p^2 \ln z + n \sum p^2 (\ln z)^2 = \sum p^2 \ln p \ln z \tag{4.7}$$

$$\ln k_{eq} \sum p^2 + n \sum p^2 \ln z = \sum p^2 \ln p \tag{4.8}$$

Solving these two equations simultaneously, one can obtain the best fitting values of n and $\ln k_{eq}$ for a particular set of data:

$$n = \frac{\sum p^2 \sum p^2 \ln p \ln z - \sum p^2 \ln p \sum p^2 \ln z}{\sum p^2 \sum p^2 (\ln z)^2 - (\sum p^2 \ln z)^2} \tag{4.9}$$

$$\ln k_{eq} = \frac{\sum p^2 \ln p - n \sum p^2 \ln z}{\sum p^2} \tag{4.10}$$

It should be mentioned that in practice the value of n calculated from the two sets of data obtained using two different sizes of plates are seldom identical, even under controlled laboratory conditions. Since the Bekker equation implies that for a particular type of terrain there is a unique value of n, an average value, n_{av}, is taken:

$$n_{av} = \frac{(n)_{b=b_1} + (n)_{b=b_2}}{2} \tag{4.11}$$

where $(n)_{b=b_1}$ and $(n)_{b=b_2}$ represent the values of n obtained from the two sets of data using plates with widths b_1 and b_2, respectively.

The value of n_{av} is then used in place of the value of n to calculate the value of k_{eq} for a particular set of data:

$$\ln k_{eq} = \frac{\sum p^2 \ln p - n_{av} \sum p^2 \ln z}{\sum p^2} \tag{4.12}$$

From the two sets of test data, two values for k_{eq}, one from the data obtained using a plate with width b_1, $(k_{eq})_{b=b_1}$, and the other from the data obtained using a plate with width b_2, $(k_{eq})_{b=b_2}$, can be derived. Thus the values of k_c and k_ϕ can be calculated by

$$k_c = \frac{(k_{eq})_{b=b_1} - (k_{eq})_{b=b_2}}{b_2 - b_1} b_1 b_2 \tag{4.13}$$

$$k_\phi = (k_{eq})_{b=b_1} - \frac{(k_{eq})_{b=b_1} - (k_{eq})_{b=b_2}}{b_2 - b_1} b_2 \tag{4.14}$$

The above procedures have been computerized and incorporated into the automatic data acquisition and processing system. Upon completion of a set of tests, values of k_c, k_ϕ and n are printed out (Wong, 1980; Wong et al., 1981).

To evaluate the goodness-of-fit using the above procedures, the following parameter ε is used:

$$\varepsilon = 1 - \frac{\sqrt{\sum (p_m - p_{c\ell})^2 / (N - 2)}}{\sum p_m / N} \tag{4.15}$$

where p_m is the measured pressure, $p_{c\ell}$ is the calculated pressure using the procedures described above, and N is the number of data points used for the curve fitting. When ε is 1, the fit is perfect. The calculation of ε for each set of data is computerized and incorporated into the automatic data acquisition and processing system. Its value is printed out upon completion of a set of tests.

Upon completion of testing in a particular area, the mean values of k_c, k_ϕ and n together with their standard deviations are calculated and printed. Thus, a complete statistical description of the response of the terrain to normal load is given.

An approach similar to that described above may be followed to fit the pressure–sinkage data to the Reece equation, Eqn (4.2). In this equation, c, γ_s, k_c' and k_ϕ' are soil constants. Thus from the curve fitting point of view, some of these constants can be combined and Eqn (4.2) can be written as

$$p = (k_c'' + bk_\phi'')(z/b)^n \tag{4.16}$$

where

$$k_c'' = ck_c' \quad \text{and} \quad k_\phi'' = \gamma_s k_\phi'$$

Employing the weighted least squares method described previously, one obtains the best fitting values of n and k_{eq}' (which is equal to $k_c'' + bk_\phi''$) for a particular set of experimental data:

$$n = \frac{\sum p^2 \sum p^2 \ln p \ln(z/b) - \sum p^2 \ln p \sum p^2 \ln(z/b)}{\sum p^2 \sum p^2 (\ln z/b)^2 - \left[\sum p^2 \ln(z/b)\right]^2} \tag{4.17}$$

$$\ln k_{eq}' = \frac{\sum p^2 \ln p - n \sum p^2 \ln(z/b)}{\sum p^2} \tag{4.18}$$

As pointed out previously, the values of n calculated from the two sets of data obtained using two different sizes of plates are seldom identical. Therefore, an average value, n_{av}, is taken:

$$n_{av} = \frac{(n)_{b=b_1} + (n)_{b=b_2}}{2} \tag{4.19}$$

This value is then used to obtain the value of k_{eq}' for a particular set of data:

$$\ln k_{eq}' = \frac{\sum p^2 \ln p - n_{av} \sum p^2 \ln(z/b)}{\sum p^2} \tag{4.20}$$

From two sets of test data, two values for k'_{eq}, one from the data obtained using a plate with width b_1, $(k'_{eq})_{b=b_1}$, and the other from the data obtained using a plate with width b_2, $(k'_{eq})_{b=b_2}$, can be derived. Thus the values of k''_c and k''_ϕ can be calculated by

$$k''_c = (k'_{eq})_{b=b_1} - \frac{(k'_{eq})_{b=b_1} - (k'_{eq})_{b=b_2}}{b_1 - b_2} b_1 \tag{4.21}$$

$$k''_\phi = \frac{(k'_{eq})_{b=b_1} - (k'_{eq})_{b=b_2}}{b_1 - b_2} \tag{4.22}$$

This procedure for fitting pressure–sinkage data with the Reece equation has been computerized and incorporated into the automatic data acquisition and processing system.

It is interesting to point out that fitting the same set of pressure–sinkage data with either the Bekker or the Reece equation results in the same goodness-of-fit. Furthermore, the values of n in both cases are identical. This is because the value of b (the smaller dimension or the radius of the plate) is a constant for a particular set of data. Consequently, both the Bekker and the Reece equations are of similar form. This does not mean, however, that the Bekker and the Reece equations will give the same prediction of pressure for any particular loading plate at any specific sinkage.

Table 4.1 summarizes the mean values of k_c, k_ϕ, k''_c, k''_ϕ and n for the various mineral terrains tested and the goodness-of-fit of the Bekker and Reece equations to the measured data.

Based on the results shown, it appears that both the Bekker and the Reece equations can be used to characterize the pressure–sinkage relationships of the mineral terrains tested, and that the weighted least squares method can provide a rational basis for processing pressure–sinkage data obtained using the bevameter. The two equations appear to provide an acceptable fit to the measured data of the mineral terrains.

4.1.2 Response to Repetitive Loading

When a vehicle is travelling over an unprepared terrain, an element of the terrain under the running gear is first subject to the normal load applied by the leading wheel (or roadwheel in a track system). When the leading wheel (or roadwheel) has passed, the load on the terrain element is reduced. Load is reapplied as a succeeding wheel (or roadwheel) rolls over it. A terrain element is thus subject to the repetitive loading of the consecutive wheels of a multi-axle wheeled vehicle or the roadwheels in a track system. The loading–unloading–reloading cycle continues until the rear wheel (or roadwheel) of the vehicle has passed over it. To

Table 4.1: Mean values of the parameters characterizing the pressure–sinkage relations of various mineral terrains

Terrain type	Constants for Bekker's equation			Constants for Reece's equation			Goodness-of-fit %	Wet density (kg/m^3)	Moisture content %
	n	k_c (kN/m^{n+1})	k_ϕ (kN/m^{n+2})	n	k_c'' (kN/m^2)	k_ϕ'' (kN/m^3)			
LETE sand	0.705	6.94	505.8	0.705	39.1	779.8	95.3	~1600	
	0.611	1.16	475.0	0.611	28.2	1066	94.5		
	0.804	3.93	599.5	0.804	16.9	879.6	93.8		
	0.728		1348	0.728	18.3	2393	88.8		
	0.578	9.08	2166	0.578	197	4365	89.2		
	0.781	47.8	6076	0.781	229.7	8940	89.8		
	0.806	155.9	4526	0.806	413.5	5420	88.1		
Upland sandy loam	1.10	74.6	2080	1.10	42.0	1833	87.7	1557	51.6
	0.97	65.5	1418	0.97	77.4	1464	92.0	1542	49.2
	1.00	5.7	2293	1.00	5.3	2283	94.8	1570	49.1
	0.74	26.8	1522	0.74	121.7	2092	95.1	1519	44.3
	1.74	259.0	1643	1.74	−0.9	763	86.0	1696	50.0
	0.85	3.3	2529	0.85	42.4	3270	87.5	1471	28.6
	0.72	59.1	1856	0.72	231.4	2323	84.2	1592	34.3
	0.77	58.4	2761	0.77	214.1	3626	86.6	1559	35.1
	1.09	24.9	3573	1.09	6.7	2982	91.9	1716	31.2
	0.70	70.6	1426	0.70	279.3	1317	94.3	1470	27.3
	0.75	55.7	2464	0.75	213.6	3244	89.4	1526	32.6
Rubicon sandy loam	0.66	6.9	752	0.66	63.3	1176	92.6	1561	43.3
	0.65	10.5	880	0.65	88.2	1358	97.0	1588	44.2
North Gower clayey loam	0.73	41.6	2471	0.73	121.2	−4.2	88.8	1681	45.8
	0.85	6.8	1134	0.85	27.0	1430	90.0	1597	52.0
Grenville loam	1.01	0.06	5880	1.01	−1.3	5814	87.4	1326	24.1
	1.02	66.0	4486	1.02	55.3	4292	89.1	1339	18.2

predict the normal pressure distribution under a moving vehicle and hence its sinkage and motion resistance, the response of the terrain to repetitive normal load must be measured, in addition to the pressure–sinkage relationship described above.

Figure 4.11 shows a typical response of a mineral terrain to repetitive loading (Wong et al., 1984). It can be seen that the pressure initially increases with sinkage along curve OA. However, when the load applied to the terrain by the plate is reduced at A, the pressure–sinkage relationship follows line AB. When the load is reapplied at B, the pressure–sinkage relationship follows, more or less, the same path as that during unloading. When the reapplied

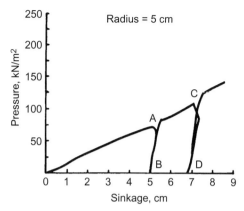

Figure 4.11: Response to repetitive normal load of a mineral terrain

load exceeds that at which the preceding unloading–reloading cycle begins (i.e. point A), additional sinkage results. With further increase of load beyond that corresponding to A, the pressure–sinkage relation follows the original pressure–sinkage curve OAC. The characteristics of the subsequent unloading–reloading cycles, such as CD, are quite similar to those of the first one (i.e. AB).

Based on the observations described above, the response of the mineral terrain to repetitive loading may be idealized, and the pressure–sinkage relationship during both unloading and reloading, such as AB and BA shown in Figure 3.11, may be described by (Wong et al., 1984)

$$\begin{aligned} p &= \left(\frac{k_c}{b} + k_\phi\right) z_u^n - k_u(z_u - z) \\ &= k_{eq} z_u^n - k_u(z_u - z) \\ &= p_u - k_u(z_u - z) \end{aligned} \quad (4.23)$$

where p_u and z_u are the pressure and sinkage, respectively, when unloading begins; and k_u is the pressure–sinkage parameter representing the average slope of the unloading–reloading line AB.

It is found that the terrain stiffness k_u during unloading or reloading is a function of z_u, which is the sinkage when unloading begins. As a first approximation, their relationship may be described by

$$k_u = k_o + A_u z_u \quad (4.24)$$

where k_o and A_u are parameters, the values of which can be derived from experimental data.

Figure 4.12 shows the relationship between k_u and z_u for the LETE sand mentioned previously. From the data shown it can be determined that $k_o = 0$ and $A_u = 503\,000\,\text{kN/m}^4$.

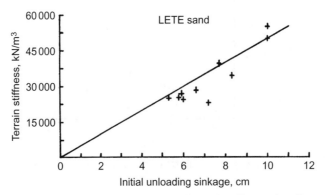

Figure 4.12: Variation of terrain stiffness during the unloading–reloading cycle with initial unloading sinkage for LETE sand

4.2 Response of Muskegs

The responses of two muskegs in the Petawawa area, Ontario, Canada, to normal load and to repetitive loading were measured using the bevameter together with the automatic data acquisition and processing system described in Chapter 3 (Wong et al., 1982). The physical features of the muskegs tested, the measured responses and the approach to characterizing their behaviour are described below.

4.2.1 Physical Features of the Muskegs Tested

Tests were carried out at two muskeg sites (Sites A and B) in the Petawawa area, about 180 km northwest of Ottawa, Ontario, Canada.

Site A was in a small but deep depression. Sedges and cat-tails were the dominant vegetation. Within the test site, there was almost pure sedge cover. The predominant cover class was F as defined by the Radforth classification system (MacFarlane, 1969). The ground was saturated (rubber boots were required for walking, water to ankles). Surface vegetation was all non-woody and fairly fragile, as it was easily trampled. The mat thickness was 5–10 cm, and the peat depth was 5.3 m. The upper 2.5 m was a non-woody fine fibrous peat with occasional woody fragments. The boundary between the mat and peat was not distinct from visual observation. However, the breaking of the surface mat could still be clearly identified in most cases from the measured load–sinkage curves, which are discussed in detail later. Above the ooze and silt at the bottom of the deposit (at 5.3 m) was a pure amorphous granular peat, dark brown with light brown lenses. There was neither wood nor fibrous peat, and the peat had a pasty consistency. The dominant features of this site could be described as high water content, small particle size and soft, weak underlying peat. This site was a textbook example of low-trafficability muskeg with a fragile surface easily disturbed by vehicles.

A set of undisturbed samples was taken with a piston sampler at three locations near the centre of the test site. Samples were taken from the top 240 cm and the bottom 80 cm of the deposit. The samples were retained in sealed containers for laboratory analysis of mat thickness, water content, bulk density, and degree of humification. The results are shown in Figure 4.13. The water content of the peat is taken as the difference between the mass of the wet sample and that of the oven-dried sample divided by the mass of the wet sample. The bulk density is defined as the ratio of the mass of the oven-dried sample to the volume of the wet sample as taken from the field. The degree of humification is defined using the von Post scale of 1–10.

Site B was in a natural depression in mineral terrain. Surface vegetation included low woody shrubs, grasses, sedges, and mosses. There were occasional small white pine trees near the north end of the site, indicating the relative dryness of the ground (rubber boots not required for walking). The sampling area within the site was in the predominantly low woody and non-woody short vegetation family (cover class EFI in the Radforth classification system) but was surrounded by medium vegetation (cover class DFI in the Radforth classification system) including noticeably higher shrubs. Occasional tree stumps were evidence of former tree growth on the site. The ground surface was very uneven and relatively firm. The woody surface vegetation (cover class E in the Radforth classification system), the mat of coarse woody fibres with a dense matrix of non-woody fine fibres, and the relative dryness combined to make a fairly stiff and strong material. The water table, revealed by cutting a hole with a chain saw, was about 20 cm beneath the ground surface at the time of testing.

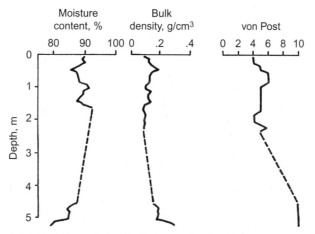

Figure 4.13: Profiles of physical properties for Petawawa Muskeg A

The predominant cover class for Site B was EFI, according to the Radforth classification system. The mat thickness was approximately 15–20 cm with a peat depth of 4.5 m. Peat in the upper 2.5 m of the deposit was found to be woody coarse fibrous, non-woody fine fibrous and amorphous granular with frequent large wood fragments. There was no particular visible, orientation of the fibrous components. The base of the deposit at 4–4.5 m contained highly decomposed sedge fragments and amorphous granular peat. The moisture content, bulk density and humification of the muskeg as a function of depth at Site B are shown in Figure 4.14.

To determine the mat density and porosity of the muskeg at Site B, a chainsaw was used to remove an undisturbed sample block of the mat and surface vegetation with dimensions of about $30 \times 20 \times 20$ cm. This sample was retained in a waterproof container and later transported to a laboratory for analysis. The total porosity of the mat, defined as the ratio of the volume of water removed from the saturated sample to that of the saturated sample, was found to be 91%. The mat density, defined as the mass of the oven-dried mat sample divided by the volume of the wet sample, was determined as 0.051 g/cm^3 for Site B.

It should be mentioned that the quantity of samples from each test site and the quantitative information obtained from them are only intended to illustrate the representative physical characteristics of the test sites, and that the data are not intended for statistical analysis.

4.2.2 Pressure–Sinkage Relationship

For pressure–sinkage tests, rectangular plates of 3.75×22.5, 5×30 and 7.5×45 cm, and circular plates with diameters of 10, 15 and 20 cm were used. To examine the effect of sinkage rate on the response of the muskeg to loading, tests were carried out at two penetration rates of approximately 2.5 and 10 cm/s.

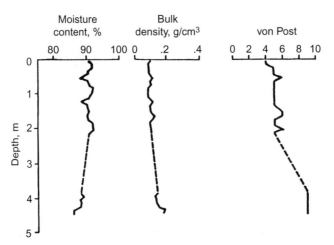

Figure 4.14: Profiles of physical properties for Petawawa Muskeg B

Representative pressure–sinkage characteristics of the muskeg at Site A obtained using a rectangular and a circular plate at a penetration rate of approximately 2.5 cm/s are shown in Figures 4.15 and 4.16, respectively. As can be seen, a well-defined critical pressure at which the surface mat of the muskeg is broken can be identified (Wong et al., 1982). The sinkage where the surface mat is broken is called the critical sinkage. Beyond the critical sinkage, the resistance to penetration of the muskeg at Site A decreased, as the load was mainly carried by the underlying peat which had a relatively low bearing strength. The pressure–sinkage curves of the muskeg shown in the figures are similar to those obtained in Michigan by Niemi and Bayer and reported by Bekker (1969).

In examining the bearing capacity of muskeg, Bekker (1969) assumed that the failure of the surface mat was due to shearing along the area delineated by the circumference of the loading area as shown in Figure 4.17 (Bekker, 1969). Therefore, the punching pressure p_p for a circular plate is determined by

$$p_p = \frac{2\pi b t_o \tau_r + \pi b^2 p''}{\pi b^2} \tag{4.25}$$

where b is the radius of the plate, t_o is the thickness of the surface mat, τ_r is the shear strength of the mat, and p'' is the reaction of the sublayer and is assumed to be a constant.

It has been reported that normally the tensile strength of the surface mat is much lower than its shear strength. Therefore, when the muskeg is subject to an applied vertical load, failure of the surface mat due to tension is a distinct possibility. Following is a proposed mathematical model

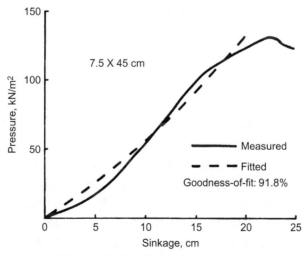

Figure 4.15: Pressure–sinkage relationship measured using a rectangular plate for Petawawa Muskeg A

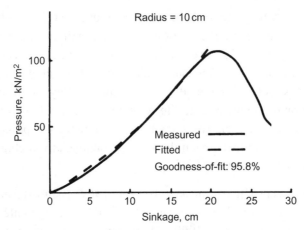

Figure 4.16: Pressure–sinkage relationship measured using a circular plate for Petawawa Muskeg A

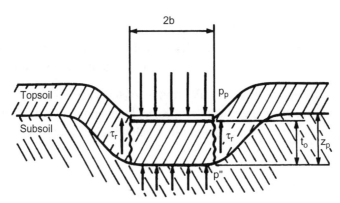

Figure 4.17: Failure mechanism of surface mat proposed by Bekker (Reprinted by permission from M.G. Bekker, *Introduction to Terrain–Vehicle Systems*, University of Michigan Press, 1969, copyright © by the University of Michigan)

for the failure of the surface mat due to tension (Wong et al., 1979). Based on this model, the punching load can be predicted.

In formulating the model, it is assumed that the muskeg consists of two layers; one is the surface mat of fibrous composition and the other is the underlying non-fibrous peat. The surface mat is further idealized as a membrane-like structure. This means that it can only sustain a force of tension directed along the tangent to the surface and cannot offer any resistance to bending. The underlying peat deposit is assumed to be a medium that offers a resistance proportional to its deformation in the vertical direction. This assumption regarding the characteristics of the underlying peat appears to have gained wide acceptance. The proposed model for the response of muskeg to vertical load is shown schematically in Figure 4.18 (Wong et al., 1979).

Based on this model, a theory for describing the load–sinkage relationship up to the critical sinkage (just prior to the breaking of the surface mat) has been proposed (Wong et al., 1979). For a rectangular loading area with a length ℓ much greater than its width b, the problem may be considered as a two-dimensional one. Considering the equilibrium of an element of the surface mat, one can obtain the following equations (see Figure 4.19):

$$dH = 0 \tag{4.26}$$

and

$$dV = -q\,dx \tag{4.27}$$

where H and V are the horizontal and vertical components of the tension force T (per unit length of the loading plate) of the surface mat, respectively, and q is the pressure exerted on the mat by the underlying peat.

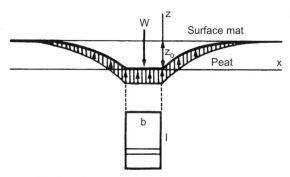

Figure 4.18: Failure mechanism for surface mat due to tension proposed by Wong et al. (Reprinted by permission of *ISTVS* from Wong et al., 1979)

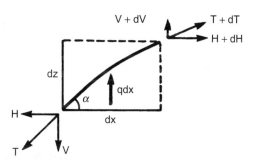

Figure 4.19: Equilibrium of an element of surface mat (Reprinted by permission of *ISTVS* from Wong et al., 1979)

Equation (4.26) indicates that the horizontal component of the tension force does not vary from point to point and that its value depends on the loading conditions.

The pressure q exerted on the mat by the underlying peat is assumed to be proportional to the deformation of the peat, that is

$$q = k(z_o - z) \tag{4.28}$$

where k is the coefficient representing the stiffness of the peat, and z_o is the sinkage of the loading plate as shown in Figure 4.18.

Since the mat can only sustain tension directed along the tangent to the surface of the mat, the ratio of the vertical to horizontal component of the tension force is given by

$$V/H = dz/dx \tag{4.29}$$

where $z(x)$ is a function which describes the profile of the mat as shown in Figure 4.18.

Combining Eqns (4.26) to (4.29), one obtains the following second order differential equation governing the profile of the muskeg when subject to a vertical load:

$$H \frac{d^2 z}{dx^2} = k(z - z_o) \tag{4.30}$$

Using non-dimensional coordinates $\bar{x} = x/z_o$ and $\bar{z} = z/z_o$, one can rewrite Eqn (4.30) in the following form:

$$\frac{H}{kz_o^2} \frac{d^2 \bar{z}}{d\bar{x}^2} = \bar{z} - 1 \tag{4.31}$$

From observations, it appears that the profiles of the mat for different sinkages resulting from various applied loads are similar. This implies that H/kz_o^2 is a constant.

Letting $H/kz_o^2 = m^2$, one can rewrite Eqn (4.31) as

$$m^2 \frac{d^2 \bar{z}}{d\bar{x}^2} = \bar{z} - 1 \tag{4.32}$$

m may be considered as a parameter characterizing the relation between the response to normal load of the mat and that of the underlying peat.

The solution to the non-dimensional differential equation, Eqn (4.32), is

$$\bar{z}(\bar{x}) = C_1 \exp(\bar{x}/m) + C_2 \exp(-\bar{x}/m) + 1 \tag{4.33}$$

where C_1 and C_2 are integration constants dependent upon boundary conditions. For the case under consideration, the boundary conditions are $\bar{x} = 0$, $\bar{z} - 0$,

$$\lim_{\bar{x} \to \infty} \bar{z} = 1$$

$$\lim_{\bar{x} \to \infty} \frac{d\bar{z}}{d\bar{x}} = 0 \tag{4.34}$$

Applying the above boundary conditions to Eqn (4.33), one obtains

$$C_1 = 0 \text{ and } C_2 = -1$$

The profile of the mat under load can therefore be described by

$$\bar{z}(\bar{x}) = 1 - \exp(-\bar{x}/m) \tag{4.35}$$

This shows that the profile of the mat is described by an exponential function. Figure 4.20 shows the profiles of mats with different values of m.

From Figure 4.19, the equilibrium equation of the loading plate with a high aspect ratio (i.e. $\ell \gg b$) is given by

$$W = b\ell q\big|_{x=0} + 2\ell V\big|_{x=0} \tag{4.36}$$

where W is the load applied to the plate.

The first term on the right-hand side of Eqn (4.36) represents the vertical force exerted on the plate by the underlying peat and the second term represents the vertical component of the tension force in the mat.

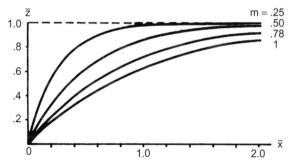

Figure 4.20: Profiles of surface mat under vertical load as a function of muskeg strength parameter m (Reprinted by permission of ISTVS from Wong et al., 1979)

From Eqn (4.28)

$$q|_{x=0} = kz_o \quad (4.37)$$

and from Eqns (4.29) and (4.35)

$$V|_{x=0} = H\frac{dz}{dx}\bigg|_{x=0} = H\frac{d\bar{z}}{d\bar{x}}\bigg|_{\bar{x}=0} = kmz_o^2 \quad (4.38)$$

Substituting the above expressions into Eqn (4.36), the relationship between the applied load W and the resulting sinkage z_o becomes

$$W = kz_o \ell b + 2mkz_o^2 \ell \quad (4.39)$$

Equation (4.39) may be rewritten to describe the pressure–sinkage relationship for muskeg prior to the breaking of the surface mat:

$$p = W/\ell b = kz_o + 2mkz_o^2/b$$
$$= kz_o + 2m_m z_o^2/b \quad (4.40)$$

where p is pressure and $m_m = mk$.

The theory described above may be extended to cover the cases where the width of the loading area b is significant in comparison with its length ℓ. In these cases, the vertical component of the tension force in the muskeg mat acting along the width b of the loading area should be taken into account. A modified load–sinkage relationship for muskeg prior to the failure of the surface mat is given by

$$W = kz_o A + 2mkz_o^2(\ell + b)$$
$$= kz_o A + m_m z_o^2 L \quad (4.41)$$

where A and L are the area and perimeter of the loading area, respectively.

Equation (4.41) may be rewritten to describe the general pressure–sinkage relationship for any loading area, including a circular one:

$$p = kz_o + m_m z_o^2 L/A$$
$$= kz_o + 4m_m z_o^2/D_h \quad (4.42)$$

where D_h is the so-called hydraulic diameter of the plate and is equal to $4A/L$.

Using the least squares method, the values of k and m_m can be derived from a measured pressure–sinkage curve. The procedure for obtaining the values of k and m_m from experimental data has been computerized and incorporated into the automatic data acquisition and processing system described in Chapter 3. Thus, immediately after the completion of a load–sinkage test, the values of k and m_m are automatically printed out. Comparisons between the measured curves and the fitted ones based on the values of k and m_m derived from the measured data are shown in Figures 4.15 and 4.16. The goodness-of-fit of Eqn (4.42) to the measured data is also shown in the figures.

Based on a large number of load–sinkage measurements, using various shapes and sizes of plates, the mean values of k, m_m and the critical sinkage z_b and their associated standard deviations for the muskeg at Site A were calculated and are given in Table 4.2.

Table 4.2: Values of k, m_m, z_b for the muskeg at Site A

Plate shape	Plate size (cm)	Penetration rate (cm/s)	k (kN/m³)		m_m (kN/m³)		z_b (cm)		Stiffness of peat measured with mat cut (kN/m³)	
			Mean value	SD	Mean value	SD	Mean value	SD	Mean value	SD
Circular	$r = 5$	2.5	407	212	97	31	20	2	410	59
		10	471	203	112	40	17	2		
	$r = 7.5$	2.5	374	108	92	30	21	1	424	95
		10	488	197	60	19	19	2		
	$r = 10$	2.5	290	107	51	16	23	2	339	83
		10	393	121	45	28	21	1		
Rectangular	3.75 × 22.5	2.5	451	17	76[1]	26[1]	23	2		
					88[2]	30[2]				
		10	612	201	48[1]	9[1]	23	3		
					56[2]	10[2]				
	5 × 30	2.5	278	91	46[1]	18[1]	25	2	490	113
					54[2]	21[2]				
	7.5 × 45	2.5	338	172	45[1]	34[1]			302	104
					53[2]	40[2]				
		10	530	48	44[1]	8[1]				
					51[2]	10[2]				

1 – obtained using Eqn (4.42); 2 – obtained using Eqn (4.40).

To directly determine the stiffness of the underlying peat and to verify the validity of the procedure described above for deriving the value of the stiffness of the peat from pressure–sinkage tests, the surface mat was cut around the perimeter of the plate to a depth of about 30 cm. Load was then applied to the muskeg surface and the resulting sinkage was measured.

Representative pressure–sinkage curves with the mat cut are shown in Figure 4.21. Using the least squares method, a linear equation was fitted to the measured curve. The straight line approximating the experimental curve (the chain line in Figure 4.21) is plotted following each test. The slope of the straight line characterizes the stiffness of the underlying peat. The values of the stiffness of the peat obtained with the surface mat cut using various shapes and sizes of plates are summarized in Table 4.2.

Based on the results of load–sinkage tests summarized in Table 4.2, the following observations may be made (Wong et al., 1982).

(a) From 19 pressure–sinkage tests with circular plates with radii of 5, 7.5 and 10 cm at a penetration rate of 2.5 cm/s, the mean values of k and m_m were found to be 357 and 80 kN/m³, respectively. Based on the results of 13 measurements with rectangular plates of 3.75 × 22.5 cm (hydraulic diameter D_h = 6.4 cm), 5 × 30 cm (D_h = 8.6 cm), and 7.5 × 45 cm (D_h = 12.9 cm), the mean values of k and m_m for Eqn (4.42) were found to be 356 and 56 kN/m³, respectively. Considering the possible variation of the muskeg structure and mat thickness from point to point in the field, the agreement between the values of k and m_m obtained using circular and rectangular plates should be regarded, in general, as reasonable. Thus, it can be concluded that as a first approximation, the proposed equation can be used to predict the pressure–sinkage relations for plates of different shapes in muskeg.

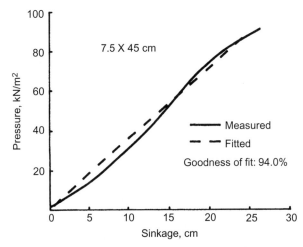

Figure 4.21: Pressure–sinkage relationship obtained with surface mat cut before tests for Petawawa Muskeg A

(b) As shown in Table 4.2, the mean values of k and m_m vary somewhat with plate size. For instance, at a penetration rate of 2.5 cm/s the mean values of k derived from test results using circular plates of radii 5, 7.5 and 10 cm are 407, 374 and 290 kN/m³, respectively. The difference between the mean values of k derived from the 5 and 10 cm plates is about 29%, while the corresponding difference between the values for the 7.5 and 10 cm plates is approximately 22%. Although these differences are not insignificant, one should bear in mind the possible natural variation of muskeg conditions from point to point in the field. In practice, a variation of 25% in this type of measurement is quite common in field work. A standard procedure for estimating the unknown population parameters by sample statistics was followed to analyse the data obtained from eight measurements using a circular plate with a radius of 5 cm. It was found that the lower and upper limits of the 95% confidence interval for the value of k are 230 and 584 kN/m³, respectively. It is noted that the mean values of k obtained using circular plates of 7.5 and 10 cm in radius fall in this range. From a theoretical viewpoint, the problem of whether the parameters k and m_m are dependent on the size of the plate used in the tests may require further study. In the meantime, for practical purposes the size of the plate used in pressure–sinkage tests for obtaining muskeg parameters should be as close as practicable to that of the contact surface of a tyre or a link of a track, in order to minimize the error in using test data for predicting the performance of full-scale vehicles.

(c) From the results summarized in Table 4.2, it can be seen that the value of k increases with the increase of penetration rate. For instance, for a circular plate with a radius of 7.5 cm, the mean value of k increases from 374 to 488 kN/m³ (equivalent to 30%), when the penetration rate increases from 2.5 to 10 cm/s. Similarly, the mean value of k for a rectangular plate of 7.5 × 45 cm increases from 338 to 530 kN/m³, when the penetration rate is increased from 2.5 to 10 cm/s.

The increase in the apparent stiffness k of the peat with the increase of penetration rate is probably due to the movement of water within the muskeg. It is also noted that the mean value of m_m, which is related to the behaviour of the surface mat, does not vary significantly with the increase in penetration rate in many cases. This indicates that as the penetration rate increases, the surface mat takes up a smaller proportion of the load.

(d) It appears that the value of the peat stiffness, k, derived from the results of pressure–sinkage tests using Eqn (4.40) or (4.42), agrees reasonably well with that obtained directly from measurements with the surface mat cut. For instance, the mean value of k derived from 19 measurements with circular plates of 5, 7.5 and 10 cm in radius at a penetration rate of 2.5 cm/s is 357 kN/m³, whereas that obtained directly from measurements with the surface mat cut is 391 kN/m³. Thus, it can be concluded that the procedure for deriving the stiffness of the peat from results of pressure–sinkage tests based on Eqn (4.40) or (4.42) is valid for all practical purposes. This also validates the basic features of the proposed theory that the load applied to the surface of a muskeg is carried partly by the peat through compression and partly by the mat through tension (Wong et al., 1979, 1982).

(e) The test results summarized in Table 4.2 indicate that the critical sinkage at which the surface mat is broken is more or less independent of the shape and size of the plate. For instance, the mean values of the critical sinkage, determined using circular plates of various sizes at a penetration rate of 2.5 cm/s, vary within a very narrow range of 20–23 cm. Similarly, the mean values of the critical sinkage measured using different sizes of rectangular plates do not differ much. The observation that the critical sinkage does not depend on the size and shape of the contact surface is consistent with the proposed theory that the failure of muskeg mat is primarily due to tension.

A similar set of pressure–sinkage tests was carried out on the muskeg at Site B, though it was not as extensive as that at Site A. To provide a basis for comparing the mechanical properties of the muskegs at the two sites, Table 4.3 summarizes some of the results of pressure–sinkage tests obtained at Site B. It can be seen that the mean values of k and m_m for the muskeg at Site B are higher than those obtained at Site A. It indicates that both the surface mat and the underlying peat at Site B were stronger (stiffer) than those at Site A. This is consistent with the observations described previously of the physical features of the muskeg at the two test sites.

4.2.3 Response to Repetitive Loading

When a vehicle is in motion, the terrain is subject to the repetitive loading of the consecutive wheels (or roadwheels on a track) of a vehicle. The knowledge of the response of muskeg to repetitive loading is, therefore, of importance to the prediction of the multipass performance of vehicle running gear.

Table 4.3: Values of k, m_m, z_b for the muskeg at Site B

Plate shape	Plate size (cm)	Penetration rate (cm/s)	k (kN/m³)		m_m (kN/m³)		z_b (cm)		Stiffness of peat measured with mat cut (kN/m³)	
			Mean value	SD	Mean value	SD	Mean value	SD	Mean value	SD
Circular	$r = 5$	2.5	954	431	99	65	21	3	860	339
		10	1243	228	99	40	22	6		
	$r = 7.5$	2.5	762	77	97	64	19	2	555	105
Rectangular	5 × 30	2.5	835	382	83[1]	57[1]				
					97[2]	67[2]				

1 – obtained using Eqn (4.42); 2 – obtained using Eqn (4.40).

The general features of the load–sinkage curve for muskeg under repetitive loading are shown in Figure 4.22. When load is first applied to the muskeg, the load–sinkage relation follows the usual pattern as OA shown in Figure 4.22. When the load is reduced at A, the load–sinkage relation follows line AB. The inclination of line AB indicates that during unloading a certain amount of elastic rebound occurs. When the load is reapplied at B, the load–sinkage relation follows a different path from that during unloading. This indicates that a certain amount of hysteresis exists during the unloading–reloading cycle. During reloading water is being squeezed out of a certain zone under the plate, while during unloading the water is flowing in. The hydrodynamic effect caused by the water flow results in the noticeable difference in the pressure-sinkage relation between unloading and reloading. When the reapplied load exceeds that at which the preceding unloading begins (point A shown in Figure 4.22) additional sinkage results. With further increase of load, the load–sinkage relation appears to follow the continuous loading path as AC shown in Figure 4.22. The characteristics of the second unloading–reloading cycle which begins at point C are quite similar to those of the first.

Based on the observations described above, the response of muskeg to repetitive loading may be idealized as follows. When the muskeg is subject to continuously increasing load, such as along line OA shown in Figure 4.22, the pressure–sinkage relation is described by Eqn (4.40) or (4.42). For the unloading–reloading cycle, the unloading path does not coincide with that of reloading. However, as a first approximation, the average pressure–sinkage relation during both the unloading and reloading (such as AB or CD in Figure 4.22) may be represented by Eqn (4.23), when the unloading sinkage z_u is less than the critical sinkage z_b.

It is observed that in muskeg the value of the pressure–sinkage constant k_u in Eqn (4.23) also varies with the sinkage z_u at which unloading begins. Based on the results of repetitive loading tests performed at Site A using circular plates, the values of k_u are plotted against z_u in

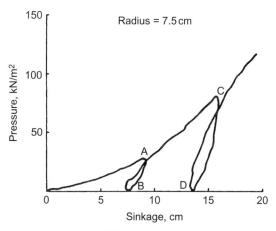

Figure 4.22: Response to repetitive normal load of Petawawa Muskeg A

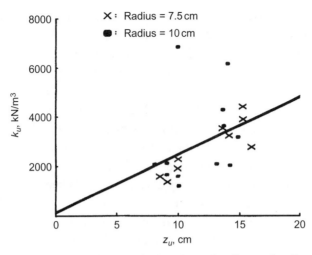

Figure 4.23: Variation of terrain stiffness during the unloading–reloading cycle with initial unloading sinkage for Petawawa Muskeg A

Figure 4.23. It should be noted that when z_u approaches zero, the value of k_u may be considered to be equal to the slope of line OA (Figure 4.22) at the origin. This is because when z_u is small (i.e. p_u is small), the muskeg can be considered as an elastic medium with a linear pressure–sinkage relation. Consequently, the value of k_u approaches that of k in Eqn (4.40) or (4.42), when z_u approaches zero. An analysis was performed on the data shown in Figure 4.23 and as a first approximation, Eqn (4.24) can be used to describe the relationship between k_u and z_u, when z_u is less than the critical sinkage z_b:

$$k_u = k_o + A_u z_u$$

where k_o and A_u are the intercept and the slope of the regression line, respectively.

Based on the data obtained using various plates, the values of k_o and A_u for Site A were calculated and are given in Table 4.4. From the table it appears that for all practical purposes the parameters characterizing the unloading and reloading characteristics of the muskeg may be regarded as independent of plate shape.

4.3 Response of Snow Covers

The mechanical properties of two snow-covered terrains in the Petawawa area, Ontario, Canada, were measured and characterized, and the results are described in this section.

In the northern temperate zones, the snow on the ground is often subject to 'melt–freeze' cycles during the winter season. Consequently, crusts (ice layers) of significant strength form at the surface of snow covers in open areas. As the accumulation of snow on top of the crusts progresses, snow covers containing ice layers are formed. Figures 4.24 and 4.25 show,

Table 4.4: Values of k_o and A_u for the muskeg at Site A

Plate	k_0 (kN/m³)	A_u (kN/m⁴)	95% confidence limits for A_u
Circular plates $r = 7.5$ and 10 cm	123	23 540	upper 23 610 lower 23 480
Rectangular plates 5×30 and 7.5×45 cm	334	25 430	upper 25 520 lower 25 340

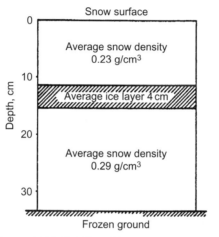

Figure 4.24: Profile of Petawawa Snow A

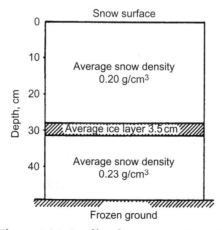

Figure 4.25: Profile of Petawawa Snow B

respectively, the profiles of two snow covers, referred to as Petawawa Snow A and B (Wong, and Preston-Thomas, 1983b).

It can be seen that a significant ice layer was present at a certain depth below the surface in these two snow covers. From the point of view of vehicle mobility over snow-covered terrain, the bearing capacity of the ice layer and the overall pressure–sinkage characteristics of the snow cover are of considerable interest. It is apparent that if the vertical load exerted by the vehicle is lower than the bearing capacity of the ice layer, the vehicle will float on top of the ice layer. On the other hand, when the load exerted by the vehicle is higher than the bearing capacity of the ice layer, the ice layer will break and a higher vehicle sinkage and consequently a higher motion resistance will result.

The compressibility characteristics of fresh and sintered snow samples obtained under laboratory conditions have been reported in the literature. However, the load–deformation characteristics of snow covers containing ice layers, as they exist in the natural environment, have not been investigated to any great extent. This section examines, from the vehicle mobility viewpoint, the bearing capacity of an ice layer in a snow cover and its overall pressure–sinkage characteristics.

4.3.1 Bearing Capacity of an Ice Layer in a Snow Cover

The pressure–sinkage curves for snow covers A and B were obtained using the bevameter and the automatic data acquisition and processing system described in Chapter 3. The mean curves, obtained using circular plates of radii 5 and 7.5 cm at a penetration rate of approximately 2.5 cm/s, for snow covers A and B are shown in Figures 4.26 and 4.27, respectively.

It can be seen from the curves that the pressure first increased gradually with sinkage as the snow within a certain boundary under the plate was deformed. It was observed that this

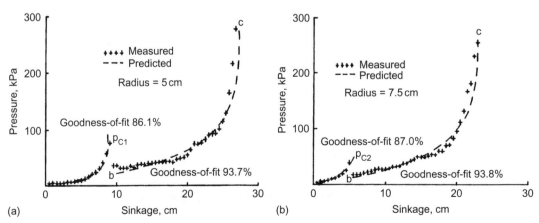

Figure 4.26: Pressure–sinkage relationships for Petawawa Snow A

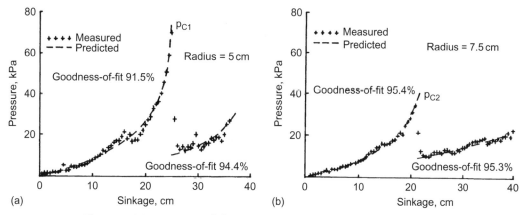

Figure 4.27: Pressure–sinkage relationships for Petawawa Snow B

deformation was more or less unidirectional. When the lower boundary of the deformation zone of the snow under the plate reached the ice layer the pressure increased rapidly with the increase of sinkage. When the applied pressure exceeded a certain level, the ice layer failed, resulting in a sudden drop in pressure. After the ice layer was fractured, further penetration of the plate produced increasing deformation of the snow beneath the ice layer. As the plate approached the frozen ground at the base of the snow cover, the pressure again increased rapidly and the pressure–sinkage curve approached an asymptote, as can be seen from Figure 4.26. This trend was not, however, apparent for the pressure–sinkage curves obtained in snow cover B shown in Figure 4.27, because the overall depth of the snow deposit was considerably greater than the maximum penetration of the plate, which was limited by the capability of the equipment used.

For snow covers with a significant ice layer, the determination of the load that causes the collapse of the ice layer is of importance to the prediction of vehicle mobility. The bearing capacity of ice sheets and methods for estimating the collapse loads under circular and strip loading areas, based on the theory of plasticity, have been discussed by Meyerhof (1960). When a light load, much lower than the ultimate, which is defined later in the section, is applied over a circular area on a large ice sheet of uniform thickness, the stress and deflection of the sheet may be predicted as for a thin, elastic and infinite plate on an elastic foundation. As the load applied to the ice sheet increases to a certain level, the stresses due to bending in the ice sheet below the loading area become equal to the flexural strength of the ice, and the sheet begins to yield leading to radial tension cracks at the bottom of the sheet, as shown in Figure 4.28. With further increase of the load, the cracks increase in length and number until the bending stresses along a circumferential section (such as CDE shown in Figure 4.28) of the sheet become equal to the flexural strength of the ice sheet, and a circumferential tension crack is formed on the top of the sheet. The ultimate bearing capacity or collapse load of the ice sheet is then reached. Based on the assumption that the ice sheet, prior to collapse, is

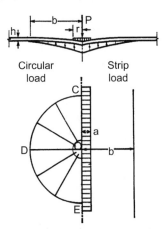

Figure 4.28: Infinite plate under circular and strip loads (Reprinted by permission of *ASCE Journal of Engineering Mechanics*, from Meyerhof, 1960)

equivalent to a thin, rigid, ideally plastic and infinite plate, the collapse load for a circular contact area, which produces circumferential tension cracks, may be expressed by (Meyerhof, 1960)

$$P_{co} = 3.3\pi(1 + 6r/b)M_o \text{ (for } 0.02 < r/b < 0.2) \tag{4.43}$$

where P_{co} is the collapse load, r is the radius of the contact area, b is the radius of the circumferential yield–hinge–circle (Figure 4.28), and M_o is the limit bending moment per unit length of the ice sheet. This bending moment per unit length is equal to $f_o h^2/4$, where f_o is the yield strength of the ice sheet in tension and h is the thickness of the ice sheet.

The radius b may be expressed in terms of the characteristic length L of the ice sheet, which is given by (Meyerhof, 1960)

$$L = \left[\frac{Eh^3}{12(1-\mu^2)k}\right]^{1/4} \tag{4.44}$$

where E and μ are the modulus of elasticity and Poisson's ratio of the ice sheet, respectively, and k is the stiffness of the material supporting the ice sheet.

It was suggested by Meyerhof (1960) that as an approximation, an effective radius b may be taken as $3.9L$. Equation (4.43) therefore may be rewritten as

$$P_{co} = 3.3\pi(1 + 3r/2L)M_o \text{ (for } 0.05 < r/L < 1) \tag{4.45}$$

Following a similar approach, the collapse load per unit length of an ice sheet for a strip loading area may be approximated as (Meyerhof, 1960)

$$P_{co} = \frac{3.6M_o}{L - a/2} \quad \text{(for } 0 < a/L < 1\text{)} \tag{4.46}$$

where a is the half-width of the loading area as shown in Figure 4.28.

When applying the above equations to the estimation of the collapse load of the ice layer in a snow cover, it should be noted that the load on the ice layer is transmitted from the plate through the compressed snow sandwiched between the plate and the ice layer. Consequently, the exact dimensions of the loading area on the ice layer may not be easy to ascertain. However, as mentioned previously, field observations showed that the snow beneath the plate underwent more or less unidirectional deformation. Therefore, as a first approximation, the loading area on the ice layer may be taken as the area of the plate. It should also be mentioned that the weight of the snow on top of the ice layer produces a pressure distributed over the surface of the ice layer, which is referred to as a surcharge. However, usually the density of undisturbed snow is relatively low. Therefore, if the snow on top of the ice layer is not very deep, the effect of the surcharge on the failure of the ice layer may be neglected in a first approximation.

It is noted that Eqns (4.45) and (4.46) contain two parameters, L and M_o. The values of these two parameters may be derived from the results of pressure–sinkage tests using two different sizes of plates, such as those shown in Figures 4.26 and 4.27. From the curves, the collapse loads P_{c1} and P_{c2} for plates with radii r_1 and r_2 corresponding to the collapse pressures p_{c1} and p_{c2} shown in Figures 4.26 and 4.27, can be defined as

$$P_{c1} = \pi r_1^2 p_{c1} \tag{4.47}$$

and

$$P_{c2} = \pi r_2^2 p_{c2} \tag{4.48}$$

Making use of Eqn (4.45), one can establish the following relations:

$$P_{c1} = 3.3\pi(1 + 3r_1/2L)M_o \tag{4.49}$$

and

$$P_{c2} = 3.3\pi(1 + 3r_2/2L)M_o \tag{4.50}$$

From these two relations, the values of L and M_o can be determined by

$$L = \frac{3}{2}\left(\frac{P_{c2}r_1 - P_{c1}r_2}{P_{c1} - P_{c2}}\right) \tag{4.51}$$

and

$$M_o = \frac{P_{c1}}{3.3[1 + r_1(P_{c1} - P_{c2})/(P_{c2}r_1 - P_{c1}r_2)]} \tag{4.52}$$

Based on the results of the pressure–sinkage tests shown in Figures 4.26 and 4.27, the values of L and M_o for the ice layers in snow covers A and B are 16.7 cm and 40.2 N, and 26.1 cm and 41.2 N, respectively.

Having determined the values of L and M_o for the ice layer in a particular snow cover, one can then use Eqn (4.45) to predict the collapse load, or the following equation to predict the corresponding collapse pressure p_{co}, for any particular circular loading area with radius r:

$$p_{co} = \frac{3.3\pi(1 + 3r/2L)M_o}{\pi r^2} \quad \text{(for } 0.05 < r/L < 1\text{)} \tag{4.53}$$

Similarly, with the values of L and M_o known, one can use Eqn (4.46) to predict the collapse load per unit length of the ice layer, or the following equation to predict the corresponding collapse pressure p_{co}, for any particular strip loading area with half-width a:

$$p_{co} = \frac{3.6M_o}{(2aL - a^2)} \quad \text{(for } 0 < a/L < 1\text{)} \tag{4.54}$$

It should be mentioned that in the above analysis the loads are assumed to be distributed over a sufficiently large area so that local shear stresses are negligible and the full, limit bending moment is developed. A small contact area may result in either local shear failure or local circumferential tension failure (see Figure 4.29) before the limit bending moment can be attained, as pointed out by Meyerhof (1960).

The ultimate load P_{us} over a circular area of radius r due to local shear failure on an ice layer with thickness much greater than r may be estimated by

$$P_{us} = 2.85\pi r^2 f_o \text{ to } 3.07\pi r^2 f_o \tag{4.55}$$

where f_o is the yield strength of the ice layer in tension.

The lower limit of Eqn (4.55) applies to a perfectly smooth contact surface, whereas the upper limit applies to a rough surface. To avoid local shearing failure, the minimum radius for the contact area with a load P is given by

$$r = 0.33(P/f_o)^{1/2} \quad \text{(for } r/h < 0.2\text{)} \tag{4.56}$$

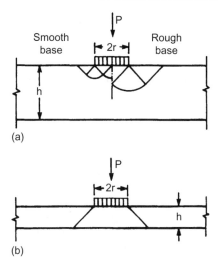

Figure 4.29: Load bearing capacity failures: (a) local shear failure, and (b) local circumferential tension failure (Reprinted by permission of *ASCE Journal of Engineering Mechanics*, from Meyerhof, 1960)

If the contact radius r is greater than about one-fifth of the thickness of the ice layer, the resistance to local shear failure decreases with the increase of the ratio r/h. At about $r = h$, local circumferential tension failure may occur and the corresponding ultimate load P_{ut} may be estimated by

$$P_{ut} = (2r + h)\pi h f_o \qquad (4.57)$$

To avoid local circumferential tension failure, the minimum radius for the contact area with a load P is given by

$$r = P/2\pi h f_o - h/2 \quad \text{(for } r \geq h\text{)} \qquad (4.58)$$

Following a similar approach, the ultimate loads due to local shear failure and local circumferential tension failure under a strip loading area can be derived.

The results of the preceding analysis indicate that to determine the actual failure mechanism of an ice layer under a given loading area, one may calculate the load P_{co} due to circumferential tension cracking and then compare it with either the ultimate load P_{us} due to local shear failure or the ultimate load P_{ut} due to local circumferential tension failure, dependent upon the ratio of r/h as mentioned previously. The smaller of the two will be the actual failure load for a given ice layer.

Based on the values of h, L and M_o for the ice layers in the two snow covers A and B described previously, it was determined that under circular loading areas with radii of 5 and

7.5 cm, the ice layers collapsed due to circumferential tension cracks and that their ultimate bearing capacities were defined by Eqn (4.45).

4.3.2 Pressure–Sinkage Relationship of Snow Covers with a Significant Ice Layer

The procedure described in the preceding section enables one to estimate the ultimate bearing capacity of an ice layer in a snow cover. However, to be able to predict the sinkage of a vehicle running gear, the overall pressure–sinkage relation of the snow cover, including that before and after the failure of the ice layer, has to be characterized (Wong and Preston-Thomas, 1983b).

Based on the results shown in Figures 4.26 and 4.27, it is proposed that the pressure–sinkage relation before as well as after the failure of the ice layer may be described by an exponential function of the following form:

$$z = z_w \left[1 - \exp(-p/p_w) \right] \quad (4.59)$$

Rearranging the above equation, one can obtain

$$p = p_w \left[-\ln(1 - z/z_w) \right] \quad (4.60)$$

where p and z are pressure and sinkage, respectively, and p_w and z_w are empirical parameters, the values of which are dependent upon the structural profile and physical properties of the snow cover. z_w defines the asymptote of the pressure–sinkage curve. It should be noted that the original snow surface is used as the origin for the sinkage z in Eqns (4.59) and (4.60).

A procedure based on the least squares principle has been developed for deriving the best values of p_w and z_w from a given pressure–sinkage curve. The chain line in Figures 4.26 and 4.27 represent the fitted curves obtained using this procedure.

It is interesting to find that Eqn (4.59) fits equally well to the stress–deformation curves of undisturbed snow obtained from confined compression tests reported in the literature.

Further examination of the experimental data shown in Figures 4.26 and 4.27 reveals that the size of the plate may have noticeable effects on the pressure–sinkage relation under certain circumstances. For example, section *bc* of the pressure–sinkage curve for a plate with a radius of 7.5 cm shown in Figure 4.26(b) approaches its asymptote at a lower sinkage than that for a plate with a radius of 5 cm shown in Figure 4.26(a). This may be explained by the fact that the lower boundary of the deformation zone of the snow under a larger plate reaches the frozen ground before that under a smaller plate at the same sinkage. As Bekker (1969) put it, the larger plate 'senses' the hard layer (or the frozen ground in the present case) sooner than the smaller plate.

Based on the results of field tests obtained using two different sizes of plate, it is proposed that as a first approximation, the empirical parameters p_w and z_w may be expressed as

$$p_w = k_{p1} + rk_{p2} \tag{4.61}$$

$$z_w = k_{z1} + k_{z2}/r \tag{4.62}$$

where r is the radius (or width) of the plate, and k_{p1}, k_{p2}, k_{z1} and k_{z2} are parameters, the values of which are dependent upon the depth and physical properties of the snow cover.

As mentioned previously, from a pressure–sinkage curve for a particular size of plate, a specific set of values of p_w and z_w can be derived. Therefore, the values of k_{p1}, k_{p2}, k_{z1} and k_{z2} can be derived from the two sets of values of p_w and z_w obtained using two different sizes of plates. The values of k_{p1}, k_{p2}, k_{z1} and k_{z2} for the pressure–sinkage curves shown in Figures 4.26 and 4.27 are given in Table 4.5.

Based on the analysis given above, the procedure for predicting the overall pressure–sinkage relation for a given loading area, such as that of a tyre or that of a track link, in a snow cover containing an ice layer may be summarized as follows:

(a) Conduct a series of pressure–sinkage tests using two different sizes of plates. The sizes of the plates used in the tests should be selected in accordance with the principles described in the preceding section, so that for commonly encountered ice layers, failure due to circumferential tension cracks will occur. This will ensure that all the basic strength parameters of the ice layer can be obtained from the results of pressure–sinkage tests.

Table 4.5: Values for k_{p1}, k_{p2}, k_{z1} and k_{z2}

Snow cover	A		B	
Section of the pressure–sinkage curve	Before failure of ice Layer	After failure of ice layer	Before failure of ice layer	After failure of ice layer
k_{p1} (kPa)	3.18	52.71	16.3	10.8
k_{p2} (kPa/cm)	2.34	−0.48	0	0
k_{z1} (cm)	0.85	14.21	24.8	41
k_{z2} (cm^2)	39.69	67.34	0	0

It should be pointed out that if the size and shape of the loading element used in the pressure–sinkage tests are very different from that of the contact area of a vehicle running gear, such as is the case when a cone is used as a loading element, the failure mode of the ice layer and the pressure–sinkage relationship will be quite different from that under a vehicle running gear. Therefore, the strength properties of the ice layer and the snow cover relevant to vehicle mobility cannot be derived from the results of such tests.

(b) Identify the failure loads of the ice layer and derive the values of the characteristic length L, the limit bending moment per unit length M_o, and hence the yield strength in tension f_o ($f_o = 4M_o/h$) of the ice layer.

(c) From experimental data, derive the values of the pressure–sinkage parameters k_{p1}, k_{p2}, k_{z1} and k_{z2} for the snow cover before and after the failure of the ice layer.

(d) Based on the values of the parameters derived from the pressure–sinkage relation for a given loading area such as that of a tyre or that of a track link, predict the pressure–sinkage relation prior to the failure of the ice layer, using either Eqn (4.59) or (4.60) and the corresponding values of k_{p1}, k_{p2}, k_{z1} and k_{z2}.

(e) Based on the ratio of the radius of the given loading area to the thickness of the ice layer, determine the possible local failure mode, that is, local shear failure or local circumferential tension failure. Calculate the appropriate ultimate load, P_{us} or P_{ut}.

(f) Calculate the collapse load P_{co} due to circumferential tension cracks for the given loading area. Compare the value of P_{co} with that of P_{us} or P_{ut} determined in step (e). The smaller of the two will be the actual failure load for the given loading area.

(g) From the failure load determined in step (f) and the pressure–sinkage relation before the failure of the ice layer determined in step (d), identify the sinkage for the given loading area at which the failure of the ice layer will occur.

(h) Predict the pressure–sinkage relation after the failure of the ice layer, using either Eqn (4.59) or (4.60) and the appropriate values of k_{p1}, k_{p2}, k_{z1} and k_{z2}.

It should be mentioned that various methods have been proposed over the years for characterizing the pressure–sinkage relationship of stratified and non-homogeneous terrains. A general discussion of this problem was given by Bekker (1969).

4.3.3 Response to Repetitive Loading

A typical response of the snow covers to repetitive loading is shown in Figure 4.30. It is quite similar to that for the mineral terrain shown in Figure 4.11. Therefore, Eqns (4.23) and (4.24)

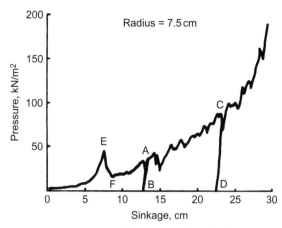

Figure 4.30: Response to repetitive normal load of Petawawa Snow A

Table 4.6: Values of the repetitive loading parameters for two snow covers

Parameters	Petawawa Snow A	Petawawa Snow B
k_o (kN/m^3)	0	0
A_u (kN/m^4)	109 600	25 923

can equally be used to describe the pressure–sinkage relationship of the snow covers during unloading–reloading cycles (Wong et al., 1984).

The values of the parameters, k_o and A_u, characterizing the behaviour of the two snow covers A and B during the unloading and reloading cycle are given in Table 4.6.

CHAPTER 5
Characterization of the Shearing Behaviour of Terrains

A vehicle applies shear load to the terrain surface through its running gear, which results in the development of thrust and associated slip. To predict the tractive performance of an off-road vehicle, it is essential to measure the shear stress–shear displacement relationship under various normal pressures and to determine the shear strength of the terrain. This chapter describes the results of the measurement and characterization of the shearing behaviour of a variety of terrains, including mineral terrain, muskeg, and snow-covered terrain. The measurements were made using a vehicle-mounted bevameter with the associated computerized data acquisition and processing system described in Chapter 3.

5.1 Characterization of the Shear Stress–Displacement Relationships

The knowledge of the shear stress–displacement relationship of the terrain is of importance to the prediction of the thrust–slip relationship of a vehicle running gear over unprepared terrain. Consequently, the characterization of the shear stress–displacement relationship has attracted considerable interest, and a number of methods have been proposed (Wong and Preston-Thomas, 1983a).

For 'brittle' soils which display a 'hump' of maximum shear stress, Bekker (1956) proposed the following equation to describe the shear stress–displacement relationship:

$$s/s_{max} = \frac{\exp\left[-K_2 + \sqrt{K_2^2 - 1}\right]K_1 j - \exp\left[-K_2 - \sqrt{K_2^2 - 1}\right]K_1 j}{\exp\left[-K_2 + \sqrt{K_2^2 - 1}\right]K_1 j_0 - \exp\left[-K_2 - \sqrt{K_2^2 - 1}\right]K_1 j_0} \quad (5.1)$$

where s and s_{max} are the shear stress and the maximum shear stress, respectively, j and j_0 are the shear displacement and the shear displacement at s_{max}, respectively, and K_1 and K_2 are empirical constants. Various methods have been proposed for deriving the values of K_1 and K_2 from measured shear curves. The derivation of the values of K_1 and K_2 from measured shear curves is, in general, a very involved process, and as pointed out by Bekker (1969), 'accurate definitions of K_1 and K_2 values are difficult without computer methods'. It should be mentioned that Eqn (5.1) is identical in form to that describing the aperiodic motion of an overdamped, single-degree-of-freedom system in free vibration. Thus, when $j \gg j_0$, s approaches zero.

For 'plastic' soils which do not exhibit a 'hump' of maximum shear stress, a modified version of Bekker's equation containing only one constant was proposed by Janosi and Hanamoto (1961) and is widely used in practice:

$$s/s_{max} = 1 - \exp(-j/K) \qquad (5.2)$$

where K is usually referred to as the shear deformation parameter and is a measure of the magnitude of the shear displacement required for the development of the maximum shear stress.

It has been found that for certain types of sand, saturated clay, fresh snow and peat, and for rubber–sand, rubber–snow, rubber–muskeg mat and rubber–peat shearing, the shear stress–displacement relationship can be described fairly well by Eqn (5.2) (Wong and Preston-Thomas, 1983a).

For the internal shearing of muskeg mat, the shear curves exhibit a 'hump' of maximum shear stress and then the shear stress decreases continually with the increase of shear displacement. It has been found that the following equation fits this type of shear curve fairly well (Wong et al., 1979; Wong and Preston-Thomas, 1983a):

$$s/s_{max} = (j/K_w)\exp(1 - j/K_w) \qquad (5.3)$$

where K_w is a constant. In physical terms, K_w is equal to the shear displacement j_0 where the shear stress peaks.

In addition to the above methods, a hypothesis has been put forward by Sela (1964) that the shear stress–displacement relationship may be described by the sum of two exponential functions, one for the cohesive component and the other for the frictional component. However, as pointed out by Sela (1964), the hypothesis has not yet been verified by experimental data.

There are shear curves that have different characteristics from those described above. They display a 'hump' of maximum shear stress at a certain shear displacement j_0. However, beyond j_0 the shear stress decreases and approaches a more or less constant level of residual stress. This type of curve was observed during field measurements of the shearing characteristics of snow covers and certain types of loam (Wong, 1983; Wong and Preston-Thomas, 1983a). A representative shear curve of this kind is shown in Figure 5.1.

It has been found that this type of shear curve cannot be described by the equations given above with sufficiently high accuracy.

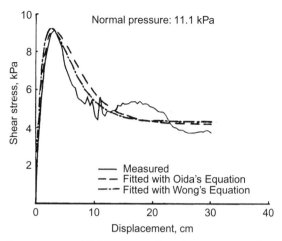

Figure 5.1: A shear curve exhibiting a peak and constant residual shear stress

Inspired by the work of Pokrovski and of Kacigin and Guskov (1968), Oida (1979) proposed the following equation for describing the type of shear stress–displacement relationship shown in Figure 5.1.

$$s/s_{max} = K_r \left[1 - \frac{\sqrt{1-K_r}\left[1+\left(\sqrt{1-K_r}-1\right)/K_r\right]^{j/K_w}}{\sqrt{1-K_r}(1-2/K_r)+2/K_r-2} \right.$$

$$\left. \times \left[1 - \left[1+\left(\sqrt{1-K_r}-1\right)/K_r\right]^{j/K_w}\right]\right]$$
(5.4)

where K_r is the ratio of the residual shear stress s_r to the maximum shear stress s_{max}, and K_w is the shear displacement where s_{max} occurs (i.e. $K_w = j_0$).

An analysis of Eqn (5.4) shows that it satisfies the following conditions:
(a) when $j = K_w$, $s = s_{max}$;
(b) when $j = K_w$, $ds/dj = 0$;
(c) when $j \gg K_w$, $s \to s_r$.

Figure 5.2 shows the relationship between s/s_{max} and j/K_w at various values of K_r based on Eqn (5.4). One major difference between Eqn (5.1) and Eqn (5.4) is that when $j \gg j_0$, s approaches zero using Eqn (5.1), whereas using Eqn (5.4), when $j \gg K_w$, s approaches s_r.

Although Eqn (5.4) has certain attractive features, it is relatively complex. Of the two constants K_w and K_r in the equation, K_w may be easily identified from the measured shear curve, since K_w represents the shear displacement where the maximum shear stress occurs

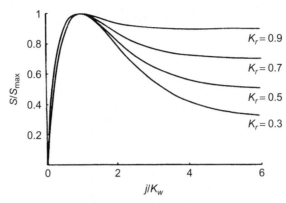

Figure 5.2: Variation of the shear stress ratio with shear displacement ratio for the Oida equation (Oida, 1979)

(i.e. $K_w = j_0$). For a 'smooth' shear curve, K_r can usually be identified directly. However, measured shear curves, particularly those obtained in the field, are seldom smooth. To minimize the error in curve fitting, an optimization procedure has to be introduced to derive an optimum value of K_r from the shear data. Owing to the non-linear nature of the relation between s/s_{max} and K_r, an iterative procedure may have to be employed to obtain an optimum value of K_r for a particular measured shear curve. It has been found that this is a relatively time-consuming procedure. Furthermore, the application of Eqn (5.4) to the prediction of the thrust–slip relationship of a vehicle running gear is quite an involved process, because of the complex form of the relation between shear stress and shear displacement described by Eqn (5.4).

In an attempt to overcome some of the drawbacks of Eqn (5.4), the following equation which, similar to Oida's equation, originates from Pokrovski's work, is suggested by Wong (1983):

$$s/s_{max} = K_r \left[1 + [1/(K_r(1 - 1/e)) - 1]\exp(1 - j/K_w) \right] [1 - \exp(-j/K_w)] \quad (5.5)$$

The definitions of K_r and K_w in Eqn (5.5) are the same as those in Eqn (5.4).

Equation (5.5) satisfies the conditions that when $j = K_w$, $s = s_{max}$; and when $j \gg K_w$, s approaches s_r. However, it can only satisfy the condition that when $j = K_w$, $ds/dj = 0$, if $K_r = 0.66$. This means that when $K_r \neq 0.66$, the value of the maximum shear stress predicted using Eqn (5.5) will be slightly different from that obtained from experiments. However, the difference in most cases ($0.3 < K_r < 0.9$) is less than 3%. It should also be pointed out that when $K_r = 0.66$ the maximum shear stress predicted using Eqn (5.5) occurs at a shear displacement slightly different from K_w. The relationship between s/s_{max} and j/K_w at various values of K_r based on Eqn (5.5) is illustrated in Figure 5.3. It has been found that Eqn (5.5) fits

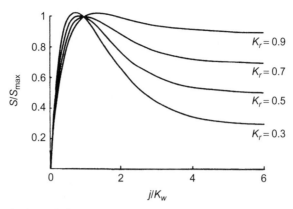

Figure 5.3: Variation of the shear stress ratio with the shear displacement ratio for the Wong equation (Wong, 1983)

well to experimental shear curves obtained from snow-covered terrains and certain types of loam (Wong, 1983; Wong and Preston-Thomas, 1983a).

In contrast with Eqn (5.4), the procedure for deriving an optimum value of K_r for Eqn (5.5) is relatively straightforward, since the relation between s/s_{max} and K_r in Eqn (5.5) is simpler than that in Eqn (5.4). Following the least squares principle, to obtain the optimum value of K_r of Eqn (5.5) for a particular shear curve, the following function F, which represents the sum of the squares of the deviations, should be minimized:

$$F = \sum\{s/s_{max} - K_r\{1 + [1/(K_r(1 - 1/e)) - 1]\exp(1 - j/K_w)\} \\ \times [1 - \exp(-j/K_w)]\}^2 \tag{5.6}$$

To minimize the function F, the first partial derivative of F with respect to K_r is taken and set to zero. This leads to the following expression for determining the optimum value of K_r for a particular set of shear data:

$$K_r = \{\sum s/s_{max}[1 - \exp(1 - j/K_w)][1 - \exp(-j/K_w)] \\ - \sum[1/(1 - 1/e)][1 - \exp(1 - j/K_w)][1 - \exp(-j/K_w)]^2 \\ \times [\exp(1 - j/K_w)]\}/\sum[1 - \exp(1 - j/K_w)]^2[1 - \exp(-j/K_w)]^2 \tag{5.7}$$

It has been found that the computing time required to obtain the optimum value of K_r for Eqn (5.5) using Eqn (5.7) is shorter than that for Eqn (5.4) using an iterative procedure. Figure 5.1 shows a comparison between the fitted curves using Eqns (5.4) and (5.5) for a particular experimental shear curve. The values of the goodness-of-fit for the Oida equation and the Wong equation are 85.4 and 86.6%, respectively. The goodness-of-fit is defined by

$$\varepsilon = 1 - \frac{\sqrt{\left[\sum (s_m - s_c)^2/(N-2)\right]}}{\sum s_m/N} \tag{5.8}$$

where s_m is the measured shear stress, s_c is the calculated shear stress at a particular shear displacement using Eqn (5.4) or Eqn (5.5), and N is the number of data points used in curve fitting.

It is interesting to point out that Eqns (5.2), (5.3) and (5.5) belong to the same family of exponential functions. In fact, Eqns (5.2) and (5.3) may be regarded as special forms of Eqn (5.5), since it contains the exponential terms $\exp(1 - j/K_w)$ used in Eqn (5.3), and $[1 - \exp(-j/K_w)]$ similar to that used in Eqn (5.2).

Based on the analysis given above and on the results of measurements of the shearing behaviour of a variety of terrains including mineral terrain, muskeg and snow-covered terrain, it can be summarized that the shear stress–displacement relationship of terrain may be characterized by one of the following three methods.

(a) If a shear curve does not display a 'hump' of maximum shear stress, but rather the shear stress increases with shear displacement and approaches a constant value, such as that shown in Figure 5.4, then Eqn (5.2) may be used to approximate the measured shear stress–shear displacement relationship:

$$s/s_{max} = 1 - \exp(-j/K)$$

It is interesting to note that the shear curve of simple exponential form shown in Figure 5.4 and described by Eqn (5.2) is similar in shape to the curve 'b' in Figure 2.24

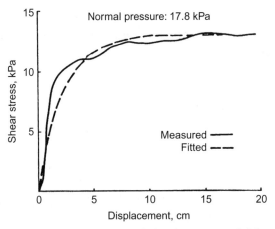

Figure 5.4: A shear curve of simple exponential form

describing the stress–strain relationship for soil in a loose state based on the critical state soil mechanics.

To obtain the best value of K for a given set of measured data, the following procedure based on the weighted least squares method is suggested. Rearranging Eqn (5.2) and taking logarithms, one obtains

$$\ln(1 - s/s_{max}) = -j/K \tag{5.9}$$

If an equal liability to error for all observations is assumed, then the appropriate value of K can be obtained by minimizing the value of the following function using a weighting factor of $(1 - s/s_{max})^2$:

$$F = \sum (1 - s/s_{max})^2 \left[\ln(1 - s/s_{max}) + j/K\right]^2 \tag{5.10}$$

To obtain the value of K that minimizes the above function, the first partial derivative of F with respect to K is taken and set to zero. This leads to the following condition:

$$\sum (1 - s/s_{max})^2 j\left[\ln(1 - s/s_{max}) + j/K\right] = 0 \tag{5.11}$$

From the above equation, the appropriate value of K that minimizes the error in curve fitting is given by

$$K = -\frac{\sum (1 - s/s_{max})^2 j^2}{\sum (1 - s/s_{max})^2 j\left[\ln(1 - s/s_{max})\right]} \tag{5.12}$$

It should be mentioned that an appropriate value of K can generally be obtained using Eqn (5.12) for a relatively 'smooth' exponential curve. For measured shear curves that are not so 'smooth', however, an appropriate value of s_{max} may be difficult to identify. Therefore, to obtain an overall good fit for the measured shear curve, both K and s_{max} may have to be regarded as parameters to be optimized. Based on the least squares principle, an iterative procedure has been developed for deriving the optimum values of K and s_{max} for a measured shear curve that is not of a 'smooth' exponential form.

(b) If a shear curve exhibits a 'hump' of maximum shear stress and then the shear stress decreases continually with the increase of shear displacement, such as that observed in the shearing of muskeg mat shown in Figure 5.5, then Eqn (5.3) may be used (Wong et al., 1979; Wong and Preston-Thomas, 1983a):

$$s/s_{max} = (j/K_w)\exp(1 - j/K_w)$$

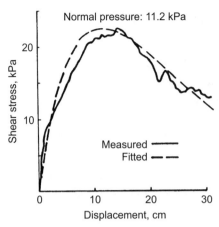

Figure 5.5: A shear curve exhibiting a peak and decreasing residual shear stress ('hump' shear curve)

To obtain the most appropriate value of K_w, the weighted least squares method described previously may be employed. Rearranging Eqn (5.3) and taking the logarithm of both sides, one obtains

$$\ln(s/s_{max}) = 1 + \ln(j/K_w) - j/K_w \qquad (5.13)$$

As mentioned previously, the conventional least squares method cannot be directly applied to Eqn (5.13) for determining the best value of K_w, because (s/s_{max}) is the measured quantity and not $\ln(s/s_{max})$. Applying the conventional least squares principle to the logarithms of the deviations will give excessive weight to deviations at low shear stresses which are usually the least significant. However, one can approach the curve fitting problem by minimizing the value of the following function using $(s/s_{max})^2$ as a weighting factor:

$$F = \sum (s/s_{max})^2 \left[\ln(s/s_{max}) - (1 + \ln(j/K_w) - j/K_w)\right]^2 \qquad (5.14)$$

To obtain the value of K_w which minimizes the above function, the first partial derivative of F with respect to K_w is taken and set to zero. This leads to the following condition:

$$\sum (s/s_{max})^2 \left[\ln(s/s_{max}) - (1 + \ln(j/K_w) - j/K_w)\right]\left[K_w - j\right] = 0 \qquad (5.15)$$

The value of K_w that satisfies the above condition is the desired value.

In certain cases where the measured shear curve is not so 'smooth', to obtain an overall good fit, both K_w and s_{max} have to be regarded as parameters to be optimized. Based on the least squares principle, an iterative procedure has been developed for deriving the optimum values of K_w and s_{max} for an experimental shear curve that is not so 'smooth'.

(c) If a shear curve displays a 'hump' of maximum shear stress and then decreases with the increase of shear displacement to a constant value of residual stress, such as that shown in Figure 5.1, then Eqn (5.5) may be used:

$$s/s_{max} = K_r\left[1 + \left[1/(K_r(1 - 1/e)) - 1\right]\exp(1 - j/K_w)\right]\left[1 - \exp(-j/K_w)\right]$$

As mentioned previously, the optimum value of K_r which minimizes the error in curve fitting can be obtained by Eqn (5.7).

It is interesting to note that the shear curve shown in Figure 5.1 and described by Eqn (5.5) is similar in shape to the curve '*a*' in Fig. 2.24 describing the stress–strain relationship for soil in a dense state based on the critical state soil mechanics.

Based on the shear stress–shear displacement curves measured under different normal pressures, the relation between the maximum shear stress and applied normal pressure can be derived. It has been found that the Mohr-Coulomb equation describes this relation adequately in many cases:

$$s_{max} = c + p\tan\phi \tag{5.16}$$

In this equation, s_{max} is the maximum shear stress, p is the normal stress, and c and ϕ are the cohesion (or adhesion) and angle of shearing resistance, respectively. In deriving the values of c and ϕ from measured shear data, a procedure proposed by Reece (1964) in which the effect of grouser height of the shear ring is taken into consideration should be followed.

It should be mentioned that during shear tests the shear head may have additional sinkage due to the rotation of the shear head. This additional sinkage is usually referred to as 'slip–sinkage' and should normally be monitored.

5.2 Shearing Behaviour of Various Types of Terrain

Using the methods for characterizing the shear stress–displacement relationship and for the shear strength of terrain described in Section 5.1, the shear data obtained from field measurements on mineral terrains, muskegs and snow-covered terrains were analysed. The parameters characterizing the shearing behaviour of various types of terrain are summarized in Tables 5.1, 5.2, 5.3 and 5.4.

5.3 Behaviour of Terrain under Repetitive Shear Loading

When a multi-axle wheeled vehicle (or a tracked vehicle) is in straight line motion over an unprepared terrain, an element of the terrain is subject to the repetitive shearing of the

Table 5.1: Parameters for characterizing the shearing behaviour of a sandy terrain

Terrain type	Type of shearing	Rate of shearing cm/s	Cohesion c (kPa)	Angle of shearing resistance ϕ (deg.)	Shear deformation parameter K (cm)
LETE sand	Internal	2.5	1.15	31.5	1.15
		2.5	1.39	30.6	1.13
	Rubber	2.5	0.96	27.3	1.14
		2.5	0.36	27.6	0.75

consecutive wheels (or roadwheels). To predict the shear stress distribution on the vehicle–terrain interface more realistically, the response to repetitive shear loading of the terrain should be known. Figure 5.6 shows the response of a frictional terrain (a dry sand) to repetitive shear loading under a constant normal load (Wong et al., 1984). It indicates that when the shear loading is reduced from B to zero and is then reapplied at C, the shear stress–displacement relationship during reshearing, such as CDE, is similar to that when the terrain is being sheared in its virgin state, such as OAB. This means that when reshearing takes place after the completion of a loading–unloading cycle, the shear stress does not instantaneously reach its maximum value for a given normal pressure. Rather a certain amount of additional shear displacement must take place before the maximum shear stress can be developed, similar to that when the frictional medium is being sheared in its virgin state.

Results of an investigation by Keira (1979) on the shearing force developed beneath a shear plate under a cyclic normal load lead to similar conclusions. Figure 5.7 shows the variation of the shearing force beneath a rectangular shear plate on a dry sand subject to a vertical harmonic load with a frequency of 10.3 Hz. It indicates that during the loading portion of each cycle, the shearing force S does not reach its maximum value instantaneously ($S_{max} = P \tan \phi$, where P is the instantaneous value of the normal load and ϕ is the angle of shearing resistance). This is demonstrated by the fact that the slope of the normal load curve is steeper than that of the shearing force curve. During the unloading portion of the cycle, however, the shearing force decreases in proportion to the instantaneous value of the normal load.

The response of the terrain under repetitive shear loading has a significant effect on the development of shear stress on the track–terrain interface, and on the prediction of the thrust developed by a track. Figure 5.8 illustrates how the development of the shear stress may be affected by the response of the terrain to repetitive shear loading for an idealized case. If the response of the terrain to repetitive shear loading is not taken into consideration, the shear stress distribution under the track with constant normal pressure beneath the roadwheels is shown in Figure 5.8(a). In the analysis, it is assumed that the shear stress–displacement relationship of the terrain is of the simple exponential form described by Eqn (5.2). Under

Table 5.2: Parameters for characterizing the shearing behaviour of loams

Terrain type	Type of shearing	Rate of shearing (cm/s)	Cohesion c (kN/m^2)	Angle of shearing resistance ϕ (deg)	Shear deformation parameter K_w (cm)	K_r Oida's equation	K_r Wong's equation	Wet density (kg/m^3)	Moisture content %
Upland sandy loam	Internal	2.5	2.2	39.4	6.1	0.655	0.659	1468	49.4
			3.3	33.7	9.3	0.832	0.835	1549	50.1
			2.8	33.4	7.1	0.673	0.724	1497	62.2
			1.1	33.5	3.6	0.435	0.497	1479	53.9
			3.4	24.1	4.9	0.457	0.557	1646	54.2
			2.6	29.1	4.8	0.366	0.481	1641	58.1
			5.1	25.6	3.6	0.392	0.448	1445	34.0
			4.3	22.7	3.5	0.420	0.462	1459	39.3
			2.7	26.1	4.2	0.411	0.456	1441	41.3
			2.5	28.2	4.1	0.323	0.379	1384	30.0
	Rubber	2.5	3.1	22.4	3.6	0.779	0.787	1471	31.3
			0.6	31.4	4.5	0.784	0.787	1471	31.3
			0.3	33.1	1.9	0.710	0.714	1623	45.1
			3.0	28.5	2.7	0.637	0.645	1612	47.9
Rubicon sandy loam	Internal	2.5	3.7	29.8	4.1	0.300	0.382	1551	48.5
			3.2	30.5	4.2	0.359	0.413	1551	48.5
North Gower clayey loam	Internal	2.5	6.1	26.6	4.5	0.325	0.398	1664	47.9
			6.2	30.3	4.7	0.349	0.419	1653	34.8
Grenville loam	Internal	2.5	3.1	29.8	3.8	0.396	0.439	1190	30.5

Table 5.3: Parameters characterizing the shearing behaviour of muskegs

Terrain type	Type of shearing	Rate of shearing (cm/s)	Cohesion c (kPa)	Angle of shearing resistance ϕ (degree)	Shear deformation parameter K_w (cm)	K (cm)
Petawawa Muskeg A	Internal shearing of mat	2.5 10	5.7 5.9	51.6 45.3	16 16.4	
	Internal shearing of peat	2.5 10	2.8 2.1	39.4 34.4		3.1 1.9
	Rubber-mat	2.5 10	1.2 1.7	40.7 40.3		1.5 0.9
	Rubber-peat	2.5	1.5	27.9		0.8
Petawawa Muskeg B	Internal shearing of mat	2.5	2.3	54.9	14.4	
	Internal shearing of peat	2.5	2.5	39.2		3.1
	Rubber-mat	2.5	1.6	40.9		0.8
	Rubber-peat	2.5	0.5	32.9		0.7

Table 5.4: Parameters for characterizing the shearing behaviour of snow covers

Terrain type	Type of shearing	Rate of shearing (cm/s)	Cohesion c (kPa)	Angle of shearing resistance ϕ (deg)	Shear deformation parameter K_w (cm)	K_r Oida's equation	K_r Wong's equation	Shear deformation parameter K (cm)
Petawawa Snow B	Internal	2.5	0.9 0 0 0.6	20.1 25.8 29.7 20.3	2.2 2.4 2.0 2.1	0.6 0.64 0.66 0.69	0.61 0.65 0.68 0.68	
	Rubber	2.5	0 0.3 0.16 0	16.1 13.5 17.0 18.9				0.35 0.46 0.34 0.43

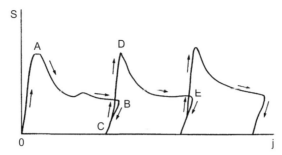

Figure 5.6: Response to repetitive shear loading of dry sand

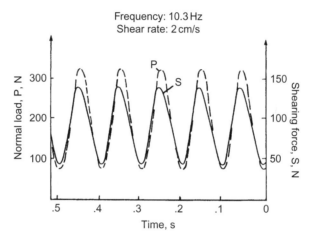

Figure 5.7: Shear force developed under a rectangular shear plate subject to cyclic normal load on dry sand (Keira, 1979)

the front roadwheel, the shear stress increases from zero following the initial part of the shear stress–displacement curve shown in Figure 5.4. Under the second or subsequent roadwheel, the shear stress would be higher than or close to that at the rear contact point of the front roadwheel dependent on the shear deformation parameter K, as the shear displacement is higher. However, if the response of the terrain to repetitive shear loading shown in Figure 5.6 is taken into account, the shear stress under the second or subsequent roadwheel will always start from zero at its front contact point and follow the initial part of the shear curve shown in Figure 5.4, similar to that under the front roadwheel illustrated in Figure 5.8(b). It should be noted that the thrust developed by a track is determined by the integration of the shear stress over the track contact area. It is, therefore, proportional to the sum of the areas enclosed by the shear stress shown in Figure 5.8(a) or (b). This indicates that if the response of the terrain to repetitive shear loading is taken into account, the predicted thrust developed by a track will be lower than that if it is not taken into consideration.

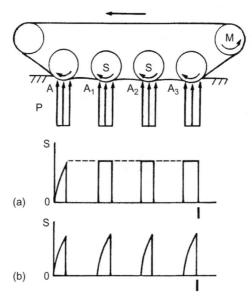

Figure 5.8: Development of shear stress under a track on a frictional terrain predicted by (a) the conventional method, and (b) the improved method taking into account the response of terrain to repetitive shear loading

It should be mentioned that the shearing characteristics of the mineral terrains, muskegs and snow-covered terrains described in Section 5.2 are predominantly frictional. Consequently, the behaviour of the terrain under repetitive shear loading described above is applicable to these terrains.

CHAPTER 6
Performance of Off-Road Vehicles

Performance of an off-road vehicle refers to its ability to overcome motion resistance, to develop drawbar pull, to negotiate grades, or to accelerate in straight-line motion. Dependent upon their intended use, different criteria are employed to evaluate the performance of various types of off-road vehicle. For tractors and the like, their prime function is to provide adequate draft to pull (or push) implements such as ploughs or bulldozer blades; the drawbar performance is, therefore, of importance. It may be characterized by the ratio of drawbar pull to tractor weight (drawbar pull coefficient), drawbar power (product of drawbar pull and vehicle forward speed), and tractive (drawbar) efficiency (the ratio of drawbar power to the corresponding power input). For off-road transport vehicles, such as trucks used in mining or logging industry, the time rate of transporting payload from the origin to destination is of interest. It may be characterized by the transport productivity (the product of payload and average vehicle operating speed from origin to destination) and the transport efficiency (the ratio of transport productivity to the corresponding power input). For military vehicles, the maximum feasible speed between two specific locations within a given operating theatre may be employed as a major criterion for evaluating their performance.

In this chapter, the performance criteria for various types of off-road vehicle are discussed.

6.1 Factors Affecting Off-Road Vehicle Performance

The major external forces acting on a tracked vehicle and on an off-road wheeled vehicle are shown in Figure 6.1(a) and (b), respectively. In the longitudinal direction, they include the aerodynamic resistance (drag) R_a, the motion resistance acting on the vehicle running gear R_v (in the case of a two-axle off-road wheeled vehicle shown in Figure 6.1(b), it is the sum of the motion resistances acting on the front and rear wheels, $R_v = R_{vf} + R_{vr}$), thrust (tractive effort or propelling force) F (in the case of a two-axle, all-wheel-drive off-road vehicle, it is the sum of the thrusts developed by the front and rear wheels, $F = F_f + F_r$), and drawbar pull F_d.

If the vehicle operates on a slope at an angle θ_s to the horizontal, the equation of motion along the longitudinal direction (or x-axis) is expressed by

$$m\frac{d^2x}{dt^2} = \frac{W}{g}a = F - R_a - R_v - F_d \mp W\sin\theta_s = F - R_a - R_v - F_d \mp R_g \quad (6.1)$$

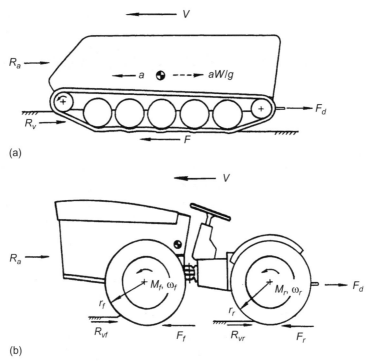

Figure 6.1: Major external forces acting on (a) a tracked vehicle and (b) a wheeled vehicle

where m and W are vehicle mass and weight, respectively, d^2x/dt^2 or a is the linear acceleration of the mass centre along the longitudinal direction, and g is the acceleration due to gravity, $W \sin \theta_s$ or R_g is the grade resistance. When the vehicle climbs up a slope, the negative sign for $W \sin \theta_s$ or R_g in Eqn (6.1) is used, and when it runs down a slope, the positive sign is used.

By introducing the concept of inertia force, Eqn (6.1) may be rewritten as

$$F = (aW/g) + R_a + R_v + F_d \pm R_g \qquad (6.2)$$

Under steady-state operating conditions, the inertia force (aW/g) is zero and Eqn (6.2) is simplified to

$$F = R_a + R_v + F_d \pm R_g \qquad (6.3)$$

6.1.1 Aerodynamic Resistance

Aerodynamic resistance R_a is usually expressed in the following form:

$$R_a = \frac{\rho}{2} C_D A_f V_r^2 \qquad (6.4)$$

where ρ is mass density of the air, C_D is the non-dimensional aerodynamic resistance (drag) coefficient and is mainly dependent on the shape of the vehicle, A_f is the frontal area of the vehicle, and V_r is the vehicle speed relative to wind speed.

Aerodynamic resistance is usually not a significant factor that affects the performance of off-road vehicles operating at speeds below 48 km/h. For vehicles designed for higher speed operations, such as military fighting vehicles, it may have to be taken into account in performance evaluation. For heavy fighting vehicles, such as tanks, the value of C_D is around 1 and the frontal area is in the range of 6 to 8 m². For instance, for a tank weighing 50 tonnes, with $C_D = 1.17$, frontal area of 6.5 m², air mass density of 1.225 kg/m³, vehicle speed of 48 km/h, and zero wind speed, the aerodynamic resistance amounts to 0.828 kN. The power required to overcome the aerodynamic resistance is approximately 11 kW.

6.1.2 Motion Resistance of Vehicle Running Gear

In off-road operations, the motion resistance R_v may include obstacle resistance R_{ob}, internal resistance of the running gear R_{in}, and resistance arising from vehicle running gear (track or wheel)–terrain interaction R_t.

A. Obstacle resistance

The obstacle resistance R_{ob} arises when the vehicle running gear encounters stumps, rocks and the like. Its magnitude depends on the height, size, shape, etc. of the obstacle and is determined experimentally.

B. Internal resistance of the running gear

The internal resistance of the running gear R_{in} is dependent on its type (track or wheel) and its design and operational factors. For a track system, the internal resistance is caused by the roadwheels rolling on the track, mechanical losses in the connecting pins when track links are rotating relative to each other (or losses caused by the flexing of rubber belt tracks), and other factors. The magnitude of track internal resistance is determined experimentally.

For metal link tracks with dry pins, commonly used in agricultural or industrial tractors, the coefficient of internal resistance f_{in}, which is the ratio of the internal motion resistance to vehicle weight, may be estimated by the following empirical equation (Bekker, 1969):

$$f_{in} = 0.06 + 0.009\,V \tag{6.5}$$

where V is vehicle speed in km/h.

For military tracked vehicles operating on hard, smooth road surfaces, the coefficient of motion resistance may be considered representative of the coefficient of internal resistance f_{in}. It may be estimated by the following empirical equation (Ogorkiewicz, 1991):

$$f_{in} = f_0 + f_s V \tag{6.6}$$

where f_0 and f_s are empirical coefficients, and V is vehicle speed in km/h. For tracks having double, rubber-bushed pins and rubber pads, the value of f_0 is typically 0.03; for all-steel, single-pin tracks, it is 0.025; and for tracks having sealed, lubricated pin joints with needle bearings, it can be as low as 0.015. The value of f_s varies with the type of track, and as a first approximation, it may be taken to be 0.00015.

When a pneumatic tyre rolls, energy is dissipated primarily through hysteresis of tyre material caused by carcass flexing. It appears as a resisting force acting against the motion of the tyre. This may be considered as the internal resistance of a pneumatic tyre R_{in}. Its magnitude depends on a number of design and operational factors, including tyre construction (bias or radial), carcass material, inflation pressure, speed, etc. A detailed discussion on the motion resistance of pneumatic tyres is presented in the reference (Wong, 2008). The internal resistance of off-road tyres will be discussed in Section 11.2.2 of Chapter 11.

C. Resistance due to vehicle running gear–terrain interaction

On unprepared terrain, the resistance R_t caused by vehicle running gear–terrain interaction, such as that caused by the sinkage of the running gear, is usually the most significant. Analysis and prediction of the resistance due to track–terrain interaction and that due to wheel–terrain interaction are presented in subsequent chapters.

On soft terrain, such as deep snow, the sinkage of the running gear may exceed vehicle ground clearance and the vehicle belly may be in contact with the terrain surface. This induces an additional resisting force, commonly known as belly drag. It is caused by the sliding of the belly on the terrain surface and by the bulldozing effect of the belly, if it is at an angle with the direction of motion of the vehicle. Belly drag is discussed in detail later in Chapters 8 and 9.

6.1.3 Thrust (Tractive Effort or Propelling Force)

To evaluate the performance potential of off-road vehicles, the thrust that the vehicle running gear (track or wheel) develops has to be determined. There are two limiting factors to the thrust developed by an off-road vehicle. One is related to the characteristics of the power plant and transmission, and the other is related to the characteristics of shearing between the vehicle running gear and the terrain described in Chapter 5.

The thrust F as determined by the characteristics of the power plant and transmission is expressed by

$$F = \frac{M_e \xi \eta_t}{r} \qquad (6.7)$$

where M_e is the engine torque, ξ is the overall gear reduction ratio of the transmission, including the gear ratio of the gearbox and that of the drive axle, η_t is the mechanical efficiency of the transmission, and r is the pitch radius of the drive sprocket of a tracked vehicle or the radius of the driven tyre of a wheeled vehicle.

The corresponding vehicle speed V is determined by

$$V = \frac{n_e r}{\xi}(1 - i) \qquad (6.8)$$

where n_e is engine speed, and i is slip of vehicle running gear (wheel or track). The definition of slip i is given below in Eqn (6.10).

The thrust as determined by the characteristics of shearing between the track and the terrain or that between the wheel and the terrain is discussed in detail in subsequent chapters. For operations over unprepared terrain, the performance potential of an off-road vehicle is quite often limited by the thrust as determined by the characteristics of shearing between the track (or wheel) and the terrain.

6.2 Drawbar Performance

6.2.1 Drawbar Pull Coefficient and Drawbar Power

For tractors, their prime function is to develop adequate drawbar pull (net force) to pull (or push) implements or working machinery, such as ploughs, construction or earthmoving equipment. The drawbar pull F_d is the net force available at the drawbar hitch. From Eqn (6.3), under steady-state operating conditions, the drawbar pull F_d is the difference between the thrust F and the sum of all resisting forces ΣR acting on the tractor, and is expressed by

$$F_d = F - \Sigma R \qquad (6.9)$$

It should be noted that in the development of the thrust F, the shearing action of the running gear on the terrain surface causes it to slip. Slip i of the vehicle (or its running gear) is defined as follows:

$$i = 1 - \frac{V}{V_t} = 1 - \frac{V}{r\omega} \qquad (6.10)$$

where V is actual forward speed of the vehicle (or its running gear), and V_t is the theoretical speed of the vehicle (or its running gear), which is equal to the product of angular speed ω and radius r of the sprocket or wheel.

For tracked vehicles, r in Eqn (6.10) is the pitch radius of the sprocket. For a smooth rigid wheel, r is the radius of the rim. However, for a flexible wheel (such as a pneumatic tyre) with significant deflection on deformable terrain, there is some uncertainty in determining an appropriate value of r in Eqn (6.10) for defining the slip i.

A method that has been employed in practice is to use the rolling radius (i.e. the ratio of the distance the wheel centre travels per revolution (2π) of wheel rotation) of the wheel (or tyre) measured under steady-state, self-propelled condition (i.e. the wheel is supplied with a driving torque to propel itself across a horizontal operating surface, while delivering zero net force or drawbar pull) as the value of r in Eqn (6.10) for defining the slip i. In essence, it means that when the wheel is under self-propelled condition with zero drawbar pull, the slip of the wheel is taken to be zero. It should be noted that under self-propelled condition, the thrust on the wheel–terrain interface is generated to overcome the motion resistance of the wheel. The presence of the thrust indicates that shearing action takes place on the wheel–terrain interface. As described in Chapter 5, this shearing action results in shear displacement which implies that wheel slip occurs. This means that under steady-state, self-propelled condition, while no net force or drawbar pull is generated, the presence of the thrust indicates the existence of wheel slip (Onafeko and Reece, 1967; Wong and Reece, 1967a). This implies that the actual wheel slip is not zero under self-propelled condition.

Another method that has been used in practice is to use the rolling radius of the wheel measured under steady-state, towed condition (i.e. the wheel is supplied with a force at the axle in the direction of travel while no driving torque is supplied to the wheel) as the value of r in Eqn (6.10) for defining the slip i. In essence, it means that when the wheel is under steady-state, towed condition, the wheel slip is taken to be zero. It should be pointed out that under towed condition, wheel skid (or negative wheel slip) occurs, particularly on deformable terrain (Onafeko and Reece, 1967; Wong and Reece, 1967b). This means that the actual wheel slip (in this case, negative wheel slip) is not zero under towed condition.

From the discussions presented above, it appears that for a flexible wheel (or tyre) with significant deflection operating on a deformable terrain, further detailed investigations are needed to establish a generally accepted method for determining the appropriate value of the radius r in Eqn (6.10) for defining the theoretical speed or the zero-slip condition (Schreiber and Kutzbach, 2007). Meanwhile, whatever method is selected for determining the wheel radius r in the analysis of wheel slip, it should be clearly stated and used in a consistent manner. The value of slip so determined should be viewed as a parameter primarily for comparing vehicle (or wheel) performance on a relative basis.

Since the thrust is related to slip, the drawbar pull is also a function of slip. The relationships between the thrust (or drawbar pull) and slip of a vehicle with tracks or wheels and the various methods for predicting them are discussed in detail in subsequent chapters.

To compare the performance of off-road vehicles of different designs, the ratio of the drawbar pull to vehicle weight (F_d/W), usually referred to as the drawbar pull coefficient, is used. It represents the extent to which the vehicle weight is utilized in generating drawbar pull. As mentioned previously, drawbar pull is a function of slip; to compare the performances of off-road vehicles on a common basis, the drawbar pull coefficient at 20% slip is usually used as one of the major performance parameters.

The product of the drawbar pull F_d and the actual forward speed of the vehicle V is referred to as the drawbar power P_d. It represents the potential productivity of the vehicle, that is, the rate at which productive work, such as ploughing or bulldozing, may be accomplished. The drawbar power P_d is expressed by

$$P_d = F_d V = (F - \Sigma R) V_t (1 - i) \tag{6.11}$$

It is shown that in Eqn (6.11) the drawbar power P_d is a function of slip i. Similar to using the drawbar pull coefficient at 20% slip as a performance parameter, the drawbar power at 20% slip is also a widely used performance parameter. The reason for selecting the drawbar pull coefficient or drawbar power at 20% slip as a basis for comparing the drawbar performances of off-road vehicles is that the operating efficiency at slip around 20% is generally satisfactory.

To illustrate the performance of an off-road vehicle, a drawbar performance diagram may be used. In the diagram, the slip i, actual forward speed V, and drawbar power P_d are plotted against the drawbar pull F_d, as shown in Figure 6.2.

6.2.2 Tractive (Drawbar) Efficiency

To characterize the efficiency of a tractor in converting engine power to the power available at the drawbar hitch to perform productive work, the tractive (drawbar) efficiency is used. The overall tractive efficiency η_{do} is defined as the ratio of drawbar power P_d to the corresponding engine power P_e and is expressed by

$$\eta_{do} = \frac{P_d}{P_e} = \frac{F_d V}{P_e} = \frac{(F - \Sigma R) V_t (1 - i)}{P_e} \tag{6.12}$$

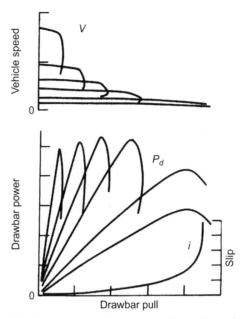

Figure 6.2: Drawbar performance of an off-road vehicle

The engine power P_e corresponding to the drawbar power P_d may be expressed in terms of the power available at the drive sprockets of a tracked vehicle or that at the driven wheels of a wheeled vehicle and the mechanical efficiency of the transmission η_t:

$$P_e = \frac{FV_t}{\eta_t} \tag{6.13}$$

Substituting for P_e from Eqn (6.13) into Eqn (6.12), the overall tractive efficiency η_{do} is expressed by

$$\eta_{do} = \frac{F - \Sigma R}{F}(1-i)\eta_t = \frac{F_d}{F}(1-i)\eta_t = \eta_m \eta_s \eta_t \tag{6.14}$$

where η_m is referred to as the efficiency of motion and is equal to F_d/F, and η_s is referred to as the efficiency of slip and is equal to $(1 - i)$.

The efficiency of motion η_m indicates the losses in converting the thrust developed by the running gear to the pull at the drawbar hitch to produce a net force to pull (or push) implements or equipment. For the motion resistance having a steady value, the efficiency of motion generally increases with the increase of drawbar pull, as shown in Figure 6.3.

The efficiency of slip characterizes the power losses and also the reduction in speed of the vehicle due to slip of the running gear. Since slip increases with an increase in thrust and

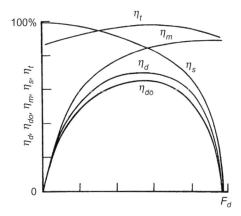

Figure 6.3: Tractive (drawbar) efficiency as a function of drawbar pull

drawbar pull, the efficiency of slip η_s decreases with the increase of drawbar pull, as shown in Figure 6.3. On unprepared terrain, slip of the running gear is a major source of power loss. Minimizing the slip for a given drawbar pull is, therefore, of practical significance in improving the operational efficiency of tractors.

As is seen from Eqn (6.14), the overall tractive efficiency η_{do} is the product of the mechanical efficiency of the transmission η_t, efficiency of motion η_m, and efficiency of slip η_s; and they vary with drawbar pull F_d, as shown in Figure 6.3. The overall tractive efficiency η_{do} exhibits a peak at an intermediate value of drawbar pull, as indicated in the figure.

Another frequently used definition of tractive (drawbar) efficiency η_d is the ratio of the drawbar power to the power input to the drive sprockets of a tracked vehicle or that to the driven wheels of a wheeled vehicle. In this case, the mechanical efficiency of the transmission η_t is not included and η_d is expressed by

$$\eta_d = \frac{F - \Sigma R}{F}(1 - i) = \frac{F_d}{F}(1 - i) = \eta_m \eta_s \qquad (6.15)$$

The variation of the tractive efficiency η_d with drawbar pull is also shown in Figure 6.3.

It should be noted that in the above analysis, the slip of all the running gear (tracks or wheels) in a vehicle is assumed to be the same. This means, for instance, that in a four-wheel-drive vehicle, all the driven wheels develop the same slip. In practice, dependent upon the type of inter-axle coupling between the drive axles and operating conditions, the front and rear wheels may not have the same slip. Consequently, the efficiency of slip η_s is not simply equal to $(1-i)$. The efficiency of slip for four-wheel-drive (or all-wheel-drive) off-road vehicles is discussed in the next section.

Example 6.1

A tracked vehicle operates on a wet clayey soil with thrust–slip relationship shown in Table 6.1. The pitch radius of its sprocket is 0.214 m. The total motion resistance of the vehicle is 6.0 kN. The transmission efficiency is 88%. Determine the drawbar performance of the vehicle in a gear with total reduction ratio of 3.53; its engine operates at a speed of 2000 rpm and develops a torque of 746 N-m.

Solution

The thrust F can be calculated using Eqn (6.7):

$$F = \frac{M_e \xi \eta_t}{r} = \frac{746 \times 3.53 \times 0.88}{0.214} = 10.829 \, kN$$

From Table 6.1, a thrust–slip curve for the vehicle can be plotted. From the curve, it can be determined that for a thrust of 10.829 kN, the slip is approximately 2.9%. The vehicle forward speed V can be calculated using Eqn (6.8):

$$V = \frac{n_e r}{\xi}(1-i) = \frac{(2000 \times 2\pi/60) \times 0.214}{3.53}(1 - 0.029) = 12.33 \, m/s = 43.39 \, km/h$$

The drawbar pull F_d is determined by

$$F_d = F - \Sigma R = 10.829 - 6.0 = 4.829 \, kN$$

The drawbar power P_d is calculated by

$$P_d = F_d \times V = 4.829 \times 12.33 = 59.54 \, kW = 79.83 \, hp$$

The efficiency of motion η_m is given by

$$\eta_m = F_d/F = 4.829/10.829 = 0.4459 = 44.59\%$$

The efficiency of slip η_s is expressed by

$$\eta_s = 1 - i = 1 - 0.029 = 0.971 = 97.1\%$$

The overall tractive efficiency η_{do} is given by Eqn (6.14):

$$\eta_{do} = \eta_m \, \eta_s \, \eta_t = 0.4459 \times 0.971 \times 0.88 = 0.3810 = 38.10\%$$

Table 6.1: Thrust–slip relationship

Slip, %	1	3	5	10	20	40
Thrust, kN	8.68	11.10	12.84	15.93	19.51	22.64

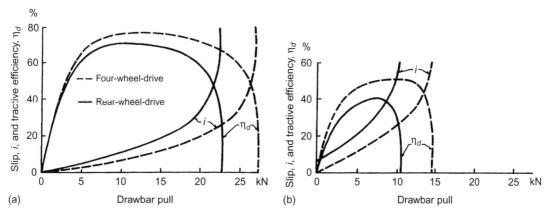

Figure 6.4: Comparison of the drawbar performances of a four-wheel-drive tractor with those of a rear-wheel-drive tractor on (a) dry loam, stubble field and (b) wet clayey loam (Reprinted by permission of *ISTVS* from Sohne, 1968)

6.2.3 Four-Wheel-Drive

In recent years, there has been an increasing demand for higher productivity in farming, logging, construction and other industries. As a result, large tractors with high engine power have been introduced. To fully utilize their potential for developing drawbar pull with improved operating efficiency, four-wheel-drive is widely used. For four-wheel-drive tractors, their total weight is utilized in developing traction, whereas for rear-wheel-drive tractors, only approximately 60–70% of their total weight is applied to the driven wheels for developing tractive effort. For a four-wheel-drive tractor with wheels of equal size, the rear wheels run in the ruts that have been compacted by the front wheels. This in general reduces the overall motion resistance of the tractor. Figure 6.4(a) and (b) show a comparison of the drawbar performances of a four-wheel-drive tractor with those of a rear-wheel-drive tractor of the same total weight on a dry loam, stubble field and on a wet clayey loam, respectively (Sohne, 1968). The figure shows that the four-wheel-drive tractor has a higher maximum drawbar pull, higher tractive efficiency and lower slip at the same drawbar pull than those of the rear-wheel-drive tractor. For instance, on the dry loam, stubble field, the drawbar pull of the four-wheel-drive tractor at 20% slip is 27% higher than that of the rear-wheel-drive tractor. The peak tractive efficiency of the four-wheel-drive tractor and that of the rear-wheel-drive tractor are 77 and 70%, respectively. This indicates that the peak tractive efficiency of the four-wheel-drive tractor is 10% higher than that of the rear-wheel-drive tractor. On the wet clayey loam, the drawbar pull of the four-wheel-drive tractor at 20% slip is approximately 62% higher than that of the rear-wheel-drive tractor. The peak value of the tractive efficiency of the four-wheel-drive tractor and that of the rear-wheel-drive tractor are 51 and 40%, respectively. This indicates that the peak tractive efficiency of the four-wheel-drive tractor is 27.5% higher than that of the rear-wheel-drive tractor.

The drawbar performance of a four-wheel-drive off-road vehicle has certain unique characteristics. For most four-wheel-drive tractors currently in use, the front and rear drive axles are rigidly coupled. This means the ratio of the angular speed of the front wheel to that of the rear wheel is fixed. However, owing to the variations of the radii of the front and rear tyres caused by mismatching of tyre sizes or by the difference in inflation pressure between the front and rear tyres, uneven wear of the front and rear tyres, or load transfer between the front and rear tyres caused by drawbar pull, the theoretical speed (the product of angular speed and wheel radius noted previously) of the front tyres may be different from that of the rear tyres. It is estimated that the difference in theoretical speed between the front and rear tyres may be as much as 10% in practice. This causes the front and rear tyres to slip at different rates. The slip efficiency η_{s4} for a four-wheel-drive tractor can no longer be simply expressed as $(1 - i)$, as shown in Eqn (6.14) or Eqn (6.15).

For a four-wheel-drive off-road vehicle the power losses due to slip occur at both the front and rear wheels. Assuming that the performances of the two front wheels are the same, and so are those of the two rear wheels, the slip efficiency η_{s4} of a four-wheel-drive vehicle is expressed by (Wong, 1970)

$$\eta_{s4} = 1 - \frac{i_f M_f \omega_f + i_r M_r \omega_r}{M_f \omega_f + M_r \omega_r} = 1 - \frac{i_f V_{tf} F_f + i_r V_{tr} F_r}{V_{tf} F_f + V_{tr} F_r} \tag{6.16}$$

where F_f and F_r are the thrusts of the front and rear wheels, respectively, i_f and i_r are the slips of the front and rear wheels, respectively, M_f and M_r are the driving torques of the front and rear wheels, respectively, V_{tf} and V_{tr} are the theoretical speeds of the front and rear wheels, respectively, and ω_f and ω_r are the angular speeds of the front and rear wheels, respectively.

When a four-wheel-drive vehicle is in straight line motion, the translatory (or actual forward) speed of the front wheels and that of the rear wheels must be the same, as the front and rear wheels are connected to the same vehicle frame. This kinematic relationship is expressed by

$$V_{tf}(1 - i_f) = V_{tr}(1 - i_r) = V \text{ or } V_{tf} = V_{tr}(1 - i_r)/(1 - i_f) \tag{6.17}$$

and

$$K_v = \frac{V_{tf}}{V_{tr}} = \frac{\omega_f r_f}{\omega_r r_r} = \frac{(1 - i_r)}{(1 - i_f)} \tag{6.18}$$

where K_v is the ratio of theoretical speed of the front wheels to that of the rear wheels, which is referred to as the theoretical speed ratio, and r_f and r_r are the radii of the front and rear wheels, respectively, as discussed previously.

From Eqn (6.18), the slip of the rear wheel i_r may be expressed as follows:

$$i_r = 1 - K_v(1 - i_f) \quad (6.19)$$

Equation (6.19) indicates that, for instance, if the value of K_v is equal to 0.85 (i.e. the theoretical speed of the front wheel is 85% of that of the rear wheel) and i_r is less than 15%, the slip of the front wheel i_f is negative, or the front wheel skids and develops a braking force opposite to the direction of motion of the vehicle. In this case, a phenomenon commonly known as 'torsional wind-up' in the transmission occurs. On the other hand, if the value of K_v is equal to 1.15 (i.e. the theoretical speed of the front wheel is 115% of that of the rear wheel) and i_f is less than 13%, the slip of the rear wheel i_r is negative, or the rear wheel skids and develops a braking force opposite to the direction of motion of the vehicle. In this case, torsional wind-up in the transmission also occurs. Torsional wind-up results in an increase of stress in the components of the transmission and in a reduction of its mechanical efficiency. Furthermore, it leads to the reduction of the maximum forward thrust that the vehicle can develop.

Substituting for V_{tf} from Eqn (6.17) into Eqn (6.16), one obtains

$$\eta_{s4} = 1 - \frac{[(1-i_r)/(1-i_f)]i_f V_{tr} F_f + i_r V_{tr} F_r}{[(1-i_r)/(1-i_f)]V_{tr} F_f + V_{tr} F_r} = 1 - \frac{i_f(1-i_r) - (i_f - i_r)K_d}{(1-i_r) - (i_f - i_r)K_d} \quad (6.20)$$

where K_d is referred to as the coefficient of thrust distribution and is equal to $F_r/(F_f + F_r)$.

Equation (6.20) indicates that, in general, the efficiency of slip of a four-wheel-drive off-road vehicle depends not only on the slips of the front and rear wheels, but also on the distribution of the thrust between them (i.e. the coefficient of thrust distribution K_d). It is, therefore, of interest to explore whether there is an optimum coefficient of thrust distribution K_d that enables the efficiency of slip to reach its peak under a given operating condition. To find this optimum coefficient of thrust distribution, the first partial derivative of η_{s4} with respect to K_d in Eqn (6.20) is taken and is set to zero:

$$\frac{\partial \eta_{s4}}{\partial K_d} = \frac{(1-i_f)(1-i_r)(i_f - i_r)}{[(1-i_r) - (i_f - i_r)K_d]^2} = 0 \quad (6.21)$$

This condition will be satisfied if the slip of the front wheels or that of the rear wheels is 100%, or the slip of the front wheels is equal to that of the rear wheels. When the slip of either the front or the rear wheels is 100%, the vehicle cannot move forward and the efficiency of slip becomes zero. Therefore, under normal operating conditions, only when the slip of the front wheels is equal to that of the rear wheels will the first partial derivative given by Eqn (6.21) be zero.

The results of a detailed analysis indicate that the slips of the front and rear wheels being equal is the necessary, as well as the sufficient, condition for achieving the maximum efficiency of slip under a given operating condition (such as operating on a particular terrain at a given vehicle total thrust to weight ratio), if either of the following two conditions is satisfied (Wong, 2008):

1. The relationship between the thrust and slip for the front wheels and that for the rear wheels are the same;
2. Their relationships are different, but the relationship between the thrust and slip for the front wheels and that for the rear wheels are linear.

The efficiency of slip η_{s4} may be expressed in an alternate form by substituting the theoretical speed ratio K_v for $(1 - i_r)/(1 - i_f)$ in Eqn (6.20):

$$\eta_{s4} = 1 - \frac{i_f K_v - (K_v - 1) K_d}{K_v - (K_v - 1) K_d} \tag{6.22}$$

Similarly, to find the optimum coefficient of thrust distribution, the first partial derivative of η_{s4} with respect to K_d in Eqn (6.22) is taken and is set to zero:

$$\frac{\partial \eta_{s4}}{\partial K_d} = \frac{(K_v - 1) K_v (1 - i_f)}{[K_v - (K_v - 1) K_d]^2} = 0 \tag{6.23}$$

When the slip of the front wheels is 100%, the vehicle forward speed is zero. Therefore, under normal operating conditions, only when the value of the theoretical speed ratio K_v is equal to one will the first partial derivative of η_{s4} with respect to K_d be zero. The condition that K_v is equal to one is equivalent to the slip of the front wheels and that of the rear wheels being equal. It follows that similar to the conclusion reached previously, the theoretical speed ratio K_v being equal to one is the necessary, as well as the sufficient, condition for achieving the maximum efficiency of slip under a given operating condition, provided that either the relationship between the thrust and slip for the front wheels and that for the rear wheels are the same; or their relationships are different, but the relationship between the thrust and slip for the front wheels and that for the rear wheels are linear.

Under normal operating conditions of off-road vehicles, such as tractors, wheel slip is usually kept below 20%. In the slip range between 0 and 20%, the relationship between thrust (or drawbar pull) and slip is approximately linear or can be linearized without excessive error on a variety of terrains. This indicates that under normal operating conditions with slip less than 20%, the slip of the front wheels and that of the rear wheels being equal or the theoretical speed ratio being equal to one enables the efficiency of slip of a four-wheel-drive off-road vehicle reaching its peak.

The results of a detailed analysis also indicate that only if the relationship between the thrust and slip of the front wheels and that of the rear wheels are different and are non-linear will the condition for achieving the peak efficiency of slip deviate from that described above. However, even under extremely unusual operating scenarios, such as the front wheels operating on farm soil and the rear wheels operating on a surface as hard as tarmac, the optimal theoretical speed ratio K_v deviates only slightly from 1, when the ratio of vehicle thrust to weight F/W is less than 0.6. For instance, under the above-noted extremely unusual operating conditions, when F/W is 0.6 (corresponding to a wheel slip of approximately 20% on farm soil), the optimal theoretical speed ratio K_v is 1.0406, which is only 4.06% different from $K_v = 1.0$ (Besselink, 2003). The efficiency of slip η_{s4} at the optimal theoretical speed ratio $K_v = 1.0406$ is 90.84%, as against 90.44% at the theoretical speed ratio $K_v = 1.0$. The difference in efficiency of slip η_{s4} between these two cases is merely 0.4%, which is negligible for all practical purposes.

The approach to the analysis of the efficiency of slip of four-wheel-drive off-road vehicles presented above is applicable to all-wheel-drive off-road vehicles with any number of driven wheels, such as 6×6 or 8×8.

A series of tests on a farm field was conducted to examine the effects of theoretical speed ratio K_v on the slip efficiency of a four-wheel-drive tractor (Wong et al., 1998, 1999, 2000). The test vehicle was an instrumented four-wheel-drive tractor with front-wheel-assist, Case-IH Magnum. The drawbar performances of the tractor, with seven combinations of front and rear tyres of different sizes at various inflation pressures, were measured in a farm field. The seven front and rear tyre combinations were 13.6R28 and 20.8R38, 14.9R28 and 20.8R38, 16.9-26 bias and 20.8R38, 16.9R26 and 20.8R38, 16.9R28 and 20.8R38, 13.6R28 and 18.4R38, and 14.9R28 and 18.4R38. The inflation pressure of the tyres varied in the range from 82 to 193 kPa. The various combinations of front and rear tyres produced a variety of theoretical speed ratios ranging from 0.908 to 1.054. The theoretical speed ratios were calculated from the rolling radii of the tyres under towed conditions on the farm field, and the fixed gear ratio between the front and rear axles was taken into account. During field tests, the driving torques on the front and rear axles, slips of the front and rear tyres (calculated based on the rolling radii of tyres under towed condition), drawbar pull, vehicle forward speed, and fuel consumption of the tractor were monitored.

Field test data show that the thrust–slip relationships for all front tyres of various sizes and under different inflation pressures used in the tests are quite similar and so are those for the rear tyres. Thus, all performance test data of approximately 350 sets obtained with the seven sets of tyres under various inflation pressures are combined in the evaluation of the effects of theoretical speed ratio K_v on the efficiency of slip of the test tractor.

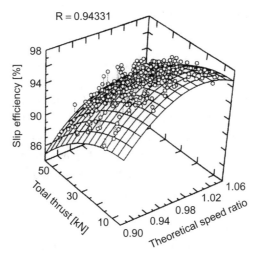

Figure 6.5: The relationship of the measured efficiency of slip, total thrust and theoretical speed ratio of a four-wheel-drive tractor on a farm field (Reprinted by permission from *SAE* paper 2000-01-2596 by Wong et al., 2000)

Figure 6.5 shows the measured relationship of efficiency of slip, defined by either Eqn (6.20) or (6.22), total thrust (i.e. the sum of the thrusts of the front and rear tyres, calculated by dividing the driving torques of the front and rear tyres by their corresponding radii under towed condition in the field), and theoretical speed ratio. A three-dimensional curved surface, representing the efficiency of slip as a quadratic function of total thrust and theoretical speed ratio, was fitted to the measured data, using the method of least squares, as shown in Figure 6.5. By slicing the curved surface along a constant total thrust plane, the relationships between the efficiency of slip and theoretical speed ratio at various values of total thrust can be obtained, as shown in Figure 6.6. It can be seen that in the range of field tests conducted with total thrust-to-weight ratio F/W up to 0.45 (or the total thrust up to 48 kN), when the theoretical speed ratio is close to or equal to 1.0, the efficiency of slip of the four-wheel-drive tractor is indeed at a maximum. Thus, the analytical finding that when the theoretical speed ratio is equal to 1.0, the efficiency of slip reaches its peak, is experimentally substantiated. It is also shown in Figure 6.6 that if the theoretical speed ratio is either higher or lower than 1.0, the efficiency of slip will be lower than the peak value. This indicates that in straight-line operation, maintaining equal slips for all driven tyres or the theoretical speed ratio $K_v = 1.0$ is a useful and practical guide for achieving high operating efficiency of all-wheel-drive, off-road vehicles.

Figure 6.7 shows the measured relationship of tractive efficiency η_d defined by Eqn (6.15), total thrust, and theoretical speed ratio. A three-dimensional curved surface, representing the tractive efficiency as a quadratic function of total thrust and theoretical speed ratio, was fitted to the measured data, using the method of least squares, as shown in Figure 6.7. By slicing the

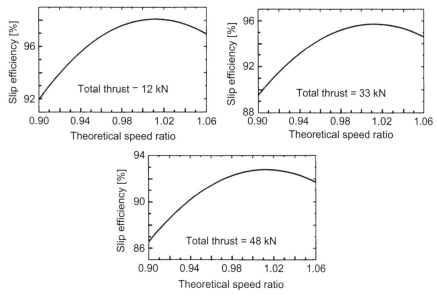

Figure 6.6: The relationships between the measured efficiency of slip and theoretical speed ratio at various values of total thrust (Reprinted by permission from *SAE* paper 2000-01-2596 by Wong et al., 2000)

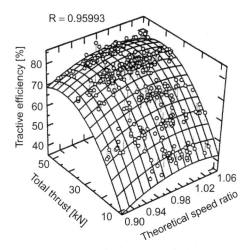

Figure 6.7: The relationship of the measured tractive efficiency, total thrust and theoretical speed ratio of a four-wheel-drive tractor on a farm field (Reprinted by permission from *SAE* paper 2000-01-2596 by Wong et al., 2000)

curved surface along a constant total thrust plane, the relationships between the tractive efficiency and theoretical speed ratio at various values of total thrust can be obtained, as shown in Figure 6.8. It can be seen that in the range of field tests conducted with total thrust-to-weight ratio F/W up to 0.45 (or the total thrust up to 48 kN), when the theoretical speed ratio is close to or equal to 1.0, the tractive efficiency of the four-wheel-drive tractor is also at a maximum.

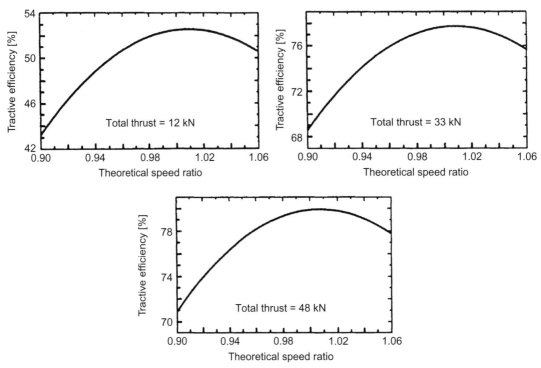

Figure 6.8: The relationships between the measured tractive efficiency and theoretical speed ratio at various values of total thrust (Reprinted by permission from *SAE* paper 2000-01-2596 by Wong et al., 2000)

■ Example 6.2

A four-wheel-drive tractor, with rigid coupling between its front and rear drive axles, operates on a farm field. The thrust–slip relationship for the two front tyres in the range of slip up to 20% may be approximated by a linear equation $F_f = 150\, i_f$ kN, where i_f is the slip of the front tyres (for instance, at 20% slip, $F_f = 150 \times 0.2 = 30$ kN). The rear tyres run in the ruts formed by the front tyres and the thrust–slip relationship for the two rear tyres in the range of slip up to 20% may be expressed by $F_r = 200\, i_r$ kN, where i_r is the slip of the rear tyres. Owing to the differences in inflation pressure and in wear between the front and rear tyres, the theoretical speed ratio K_v is 0.9. The total motion resistance of the tractor is 6.5 kN and the tractor pulls an implement with resistance of 30 kN.

(a) Determine the efficiency of slip of the four-wheel-drive tractor under the operating conditions described above.

(b) If identical tyres are installed in the front and rear drive axles, so that the theoretical speed ratio K_v is 1.0, determine the efficiency of slip of the tractor.

Solution

(a) From Eqn (6.18):

$$K_v = \frac{V_{tf}}{V_{tr}} = \frac{\omega_f r_f}{\omega_r r_r} = \frac{(1-i_r)}{(1-i_f)} = 0.9$$

and $i_r = 0.10 + 0.9\, i_f$.

Under steady-state operating conditions, the sum of the thrusts developed by the front and rear tyres is equal to the sum of the total motion resistance of the tractor 6.5 kN and the resistance of the implement 30 kN.

$$F = F_f + F_r = 150\, i_f + 200\, i_r = 150\, i_f + 200\,(0.10 + 0.9\, i_f) = 36.5\,\text{kN}.$$

Solving the above equation, one obtains

$$i_f = 0.05\,(5\%) \text{ and } i_r = 0.10 + 0.9\, i_f = 0.1 + 0.9 \times 0.05 = 0.145\,(14.5\%)$$

$$F_f = 150 \times 0.05 = 7.5\,\text{kN and } F_r = 200 \times 0.145 = 29\,\text{kN}$$

The coefficient of thrust distribution $K_d = F_r/(F_f + F_r) = 29/(7.5 + 29) = 0.7945$.

The efficiency of slip η_{s4} can be calculated using Eqn (6.22):

$$\eta_{s4} = 1 - \frac{i_f K_v - (K_v - 1)K_d}{K_v - (K_v - 1)K_d} = 1 - \frac{0.05 \times 0.9 - (0.9 - 1) \times 0.7945}{0.9 - (0.9 - 1) \times 0.7945}$$
$$= 0.873 = 87.3\%$$

(b) When the theoretical speed ratio K_v is equal to 1, the slip of the front tyres is equal to that of the rear tyres. From the thrust–slip relationships for the front and rear tyres given above, it is determined that the slips of the front and rear tyres $i_f = i_r = 36.5\,\text{kN}/350\,\text{kN} = 0.104\,(10.4\%)$.

Under these circumstances, the efficiency of slip η_{s4} is given by $\eta_{s4} = 1.0 - 0.104 = 0.896 = 89.6\%$.

This indicates that when $K_v = 1.0$, the efficiency of slip η_{s4} is 2.6% higher than that when $K_v = 0.9$.

■

6.3 Transport Productivity and Transport Efficiency

For off-road transport vehicles, such as off-road trucks used in construction, mining and logging industries, transport productivity and transport efficiency are often used as performance criteria.

Transport productivity P_{ro} is defined as the product of the payload and the average operating speed from the origin to destination on a given route. It is expressed by

$$P_{ro} = W_p V_{av} \quad (6.24)$$

where W_p is the payload of the vehicle, and V_{av} is the average operating speed of the vehicle.

Transport efficiency η_{tr} is defined as the ratio of the transport productivity P_{ro} to the corresponding power input P_{in}. It is expressed by

$$\eta_{tr} = \frac{P_{ro}}{P_{in}} = \frac{W_p V_{av}}{(\Sigma R) V_{av} / \eta_p} = \frac{W}{\Sigma R} \frac{W_p}{W} \eta_p = C_{ld} \eta_{st} \eta_p \quad (6.25)$$

where ΣR is the vehicle total motion resistance, W is the gross weight of the vehicle, C_{ld} is referred to as the lift to drag ratio and is equal to the ratio of vehicle gross weight W to total motion resistance ΣR, η_{st} is referred to as the structural efficiency and is equal to the ratio of payload W_p to vehicle gross weight W, and η_p is the propulsive efficiency, including transmission efficiency and slip efficiency, of the vehicle.

It should be pointed out that the lift to drag ratio C_{ld} can be much greater than unity. Thus, the transport efficiency η_{tr} may also be much greater than unity. Transport efficiency is quite different in concept from the mechanical efficiency which is always less than unity. It should also be noted that the transport efficiency η_{tr} is not a function of the average vehicle operating speed V_{av} from the origin to destination.

The reciprocal of transport efficiency η_{tr} expressed in terms of power consumed per unit transport productivity may also be used as a performance criterion for off-road transport vehicles (Wong, 1972a, 1975).

■ Example 6.3

A heavy mining truck has a payload of 2138 kN (with a volume capacity of 129 m³) and a gross weight of 3693 kN (including the payload). It operates on a roughly prepared trail with a coefficient of total motion resistance of 0.03 and at a slip of 5%. The transmission efficiency is 85%. Estimate the transport efficiency of the truck.

Solution

The lift to drag ratio C_{ld} can be determined from the reciprocal of the coefficient of total motion resistance:

$$C_{ld} = 1/0.03 = 33.33$$

The structural efficiency $\eta_{st} = W_p/W = 2138/3693 = 0.5789 = 57.89\%$

The propulsive efficiency $\eta_p = 0.85 \times (1 - 0.05) = 0.8075$

From Eqn (6.25), the transport efficiency η_{tr} is

$$\eta_{tr} = C_{ld}\, \eta_{st}\, \eta_p = 33.33 \times 0.5789 \times 0.8075 = 15.581 = 1{,}558.1\%$$

∎

6.4 Mobility Map and Mobility Profile

To evaluate the performance of a military vehicle, its average cross-country speed between two points in a given region is often used as a basic criterion. The average cross-country speed is a highly aggregated parameter, which represents the net results of numerous interactions among the vehicle, the driver, and the operating environment.

To predict the average cross-country speed, various computer simulation models, such as the US Army Mobility Model (AMM-75) and its successor, the NATO Reference Mobility Model (NRMM), have been developed (Nuttall et al., 1974; Jurkat et al., 1975). Because of the variation of environmental conditions in a given region, in these models the region of interest is divided into terrain units, within each of which the terrain is considered to be sufficiently uniform to permit the use of the simple, maximum, straight-line speed, referred to as the maximum feasible speed, of the vehicle to define its performance in or across that terrain unit. Dependent upon the characteristics of the terrain unit, the simulation models have three independent computational modules (Jurkat et al., 1975):

1. The areal patch module, which computes the maximum feasible speed of a single vehicle in a single areal terrain patch.
2. The linear feature module, which computes the minimum feasible time for a single vehicle, aided (such as with the assistance of winching or excavating) or unaided, to cross a uniform segment of a significant terrain feature, such as a stream, ditch, or embankment.
3. The on-road module, which computes the maximum feasible speed of a single vehicle travelling along a uniform segment of a road or trail.

All these modules draw from a common database that quantitatively describes the vehicle, the driver, and the terrain to be examined in the simulation.

When a vehicle is crossing an areal terrain patch, the maximum feasible speed may be limited by one or a combination of the following factors:

(a) The thrust (tractive effort) available for overcoming the resisting forces due to sinkage, slope, obstacles, vegetation, etc. The methods for evaluating vehicle tractive performance used in these models are empirical in nature and are discussed in some detail in Chapters 7 and 11.

(b) The driver's tolerance to ride discomfort when traversing rough terrain or to obstacle impacts.

(c) The driver's reluctance to proceed faster than the speed at which the vehicle would be able to decelerate to a stop, within the limited visibility distance prevailing in the patch.

(d) The vehicle's manoeuvrability to avoid obstacles.

(e) The acceleration and deceleration between obstacles, and speed reduction due to manoeuvring to avoid obstacles.

The speed limited by each factor is calculated and compared, and the maximum feasible speed within a particular terrain patch is then determined.

When the vehicle is traversing a linear feature segment, such as a stream, artificial ditch, canal, escarpment, railway, or highway embankment, an appropriate method is used to determine the maximum feasible speed. In the calculations, the time required to enter and cross the segment, and that required to egress from it, are taken into account. Both include allowance for engineering effort, such as winching or excavating, whenever required.

To predict the maximum feasible speed of a vehicle on roads or trails, in addition to the speed limited by the motion resistance, the speed limited by ride discomfort, by visibility, by tyre characteristics, or by road curvature has to be taken into consideration. The least of them is taken as the maximum feasible speed for the road or trail segment.

The results from these simulation models can be conveniently shown in a mobility map. As an example, Figure 6.9 shows the mobility map of a 22.24 kN (2½-ton) truck on a particular region (Jurkat et al., 1975). The numbers in the map indicate the speeds (in miles per hour) that the vehicle can achieve in various patches throughout the region under consideration. For the truck to travel from point A to point B, the optimal route indicated by the chain-line is shown in the map. It is selected based on the maximum average speed or the minimum time to travel from the origin to destination. The information shown in the mobility map may be summarized in a mobility profile. As an example, Figure 6.10 shows the mobility profile of a 44.48 kN (5-ton) truck in desert terrain (Jurkat et al., 1975). It indicates the average speed that the vehicle can sustain as a function of the percentage of the region under consideration. For instance, the intercept of the 90% point A in Figure 6.10 indicates that the truck can achieve an average speed of 21.7 km/h in 90% of the region considered. Thus, the mobility profile conveys a statistical description of vehicle performance in a particular region.

The mobility map and mobility profile are suitable formats for characterizing military vehicle performance primarily for the purpose of operational planning. However, they are not suitable for parametric analysis of vehicle design and performance. Computer-aided methods for detailed parametric analyses of design and performance of various types of off-road vehicle, intended primarily for use by vehicle design and development engineers, are discussed in Chapters 8, 9, 10 and 12.

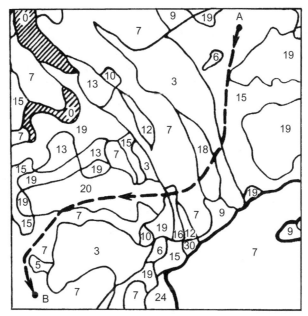

Figure 6.9: Mobility map for a 2½-ton truck. Numbers in the map designate the maximum achievable speed in miles per hour in a given terrain patch; cross-hatched areas indicate where the vehicle is immobile (Reprinted by permission of *ISTVS* from Jurkat et al., 1975)

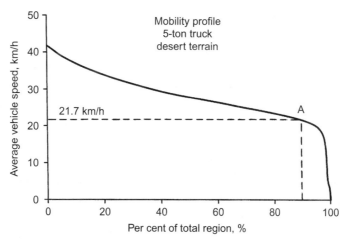

Figure 6.10: Mobility profile of a 5-ton truck in desert (Reprinted by permission of *ISTVS* from Jurkat et al., 1975)

PROBLEMS

6.1 A tracked vehicle has a total weight of 235 kN and operates on a level terrain. The coefficient of internal resistance of the track system is expressed by Eqn (6.6), where the value of f_0 is 0.03 and that of f_s is 0.00015. The coefficient of resistance due to track-terrain interaction is 0.08. The value of the aerodynamic drag coefficient is 1.15 and its frontal area is 5.95 m². The pitch radius of sprocket is 0.267 m and the transmission efficiency is 0.88. The thrust-slip relationship for the vehicle on the terrain is given in the following table. Estimate the acceleration of the vehicle in a gear with total reduction ratio (including both the gear box and final drive reduction ratios) of 13.8, when the engine operates at a speed of 2500 rpm and develops a torque of 935 N-m. In estimating the acceleration, the effect of the rotating inertia in the driveline may be neglected.

Thrust–slip relationship

Slip, %	0	3	5	10	20
Thrust, kN	0	52.32	63	80.21	97.87

6.2 An agricultural tractor with rubber tracks operates on a farm field with thrust-slip relationship shown in the following table. The pitch radius of the sprocket is 0.3 m. The total motion resistance of the tractor on the farm field is 5.5 kN. The transmission efficiency is 0.86. Determine the drawbar pull and overall tractive (drawbar) efficiency of the tractor, when its engine operates at a speed of 1500 rpm and develops a torque of 636 N-m, and a gear with total reduction ratio of 19.35 is engaged.

Thrust–slip relationship

Slip, %	0	3	5	10	20
Thrust, kN	0	21.82	26.27	34.45	40.81

6.3 A four-wheel-drive tractor, with rigid coupling between its front and rear drive axles, operates on a farm field. The thrust-slip relationship for the two front tyres in the range of slip up to 20% may be approximated by a linear equation: $F_f = 150\, i_f$ kN, where i_f is the slip of the front tyres. The rear tyres run in the ruts formed by the front tyres. The thrust-slip relationship for the two rear tyres in the range of slip below 20% may be expressed by $F_r = 175\, i_r$ kN, where i_r is the slip of the rear tyres. Owing to the differences in inflation pressure and in wear between the front and rear tyres, the theoretical speed ratio K_v is 1.1. The total motion resistance of the tractor is 6.0 kN and the tractor pulls an implement with resistance of 20 kN.

(a) Determine the tractive efficiency, as defined by Eqn (6.15), of the four-wheel-drive tractor under the operating conditions described above.

(b) If identical tyres are installed in the front and rear drive axles, so that the theoretical speed ratio K_v is 1.0, determine the tractive efficiency of the tractor.

6.4 A four-wheel-drive tractor, with rigid coupling between its front and rear drive axles, operates on a farm field. The thrust-slip relationship for the two front tyres in the range of slip up to 20% may be approximated by a linear equation: $F_f = 120\, i_f$ kN, where i_f

is the slip of the front tyres. The rear tyres run in the ruts formed by the front tyres and the thrust-slip relationship for the two rear tyres in the range of slip up to 20% may be expressed by $F_r = 135\, i_r$ kN, where i_r is the slip of the rear tyres. Owing to the differences in inflation pressure and in wear between the front and rear tyres, the theoretical speed ratio K_v is 0.9. If the total motion resistance of the tractor is 6 kN and the tractor is in a self-propelled condition (i.e. not pulling or pushing any implement or working machinery), determine whether or not torsional wind-up in the transmission will occur under these circumstances. Estimate the slip (or skid) of the front and rear tyres under these conditions.

6.5 A wheeled mining truck has a payload of 844 kN (with volume capacity of 36.4 m^3) and a gross weight of 1442 kN (including the payload). It operates on a prepared trail with an average grade of 8% and at a tyre slip of 4%. The average mechanical efficiency of the transmission is 0.86. The motion resistance coefficient is 0.035. The average speed of the truck on the given route is 25 km/h. Estimate the transport productivity and the transport efficiency of the truck operating on the route described above.

CHAPTER 7

Methods for Evaluating Tracked Vehicle Performance

The track was first conceived in the 18th century as a 'portable railway' and tracked vehicles have been used on a large-scale since the turn of the last century. However, for a long period of time, the development and design of tracked vehicles have been, by and large, guided by past experience and the 'cut and try' methodology. As economic and social conditions change with the rapid progress in technology, the 'trial and error' approach to off-road vehicle development has become extremely inefficient and prohibitively expensive. Furthermore, new requirements for greater mobility over a wider range of terrains in all seasons, and growing demands for energy conservation and environmental preservation, have emerged. This has led to the recognition of the necessity of establishing mathematical models for vehicle–terrain systems that will enable the development and design engineer, as well as the user, to evaluate a wide range of options and to select an optimum vehicle configuration for a given mission and environment. To be useful to the development and design engineer, a mathematical model for tracked vehicle performance should take into account all major vehicle design and operational parameters, as well as terrain characteristics.

A variety of models for predicting and evaluating tracked vehicle performance has been developed. A brief review of some of the empirical methods and methods for parametric analysis and for predicting static pressure distributions beneath tracks is given below.

7.1 Empirical Methods

It is generally recognized that the interaction between a tracked vehicle and terrain is very complex and is difficult to model accurately. To circumvent this difficulty, empirical models have been developed. In general, these models are based on the test results of a number of representative tracked vehicles over a range of terrains of interest. The measured vehicle performance is then empirically correlated with terrain conditions, usually identified by observations and simple measurements. This can lead to the establishment of a scale for evaluating vehicle mobility on the one hand and terrain trafficability on the other.

One of the well-known empirical methods for evaluating off-road vehicle performance is that developed by the US Army Corps of Engineers Waterways Experiment Station (WES). In developing the methods, vehicles were tested in a range of terrains, primarily fine- and coarse-grained soils. Fine-grained soils are silt or clayey soils in which 50% or more by weight of the grains are smaller than 0.074 mm in diameter. Coarse-grained soils are beach and desert soils, usually containing less than 7% of the grains smaller than 0.074 mm in diameter, or soils containing 7% or more of the grains smaller than 0.074 mm in diameter but not in a wet condition. The terrain conditions were identified using a cone penetrometer. The measured data of vehicle performance and terrain conditions were then empirically correlated, and a model known as the WES VCI model was proposed for predicting vehicle performance on fine- and coarse-grained inorganic soils (Rula and Nuttall, 1971). This model forms the basis for the subsequent developments of the AMC71 and AMM75 mobility models and the NATO Reference Mobility Model (NRMM). To illustrate the methodology of this type of empirical models, the basic features of the WES VCI model are outlined below as an example.

In the WES VCI model, an empirical equation was established to calculate first the mobility index (MI) of a vehicle in terms of certain vehicle design features. The mobility index for a tracked vehicle is defined as

$$\text{Mobility Index} = \left(\frac{\text{contact pressure factor} \times \text{weight factor}}{\text{track factor} \times \text{grouser factor}} + \text{bogie factor} - \text{clearance factor} \right) \times \text{engine factor} \times \text{transmission factor} \qquad (7.1)$$

where:

$$\text{Contact pressure factor} = \frac{\text{gross weight, lb}}{\text{area of tracks in contact with ground, in}^2}$$

Weight factor: less than 222.4 kN (50,000 lb) = 1.0

$$222.4 - 311.4\,\text{kN} \ (50{,}000\,\text{lb} - 69{,}999\,\text{lb}) = 1.2$$

$$311.4 - 444.8\,\text{kN} \ (70{,}000 - 99{,}999\,\text{lb}) = 1.4$$

444.8 kN or greater (100,000 lb) = 1.8

$$\text{Track factor} = \frac{\text{track width, in.}}{100}$$

Grouser factor: Grousers less than 3.8 cm (1.5 in.) high = 1.0

Grousers more than 3.8 cm (1.5 in.) high = 1.1

$$\text{Bogie factor} = \frac{\text{gross weight, lb, divided by 10}}{\begin{pmatrix} \text{total number of} \\ \text{bogies on tracks in} \\ \text{contact with ground} \end{pmatrix} \times \begin{pmatrix} \text{area of one} \\ \text{track shoe, in.}^2 \end{pmatrix}}$$

$$\text{Clearance factor} = \frac{\text{clearance, in.}}{10}$$

Engine factor: ≥ 8.2 kW/tonne (10 hp/ton) of vehicle weight = 1.0

< 8.2 kW/tonne (10 hp/ton) of vehicle weight = 1.05

Transmission factor: automatic = 1.0; manual = 1.05

Based on the mobility index (MI), a parameter called the vehicle cone index (VCI) is calculated. The VCI represents the minimum strength of a given soil in the critical layer which is required for a vehicle to successfully make a specific number of passes, usually one pass or 50 passes. The values of VCI for one pass and 50 passes, VCI_1 and VCI_{50}, are calculated from the mobility index (MI) using the following empirical equations (Rula and Nuttall, 1971):

$$VCI_1 = 7.0 + 0.2MI - \left(\frac{39.2}{MI + 5.6}\right) \tag{7.2}$$

$$VCI_{50} = 19.27 + 0.43MI - \left(\frac{125.79}{MI + 7.08}\right) \tag{7.3}$$

The soil strength is described in terms of either the rating cone index (RCI) for fine-grained soils or the cone index (CI) for coarse-grained soils. The critical layer referred to above varies with the type and weight of vehicle and with soil strength profile. For freely draining or clean sands, it is usually the 0–15 cm (0-6 in.) layer. For fine-grained soils and poorly drained sands with fines, it is usually the 0–15 cm (0-6 in.) layer for one pass and the 15–30 cm (0-12 in.) layer for 50 passes.

For fine-grained soils, after the VCI and soil strength values have been determined, the values of the performance parameters of a tracked vehicle, such as the net maximum drawbar pull coefficient (the ratio of drawbar pull to vehicle weight), maximum slope negotiable, and towed motion resistance coefficient (the ratio of towed motion resistance to vehicle weight), are then empirically determined as functions of vehicle type, number of passes to be completed, and the excess of RCI over VCI (i.e. RCI − VCI), also referred to as the excess soil

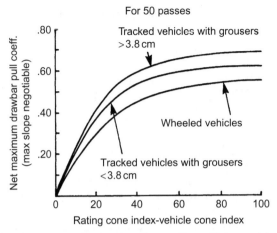

Figure 7.1: Variations of the net maximum drawbar pull coefficient with the excess of RCI over VCI for tracked and wheel vehicles for 50 passes on level, fine-grained soils (Rula and Nuttall, 1971)

strength. The methodology of the WES VCI model for predicting tracked vehicle performance on fine-grained soils has been adopted by NATO Reference Mobility Model Edition II.

Figure 7.1 shows the empirical relations between the net maximum drawbar pull coefficient (or maximum slope negotiable) and the excess of RCI over VCI for tracked vehicles with different grouser heights over fine-grained soils. The empirical relations between the first-pass towed motion resistance coefficient and the excess of RCI over VCI for tracked and wheeled vehicles operating on fine-grained soils are shown in Figure 7.2 (Rula and Nuttall, 1971).

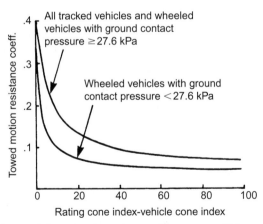

Figure 7.2: Variations of the towed motion resistance coefficient with the excess of RCI over VCI for tracked and wheeled vehicles for the first pass on level, fine-grained soils (Rula and Nuttall, 1971)

It should be noted that empirical models should only be used for estimating the performance of vehicles with design features similar to those that have been tested under similar operating conditions. Empirical relations should not be extrapolated beyond the conditions under which they were derived. It appears uncertain whether empirical equations based on test data collected years ago are still valid for predicting the performance of current and future generations of tracked vehicles. It should also be pointed out that while some vehicle design features have been taken into account in the empirical relations described above, a number of important design parameters of tracked vehicles have been omitted, such as track system configuration, initial track tension and suspension characteristics (Wong, 1986a, b, 1992a, 1995, 1997, 1999, 2007; Wong and Huang, 2005, 2006a, b, 2008; Wong and Preston-Thomas, 1986a, 1988).

Another empirical method for evaluating the mobility of tracked vehicles was suggested by Rowland (1972). He proposed to use the mean maximum pressure (MMP), which is defined as the mean value of the maxima occurring under all the roadwheel stations, as a criterion for evaluating the soft ground performance of off-road vehicles. Empirical equations were derived from vehicle test data for calculating the MMP for different types of track system.

He proposed that for link and belt tracks on rigid roadwheels:

$$MMP = \frac{1.26W}{2n_r A_l b \sqrt{t_t D}} \text{ kPa} \tag{7.4}$$

and for belt tracks on pneumatic tyred roadwheels:

$$MMP = \frac{0.5W}{2n_r b \sqrt{Df_t}} \text{ kPa} \tag{7.5}$$

where W is the vehicle weight (kN), n_r is the number of wheel stations in one track, A_l is the rigid area of link (or belt track cleat) as a proportion of $b \times t_t$, b is the track or pneumatic tyre width (m), t_t is the track pitch (m), D is the outer diameter of roadwheel or pneumatic tyre (m), and f_t is the radial deflection of pneumatic tyre under load (m).

Table 7.1 shows the values of MMP for various types of tracked vehicles calculated using the empirical formulae proposed by Rowland (1975).

To evaluate whether a particular vehicle with a specific value of MMP will have adequate mobility over a specific terrain, Rowland proposed a set of desired values of the mean maximum pressure for different terrain conditions. Table 7.2 shows the desired values of MMP suggested by Rowland (1975).

It should be pointed out that in the empirical formulae proposed by Rowland, terrain characteristics are not taken into account in the calculation of MMP. Thus, the values of MMP calculated using either Eqn (7.4) or Eqn (7.5) are independent of terrain conditions. In reality, the pressure distribution under a track, hence the actual value of MMP, is strongly influenced

Table 7.1: Values of the mean maximum pressure of some tracked vehicles

Vehicle	Track configuration	Weight (kN)	Mean maximum pressure (kPa)
Amphibious Carrier M29C Weasel	Link track	26.5	27
Armoured Personnel Carrier M113	Link track	108	119
Caterpillar D4 Tractor	Link track	59	82
Caterpillar D7 Tractor	Link track	131	80
Main Battle Tank AMX 30	Link track	370	249
Main Battle Tank Leopard I	Link track	393	198
Main Battle Tank Leopard II	Link track	514	201
Main Battle Tank M60	Link track	510–545	221–236
Main Battle Tank T62	Link track	370	242
Swedish S-Tank	Link track	370	267
Volvo BV202 All-Terrain Carrier	Belt track, Pneumatic tyre	42	33

Table 7.2: Desired values of the mean maximum pressure

Terrain	Mean maximum pressure (kPa)		
	Ideal (multipass operation or good gradability)	Satisfactory	Maximum acceptable (mostly trafficable at single-pass level)
Wet, fine-grained			
• Temperate	150	200	300
• Tropical	90	140	240
Muskeg			
Muskeg floating mat and	30	50	60
European bogs	5	10	15
Snow	10	25–30	40

by terrain characteristics. Figure 7.3(a), (b) and (c), show the measured normal pressure distributions under a tracked vehicle (an armoured personnel carrier) with total weight of 88.71 kN, over a sandy terrain (LETE sand), a muskeg (Petawawa muskeg A) and a snow-covered terrain (Petawawa snow B), respectively (Wong, 1994a). The pressure–sinkage characteristics and shearing behaviour of these three terrains are described in Chapters 4 and 5, respectively. As shown in Figure 7.3, the peak pressures under the roadwheels vary significantly with terrain conditions. For instance, on LETE sand which is dense, the track sinkage is shallow, as shown in Figure 7.3(a). The vehicle weight is primarily supported by the track links immediately beneath the roadwheels. The track links between roadwheels support relatively little load. The measured peak pressure under the rear roadwheel is approximately 427 kPa. On Petawawa muskeg A which is relatively soft, the track sinkage is much deeper than that on LETE sand, as shown in Figure 7.3(b), In this case, the vehicle weight is supported by not only the track links immediately beneath the roadwheels but also those

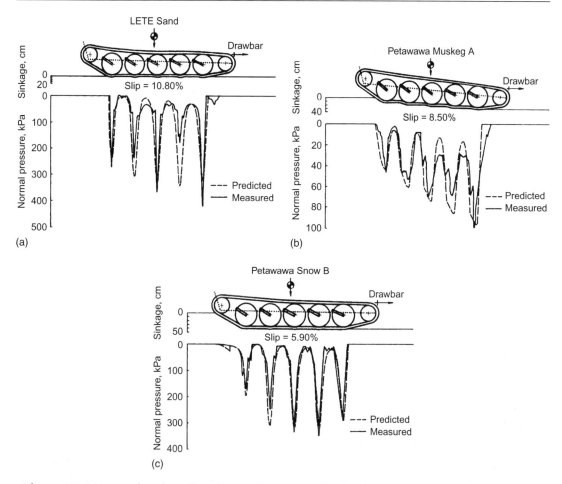

Figure 7.3: Measured and predicted normal pressure distributions by NTVPM under a tracked vehicle over (a) sandy terrain, (b) muskeg, and (c) snow-covered terrain

between roadwheels. The peak pressure under the rear roadwheel is approximately 100 kPa, which is only 23.4% of that on LETE sand.

The actual values of MMP derived from the measured peak pressures under all roadwheel stations on LETE sand, Petawawa muskeg A, and Petawawa snow B are 299, 68 and 279 kPa, respectively, as shown in Table 7.3. The calculated value of MMP using Eqn (7.4) is 98 kPa (with $W = 88.71$ kN, $n_r = 5$, $A_l = 0.992$, $b = 0.38$ m, $t_t = 0.15$ m, and $D = 0.61$ m), and is also shown in the table. This clearly indicates that the actual values of MMP are highly dependent on terrain conditions; whereas the calculated value obtained using Eqn (7.4) is the same for all types of terrain. It is also shown that there are significant differences between the calculated value of MMP using Eqn (7.4) and the actual values of MMP on the three types of terrain (Wong, 1994a). The discrepancies between the values calculated by Eqn (7.4) and the

Table 7.3: Comparison of the measured values of MMP on various terrains with those predicted using NTVPM and those calculated by Rowland's empirical formula

Terrain	MMP (kPa)		
	Measured	Predicted by NTVPM*	Calculated by Rowland's formula
LETE sand	299	310	98
Petawawa Muskeg A	68	72	98
Petawawa Snow B	279	288	98

Source: Wong (1994a).
*The values of MMP predicted by NTVPM vary with track slip. Those given in this table are for the specific slips shown in Figure 7.3(a), (b) and (c), respectively.

measured values, normalized with respect to the measured ones, are 67.2, 44.1 and 65.2% on LETE sand, Petawawa muskeg A and Petawawa snow B, respectively.

For comparison, the normal pressure distributions under the tracked vehicle on the three types of terrain predicted by the computer-aided method NTVPM are shown in Figure 7.3(a), (b) and (c). The analytical framework within which NTVPM is developed is discussed in Chapter 8. It can be seen that there is a reasonably close agreement between the measured normal pressure distributions and the predicted ones obtained using NTVPM. The values of MMP derived from the predicted normal pressure distributions by NTVPM are shown in Table 7.3. It can be seen from the table that there is a reasonably close correlation between the measured values of MMP and the predicted ones by NTVPM. The values of MMP predicted by NTVPM on LETE sand, Petawawa muskeg A, and Petawawa snow B are 310, 72 and 288 kPa, respectively. The corresponding differences between the measured values of MMP and the predicted ones by NTVPM, normalized with respect to the measured values, are 3.7, 5.9 and 3.2%, respectively. This indicates that the values of MMP predicted by NTVPM are much closer to the measured values than those calculated using Eqn (7.4).

The results of the study presented above indicate that the complete neglect of the effect of terrain conditions, hence its inability to predict the influence of terrain characteristics on the value of MMP, is one of the major deficiencies of the empirical method proposed by Rowland. It should be pointed out that the empirical equations, Eqns (7.4) and (7.5), were derived primarily from experimental data obtained prior to 1960 (Rowland, 1972). It is, therefore, uncertain whether these empirical formulae could be applied to the evaluation of the mobility of current or future generations of tracked vehicles with new design features. The empirical equations, together with Table 7.2, could only be employed to evaluate tracked vehicle mobility on a 'go/no go' basis and could not be used to quantitatively predict the tractive performance of a tracked vehicle, such as its motion resistance, thrust, drawbar pull and tractive efficiency under a given operating condition. Furthermore, Eqns (7.4) and (7.5) only take into account a limited number of vehicle design parameters. It is shown later in Chapter 9 that a number of other design parameters, such as the roadwheel system configuration, initial track tension, and roadwheel suspension characteristics, also have significant effects on tracked vehicle mobility on soft terrain.

Within the context of their intended purposes, empirical methods, such as those developed by WES and Rowland, may be useful in certain types of application, such as in the preliminary assessment of vehicle mobility. It should be pointed out, however, that empirical relations are only valid for the specific types of vehicle operating on a similar range of terrains that have been tested. They normally should not be extrapolated beyond the conditions upon which they were originally derived. Consequently, empirical models have inherent limitations. For instance, it is uncertain whether an empirical model could play a useful role in the evaluation of new design concepts or in the prediction of vehicle mobility in new operating environments. Furthermore, an entirely empirical approach to the study of vehicle mobility is only feasible where the number of variables involved in the problem is relatively small. If a large number of factors are required to define the problem, then an empirical approach may not be feasible or cost effective.

To provide the vehicle engineer with practical tools to quantitatively evaluate tracked vehicle performance or design over a wide range of operating scenarios, other methods of approaches have been developed. Some of these are outlined below and in subsequent chapters.

7.2 Methods for Parametric Analysis

One of the better known methods for parametric analysis of tracked vehicle performance is that originally developed by Bekker (1956). In this method, the track in contact with the terrain is assumed to be similar to a rigid footing. If the centre of gravity of the vehicle is located at the midpoint of the track contact area, the normal pressure distribution may be assumed as uniform, as shown in Figure 7.4. On the other hand, if the centre of gravity of the vehicle is located ahead of or behind the midpoint of the contact area, a sinkage distribution of trapezoidal form will be assumed. Based on the assumed contact pressure, and making use of the pressure–sinkage relationship obtained from the bevameter, the sinkage of the track can be predicted.

Using the Bekker pressure–sinkage equation (Eqn (4.1)), for a track with uniform contact pressure, the sinkage z_0 is given by

$$z_0 = \left(\frac{p}{k_c/b + k_\phi} \right)^{1/n} = \left(\frac{W/bl}{k_c/b + k_\phi} \right)^{1/n} \qquad (7.6)$$

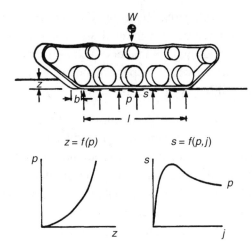

Figure 7.4: A simplified model for tracked vehicle performance

where b and l are the width and contact length of the track, respectively, and W is the normal load on the track.

The work done in compacting the terrain to a depth of z_0 by a track of width b and contact length l and with uniform contact pressure is given by

$$\text{Work done} = bl \int_0^{z_0} p\, dz = bl \int_0^{z_0} (k_c/b + k_\phi) z^n\, dz = bl(k_c/b + k_\phi) \frac{z_0^{n+1}}{n+1} \quad (7.7)$$

Substituting for z_0 from Eqn (7.6) yields

$$\text{Work done} = \frac{bl}{(n+1)(k_c/b + k_\phi)^{1/n}} \left(\frac{W}{bl}\right)^{(n+1)/n} \quad (7.8)$$

If the track is pulled a distance l in the horizontal direction, the work done by the towing force, which is equal in magnitude to the motion resistance due to compaction R_c, can be equated to the work done in making a rut of width b and length l to a depth of z_0, as expressed by Eqn (7.8):

$$R_c\, l = \frac{bl}{(n+1)(k_c/b + k_\phi)^{1/n}} \left(\frac{W}{bl}\right)^{(n+1)/n}$$

and

$$R_c = \frac{1}{(n+1)b^{1/n}(k_c/b + k_\phi)^{1/n}} \left(\frac{W}{l}\right)^{(n+1)/n} \quad (7.9)$$

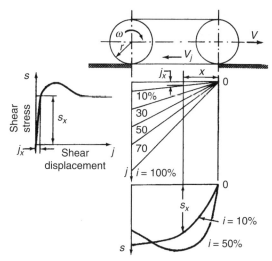

Figure 7.5: Development of shear displacement and shear stress under a rigid track (Reprinted by permission from J.Y. Wong, *Theory of Ground Vehicles*, 4th Ed., Wiley, 2008, copyright © 2008 by John Wiley)

In very soft ground where significant sinkage occurs, Bekker (1960) suggested that a bulldozing resistance should be taken into account and that it may be estimated using the retaining wall (or the passive earth pressure) theory of soil mechanics described in Section 2.2.

To predict the tractive effort of a tracked vehicle, the shear stress developed on the track–terrain interface has to be determined. As discussed in Chapter 5, the shear stress is related to shear displacement. Therefore, it is essential to analyse the development of shear displacement under a track. Figure 7.5 illustrates the approach to evaluating the shear displacement developed beneath a track having a flat contact patch. When a driving torque is applied to the sprocket, shearing action takes place on the track–terrain interface. Consequently, there is a relative movement between the track and the terrain (or shear displacement) in the horizontal direction, which results in track slip. The slip i of a track is defined by Eqn (6.10):

$$i = 1 - \frac{V}{r\omega} = 1 - \frac{V}{V_t} = \frac{V_t - V}{V_t} = \frac{V_j}{V_t}$$

and

$$V_j = r\omega i$$

where V is the actual forward speed of the track, V_t is the theoretical speed of the track which is the product of the pitch radius r and angular speed ω of the sprocket, and V_j is the slip speed of the track. It should be noted that the slip speed V_j of the track is in the opposite direction of the forward speed of the track V, as shown in Figure 7.5.

For a metal link track, such as that used in industrial or agricultural tractors, it may be assumed that it is rigid and cannot be stretched. The slip speed V_j is, therefore, the same for every point of the track in contact with the terrain. The shear displacement j at a point located at a distance x from the front of the track is given by

$$j = V_j t \tag{7.10}$$

where t is the contact time of the point in question with the terrain and is equal to x/V_t. Equation (7.10) may be rewritten as

$$j = \frac{V_j x}{V_t} = i x \tag{7.11}$$

This indicates that the shear displacement beneath a track having a flat contact surface increases linearly from zero at the front to the rear of the contact area, as shown in Figure 7.5.

If the normal pressure on the track is uniformly distributed, as shown in Figure 7.6(a), and the shear stress–shear displacement relationship is described by the simple exponential equation, Eqn (5.2), the tractive effort F of a track with a flat contact surface can be expressed by

$$\begin{aligned} F &= b \int_0^l \left(c + \frac{W}{bl} \tan\phi \right)(1 - e^{-ix/K})\,dx \\ &= (Ac + W\tan\phi)\left[1 - \frac{K}{il}(1 - e^{-il/K}) \right] \end{aligned} \tag{7.12}$$

where A, b and l are the contact area, contact width and contact length of the track, respectively, c, ϕ and K are the cohesion, angle of shearing resistance and shear deformation parameter, respectively, i is the track slip, and W is the normal load on the track.

If the normal pressure is not uniformly distributed along the track contact length, the procedure described above can still be used to predict the thrust developed by a track as a function of slip. For instance, if the normal pressure has a multi-peak sinusoidal distribution as shown in Figure 7.6(b), the normal pressure is expressed by (Wills, 1963)

$$p = \frac{W}{bl}\left(1 + \cos\frac{2n\pi x}{l}\right) \tag{7.13}$$

where n is the number of periods, as shown in Figure 7.6(b). It should be noted that in this case the peak pressure p_{max} is twice the mean pressure $p_{mean} = W/bl$. In a frictional terrain, the shear stress developed along the contact length can be expressed by

$$s = \frac{W}{bl}\tan\phi\left(1 + \cos\frac{2n\pi x}{l}\right)(1 - e^{-ix/K}) \tag{7.14}$$

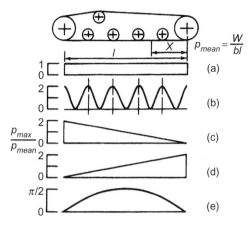

Figure 7.6: Various idealized normal pressure distributions under a track (Reprinted by permission of *J. Agri. Engng Res.* from Wills, 1963)

and hence the tractive effort as a function of slip is given by

$$F = b \int_0^l \frac{W}{bl} \tan\phi \left(1 + \cos\frac{2n\pi x}{l}\right)\left(1 - e^{-ix/K}\right) dx$$

$$= W \tan\phi \left[1 + \frac{K}{il}\left(e^{-il/K} - 1\right) + \frac{K(e^{-il/K} - 1)}{il(1 + 4n^2 K^2 \pi^2/i^2 l^2)}\right] \quad (7.15)$$

In the case where the normal pressure increases linearly from the front to the rear of the contact area as shown in Figure 7.6(c), the normal pressure distribution is described by

$$p = \frac{2W}{bl}\frac{x}{l} \quad (7.16)$$

and in a frictional terrain, the tractive effort as a function of slip is expressed by

$$F = W \tan\phi \left[1 - 2\left(\frac{K}{il}\right)^2 \left(1 - e^{-il/K} - \frac{il}{K}e^{-il/K}\right)\right] \quad (7.17)$$

If the normal pressure decreases linearly from the front to the rear of the contact area as shown in Figure 7.6(d), the normal pressure distribution is expressed by

$$p = \frac{2W}{bl}\frac{(l-x)}{l} \quad (7.18)$$

and over a frictional terrain, the tractive effort is given by

$$F = 2W\tan\phi\left[1 - \frac{K}{il}\left(1 - e^{-il/K}\right)\right]$$

$$-W\tan\phi\left[1 - 2\left(\frac{K}{il}\right)^2\left(1 - e^{-il/K} - \frac{il}{K}e^{-il/K}\right)\right] \qquad (7.19)$$

In the case where the normal pressure has a sinusoidal distribution with the maximum pressure at the centre and zero pressure at the front and rear contact points as shown in Figure 7.6(e), the normal pressure distribution is described by

$$p = \frac{W}{bl}\frac{\pi}{2}\sin\left(\frac{\pi x}{l}\right) \qquad (7.20)$$

It should be noted that in this case the maximum pressure at the midpoint, p_{max}, is equal to $(\pi/2)(W/bl)$. Over a frictional terrain, the tractive effort as a function of slip is expressed by

$$F = W\tan\phi\left[1 - \frac{\left(e^{-il/K} + 1\right)}{2\left(1 + i^2l^2/\pi^2K^2\right)}\right] \qquad (7.21)$$

Figure 7.7 shows the variations of the tractive effort with slip of a track vehicle with various types of normal pressure distribution described above (Wills, 1963). It can be seen that the normal pressure distribution beneath a rigid track affects the development of tractive effort.

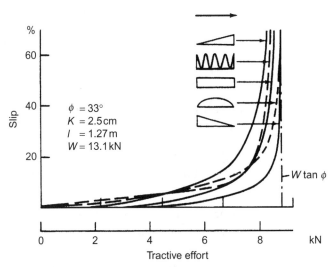

Figure 7.7: Effects of normal pressure distribution on the tractive effort (thrust) of a track on sand (Reprinted by permission of *J. Agri. Engng Res.* from Wills, 1963)

From the predicted tractive effort and motion resistance, the drawbar pull as a function of slip and the overall tractive performance of the vehicle can be determined.

Example 7.1

An industrial tractor of total weight of 329 kN is under development. Two track systems, A and B, are being considered. Each of the two tracks of system A has a width of 0.5 m and a contact length of 2.60 m. Each of the two tracks of system B has a width of 0.6 m (120% of that of track system A) and a contact length of 2.17 m (83.3% of that of track system A). Thus, the total contact areas of the two track systems are the same. Compare the drawbar performances of a vehicle with these two proposed track systems at 20% slip on a clayey soil, with pressure–sinkage parameters: $n = 1.0$, $k_c = 20.68 \text{ kN/m}^{n+1}$, and $k_\phi = 814.3 \text{ kN/m}^{n+2}$; and shear strength parameters: $c = 3.45 \text{ kPa}$, $\phi = 11°$, and $K = 2.54 \text{ cm}$. In the evaluation, uniform normal pressure distributions under the two track systems may be assumed.

Solution

(a) System A

If uniform normal pressure under the track system is assumed, then the sinkage of the track may be estimated using Eqn (7.6):

$$z_0 = \left(\frac{W/2bl}{k_c/b + k_\phi}\right)^{1/n} = \left(\frac{329/(2 \times 0.5 \times 2.60)}{20.68/0.5 + 814.3}\right) = \frac{126.538}{855.66} = 0.148 \text{ m}$$

The compaction resistance R_c may be determined from the track sinkage z_0 as follows (referring to Eqn (7.7)):

$$R_c = 2b\,(k_c/b + k_\phi)\left[\frac{z_0^{n+1}}{n+1}\right] = 2 \times 0.5 \times 855.66\,(0.148^2/2) = 9.371 \text{ kN}$$

The thrust of the track system at 20% slip may be predicted using Eqn (7.12):

$$F = (Ac + W\tan\phi)\left[1 - \frac{K}{il}(1 - e^{-il/K})\right]$$

$$= (2.6 \times 3.45 + 329 \times 0.1944)\left[1 - \frac{0.0254}{0.2 \times 2.6}(1 - e^{-20.47})\right]$$

$$= 69.365 \text{ kN}$$

The drawbar pull F_d at 20% slip (as determined by vehicle-terrain interaction) is equal to the difference between the thrust F at 20% slip and the compaction resistance R_c:

$$F_d = F - R_c = 69.365 - 9.371 = 59.994 \text{ kN}$$

(b) System B

If uniform normal pressure under the track system is assumed, then the sinkage of the track may be estimated using Eqn (7.6):

$$z_0 = \left(\frac{W/2bl}{k_c/b + k_\phi}\right)^{1/n} = \left(\frac{329/(2 \times 0.6 \times 2.167)}{20.68/0.6 + 814.3}\right) = \frac{126.519}{848.77} = 0.149 \text{ m}$$

The compaction resistance R_c may be estimated as follows:

$$R_c = 2b(k_c/b + k_\phi)\left[\frac{z_0^{n+1}}{n+1}\right] = 2 \times 0.6 \times 848.77 \, (0.149^2/2) = 11.306 \text{ kN}$$

The thrust of the track system at 20% slip may be predicted using Eqn (7.12):

$$F = (Ac + W\tan\phi)\left[1 - \frac{K}{il}\left(1 - e^{-il/K}\right)\right] = (2 \times 0.6 \times 2.167 \times 3.45 + 329 \times 0.1944)$$

$$\left[1 - \frac{0.0254}{0.2 \times 2.167}\left(1 - e^{-17.063}\right)\right]$$

$$= 68.654 \text{ kN}$$

The drawbar pull F_d at 20% slip (as determined by vehicle-terrain interaction) is equal to the difference between the thrust F at 20% slip and the compaction resistance R_c:

$$F_d = F - R_c = 68.654 - 11.306 = 57.348 \text{ kN}$$

(c) The ratio of the drawbar pull of system A to that of system B at 20% slip is 1.046. This indicates that while both systems have the same track contact area, the drawbar pull of system A at 20% slip is 4.6% higher than that of system B. This is because system B has slightly higher compaction resistance due to its wider track, although its track sinkage is essentially the same as that of system A. Furthermore system B has slightly lower thrust due to its shorter track contact length than system A.

∎

7.3 Methods for Predicting Static Pressure Distributions beneath Tracks

While the idealization of a track system as a rigid footing may not be unreasonable for tracked vehicles with low ratios of roadwheel spacing to track pitch commonly used in industrial and agricultural tractors (Rohrbach and Jackson, 1982), it is not realistic for tracked vehicles with high ratios of roadwheel spacing to track pitch designed for high speed operations (Wong et al., 1984). Ground pressure under these vehicles is usually concentrated under the roadwheels and is far from uniform. Consequently, performance predictions using the method described above will be unrealistic, particularly with respect to sinkage, motion resistance and tractive effort on soft terrain.

In an attempt to improve the prediction of ground pressure distribution under a track system with a high ratio of roadwheel spacing to track pitch, Bekker (1956) performed a pioneering theoretical study more than five decades ago. The analysis is intended for predicting static ground pressure distribution when the vehicle is at rest. The effects of vehicle weight, track width, roadwheel spacing and the pressure–sinkage relationship of the terrain were taken into account. The study was, however, limited to the analysis of the shape of the track span between two roadwheels, simplified as knife-edge supports, in a terrain with a linear pressure–sinkage relationship. The effects of roadwheel diameter, suspension characteristics and other design factors were neglected in the analysis.

In 1981, an improved method for predicting the static ground pressure distribution was developed by Garber and Wong (1981a, b). In the analysis, the major design features of a tracked vehicle, such as the dimensions, spacing and number of roadwheels, vehicle weight, track dimensions, initial track tension, and characteristics of the suspension and track tensioning device, were taken into consideration. The analysis can accommodate terrain with a linear or non-linear pressure–sinkage relationship. Based on a detailed examination of the interaction between the track and the terrain under static conditions, the interrelationships between the static ground pressure distribution and vehicle design parameters and terrain conditions were established. Figure 7.8 shows the predicted static ground pressure distributions of various tracked vehicles on different types of terrain, obtained using the analytical method developed by Garber and Wong (1981a). The design parameters of the vehicles used in the analysis are given in Table 7.4. The variations of the mean maximum pressure (MMP), the ratio of MMP to MGP (mean ground pressure, which is the normal load on the track divided by the gross contact area), and the maximum roadwheel sinkage with terrain stiffness for various types of tracked vehicle are shown in Figures 7.9, 7.10 and 7.11, respectively. It can be seen that both the track system configuration and the terrain conditions have significant influence on the static ground pressure distribution, and that the maximum static pressure under a track is generally much higher than the mean (or nominal) ground pressure.

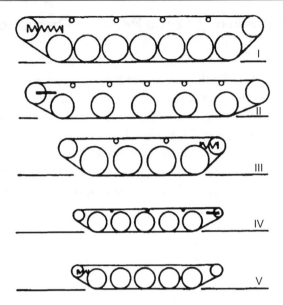

(I) Tensioning wheel in the front
(II) Tensioning wheel in the front and fixed
(III) Tensioning wheel in the rear
(IV) Tensioning wheel in the rear and fixed
(v) Tensioning wheel in the front

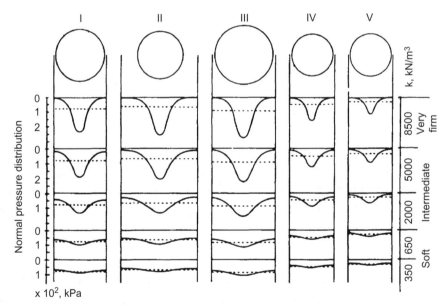

Figure 7.8: Static normal pressure distributions under various track systems over terrains with different stiffnesses

Table 7.4: Design parameters of the track systems under consideration

Design parameters	Track system I	Track system II	Track system III	Track system IV	Track system V
Weight of the vehicle, kN	514	370	370	138	78
Width of the track, m	0.65	0.57	0.67	0.50	0.43
Distance between the centres of the front and rear roadwheels, m	4.50	4.38	2.65	2.50	2.50
Radius of the roadwheel, m	0.35	0.335	0.40	0.28	0.295
Number of roadwheels	7	5	4	5	5
Angle between the horizontal surface and the track close to the sprocket	45°	22°	37°	33°	30°
Angle between the horizontal surface and the track close to the tensioning wheel	35°	28.5°	45°	28°	32°
Vertical distance between the centre of the tensioning wheel and the centre of the roadwheel, m	0.525	0.335	0.34	0.22	0.175
Radius of the sprocket, m	0.28	0.335	0.24	0.17	0.175
Radius of the tensioning wheel, m	0.30	0.285	0.22	0.11	0.175
Combined stiffness of the suspension springs of one track, kN/m	1500	1075	1250	545	215
Stiffness of the track tensioning device spring, kN/m	1000	∞^*	1250	∞^*	500
Weight of the track per unit track length, kN/m	1.2	0.9	1.1	0.55	0.3
Number of supporting rollers	4	5	2	3	0
Initial track tension per unit width of the track, kN/m	44	25	35.5	12.8	21

$^*\infty$ indicates that the tensioning wheel is fixed.

It can be seen that the analytical method for predicting the static ground pressure distribution developed by Garber and Wong (1981a) can serve a useful purpose in differentiating the potential performance of vehicles of different designs and in determining the relative significance of the effects of various vehicle design parameters on ground pressure distribution. It should be pointed out, however, that when a tracked vehicle is in straight line motion, a terrain element under the track is subject to the repetitive loading of consecutive roadwheels and track links. The response of the terrain to repetitive loading should, therefore, be taken into account in predicting the ground measure distribution under a moving vehicle. Furthermore,

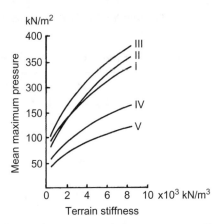

Figure 7.9: Variations of the static mean maximum pressure of various track systems with terrain stiffness

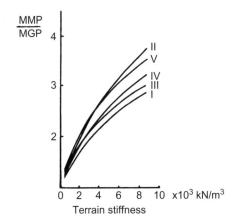

Figure 7.10: Variations of the ratio of static mean maximum pressure to mean ground pressure of various track systems with terrain stiffness

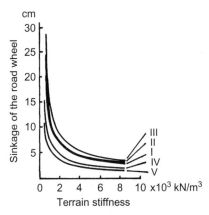

Figure 7.11: Variations of roadwheel sinkage of various track systems with terrain stiffness under static conditions

for a moving tracked vehicle, shear stresses will be developed on the track–terrain interface. To develop a comprehensive method for predicting the tractive performance of tracked vehicles, these factors should be taken into consideration.

A review of the methods for predicting tracked vehicle performance presented above indicates that some of the methods are either of empirical nature or based on assumptions that are not necessarily realistic in many cases. Therefore, there is a need for prediction methods that take into account all major design and operational parameters of the tracked vehicle and all relevant terrain characteristics. These would provide the engineer with useful tools in the development and design of tracked vehicles for given missions and operating environments. In an attempt to meet this need, two computer-aided methods for performance and design evaluation of tracked vehicles have been developed. One is called NTVPM, which is for vehicles with flexible tracks, such as link tracks with relatively short track pitch or rubber belt (band) tracks, commonly used in military fighting and logistics vehicles or in high-speed transport vehicles (Wong et al., 1984; Wong and Preston-Thomas, 1988; Wong, 1992a, 1995, 2007; Wong and Huang, 2005, 2006a, b, 2008). The other is called RTVPM for vehicles with link tracks having relatively long track pitch, commonly in use in industrial and agricultural tractors (Gao and Wong; 1994; Wong and Gao, 1994; Wong, 1998). The basic features of these two computer-aided methods, NTVPM and RTVPM, are discussed in the following chapters. Examples of their practical applications to performance and design evaluation are also presented.

PROBLEMS

7.1 An industrial tractor has a total weight of 429 kN and each of its two tracks has a ground contact length of 3.47 m and width of 0.61 m. It operates on a soil with pressure-sinkage parameters: $n = 0.15$, $k_c = 1.52 \, \text{kN/m}^{n+1}$ and $k_\phi = 119.61 \, \text{kN/m}^{n+2}$; and shear strength parameters: $c = 13.79 \, \text{kPa}$, $\phi = 11°$ and $K = 2.54 \, \text{cm}$. Estimate its drawbar pull coefficient and tractive efficiency (defined by Eqn (6.15)) at 20% slip, as determined by vehicle-terrain interaction. In the evaluation, uniform normal pressure distribution on the track-terrain interface, as shown in Figure 7.6 (a), may be assumed.

7.2 For the industrial tractor described in Problem 7.1, if it is desirable to increase its drawbar pull coefficient at 20% slip by 10% from that estimated in the above problem, what changes to the vehicle and/or track parameters would you recommend? State the reasons for your recommendations in some detail.

7.3 A tracked vehicle has a total weight of 125 kN and each of its two tracks has a ground contact length of 2.65 m and a contact width of 0.38 m. It operates on a terrain with pressure-sinkage parameters: $n = 0.9$, $k_c = 52.53 \, \text{kN/m}^{n+1}$, $k_\phi = 1127.97 \, \text{kN/m}^{n+2}$; and shear strength parameters: $c = 4.83 \, \text{kPa}$, $\phi = 20°$ and $K = 2.54 \, \text{cm}$.

(a) If the normal pressure distribution is of the multi-peak sinusoidal type, as shown in Figure 7.6 (b), estimate its drawbar pull coefficient and tractive efficiency (defined by Eqn (6.15)) at 20% slip, as determined by vehicle-terrain interaction.

(b) If the normal pressure is uniformly distributed on the track-terrain interface, estimate its drawbar pull coefficient and tractive efficiency at 20% slip and compare them with those with normal pressure distribution of the multi-peak sinusoidal type described in (a).

7.4 A tracked vehicle has the same weight and track parameters as that described in Problem 7.3 and operates on the same terrain.

(a) Owing to the rearward shift of the vehicle centre of gravity, its normal pressure on the track-terrain interface increases linearly from zero at the front to the rear, as shown in Figure 7.6 (c). Estimate its drawbar pull coefficient and tractive efficiency (defined by Eqn (6.15)) at 20% slip. Compare them with those with uniform normal pressure distribution on the track-terrain interface.

(b) Explain the reason for the difference in drawbar pull coefficient caused by the changes in the normal pressure distribution from being uniform to that shown in Figure 7.6 (c).

CHAPTER 8

Computer-Aided Method NTVPM for Evaluating the Performance of Vehicles with Flexible Tracks

To gain a competitive edge in today's globalized market, shortening the product development process and reducing the associated costs are crucial. Virtual product development (virtual prototyping) is therefore being actively pursued in the off-road vehicle industry. To successfully implement the virtual product development process, comprehensive and realistic computer-aided methods (or computer simulation models) for performance and design evaluation of tracked vehicles are of vital importance. To be useful to the engineer in the evaluation of vehicle design concepts and in the optimization of design parameters, as well as to the procurement manager in the assessment of vehicle candidates, a computer-aided method should take into account all major vehicle design features and all relevant terrain characteristics. Furthermore, to facilitate its applications, it must be user-friendly. In an attempt to meet these requirements, a computer-aided method known as NTVPM has been developed for vehicles with link tracks having relatively short track pitch or with rubber belt (band) tracks. This type of track is hereinafter referred to as the flexible track.

The short-pitch link track system commonly used in the current generation of military vehicles and off-road transport vehicles has a ratio of roadwheel diameter to track pitch in the range of 4 to 6, a ratio of roadwheel spacing to track pitch in the range of 4 to 7, and a ratio of sprocket pitch diameter to track pitch of the order of 4. The rubber belt (band) track has been used in agricultural tractors, military logistic vehicles, and off-road transport vehicles. It has also been proposed for use in future generation of fighting vehicles. The link track with relatively short track pitch or the rubber belt track is idealized as a flexible and extensible belt in the analysis of track–terrain interaction in NTVPM.

NTVPM is based on a detailed analysis of the physical nature of track–terrain interaction and on the principles of terramechanics (Wong et al., 1984). It focuses on the prediction of the normal and shear stress distributions on the track–terrain interface under steady-state operating conditions, from which the tractive performance of a tracked vehicle is evaluated. It can be used to predict the mobility of single-unit or two-unit articulated tracked vehicles

(Wong, 1992a, b, c, 1994b, 1995, 1997, 1999, 2007; Bodin, 2002; Wong and Huang, 2005, 2006a, b, 2008; Wong and Preston-Thomas, 1988).

NTVPM has undergone continual improvements and updates since its inception. In its latest version, all major design features of the vehicle have been taken into account. These include vehicle sprung weight, unsprung weight, location of the centre of gravity, number of roadwheels, roadwheel dimensions and spacing, locations of the sprocket and idler, supporting roller dimensions and spacing, track dimensions and geometry, initial track tension (i.e. the tension in the track system when the vehicle is stationary on level, hard ground), track longitudinal elasticity (stiffness), and roadwheel suspension characteristics.

The characteristics of roadwheel suspensions are fully taken into consideration in NTVPM. Pivot-arm suspensions, such as torsion bar suspensions and hydro-pneumatic suspensions, and translational spring suspensions, with linear or non-linear load–deflection characteristics, can be simulated. The non-linear behaviour of the suspension may be characterized using a polynomial up to the fifth order. On highly deformable terrain, such as deep snow, track sinkage may be greater than vehicle ground clearance. Thus, the vehicle belly (hull) may be in contact with the terrain surface. This would induce additional drag due to belly–terrain interaction. It may also reduce vehicle traction due to the belly supporting part of the vehicle weight which causes the reduction of the load applied on the track. The effects of belly–terrain interaction on vehicle performance have been taken into account. All pertinent terrain characteristics, including the pressure–sinkage and shearing characteristics and the response to repetitive loading, measured by the bevameter technique described in Chapters 3, 4 and 5, are taken into consideration. The basic features of the computer-aided method have been validated by means of full-scale vehicle tests on various types of terrain. Thus, NTVPM provides the engineer with a comprehensive and realistic tool for performance and design evaluation of vehicles with flexible tracks, from a traction perspective. It has been successfully employed by off-road vehicle manufacturers in the development of new products and by governmental agencies in the evaluation of vehicle candidates in North America, Europe and Asia.

The basic approach to the development of the computer-aided method NTVPM is outlined below.

8.1 Basic Approach to the Prediction of Normal Pressure Distribution under a Flexible Track

In the analysis of the mechanics of track–terrain interaction, the link track with relatively short track pitch or the rubber belt (band) track is idealized as a flexible and extensible belt, as noted previously. The track may elongate when subject to tension. Figure 8.1 shows the relationships between the tension and elongation of the tracks of a European main battle tank (MBT), a European infantry fighting vehicle (IFV), and a US armoured personnel carrier (APC). The elongation shown in the figure is expressed as the ratio of the longitudinal

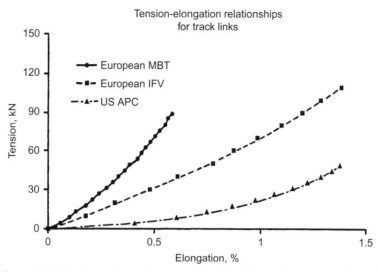

Figure 8.1: Tension–elongation relationships for various types of track

deformation to the original length of the track link. The elongation of the track links for most military vehicles currently in service is primarily due to the deformation of the rubber bushings surrounding the track pins when subject to tension. It is demonstrated later in the chapter that the idealization of the link track with relatively short track pitch or of the rubber belt track as a flexible and extensible belt is reasonable.

A schematic diagram of the flexible track–roadwheel system on a deformable terrain under steady-state forward motion is shown in Figure 8.2. When a tracked vehicle rests on a firm ground, the track lies flat on the surface. On the other hand, when the vehicle travels on a deformable terrain, the normal load applied through the track–roadwheel system causes the terrain to deform, which results in track sinkage. The track segments between roadwheels take up load and as a consequence they deflect and have a form of a curve. The actual length of that part of track in contact with the terrain between the front and rear roadwheels increases, in comparison with that when the vehicle is stationary on a firm ground. This causes a reduction in the sag of the top run of the track and a change in track tension. The elongation of the track under tension is taken into account in the analysis, as noted previously.

The deflected track in contact with the terrain may be divided into two sections: one in contact with both the roadwheel and the terrain, such as segments AC and FH shown in Figure 8.2(b); the other in contact with the terrain only, such as segment CF shown in the figure. The shape of the track segment in contact with the roadwheel, such as AC, is defined by the shape of the roadwheel, whereas the shape of the track segment in contact with the terrain only, such as CF, is determined by interaction between the track segment and the terrain. It is influenced by the track tension in the segment, spacing between two adjacent roadwheels, and the pressure–sinkage relationship and response to repetitive loading of the terrain.

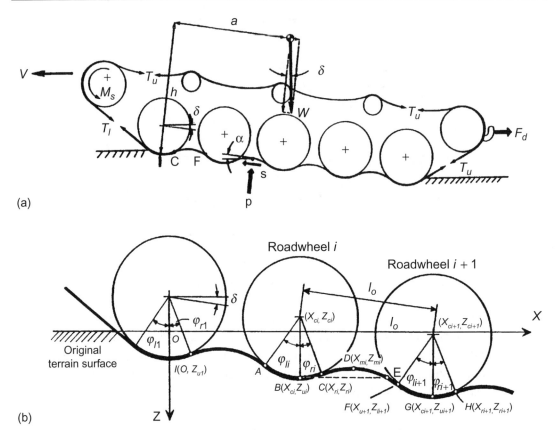

Figure 8.2: Interaction of a flexible track system with deformable terrain

Along segment AB in Figure 8.2(b), the pressure exerted on the terrain increases from A to B. From B to D the pressure decreases corresponding to the unloading portion of the repetitive loading cycle shown in Figure 4.11, 4.22 or 4.30, dependent upon the type of terrain. Along segment DE, the pressure increases again, corresponding to the reloading portion of the repetitive loading cycle. Beyond point E, which is at the same level as point B, the sinkage is higher than that at B. As a result the pressure increases and the sinkage of roadwheel $i + 1$ is greater than that of roadwheel i. Beyond point G the pressure exerted on the terrain decreases again, and another unloading–reloading cycle begins, as described in Chapter 4.

Based on the understanding of the loading exerted on the deformable terrain by consecutive roadwheels and by the track, it is apparent that to predict the normal pressure distribution under a flexible track in steady-state forward motion, the shape of the track segment between adjacent roadwheels and in contact with the terrain only should be determined. The pressure–sinkage relationship and response to repetitive loading of the terrain should also be taken into consideration. To provide a clear understanding of the approach to the development of NTVPM, the following

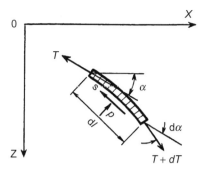

Figure 8.3: Forces acting on an element of the flexible track

analysis begins with the case where the roadwheels are fixed to the vehicle frame. Later, the analysis is expanded to include the effects of roadwheel suspensions on track–terrain interaction.

To determine the shape of the track segment between adjacent roadwheels, let us examine the equilibrium conditions of an element of the track shown in Figure 8.3. In the analysis, the pressure p exerted on the element of the track by the terrain is assumed to be normal to the track–terrain interface, and equal to that acting on a plate at same sinkage, obtained from the pressure–sinkage tests described in Chapter 4.

Considering the equilibrium of the track element along the normal and tangential directions, one obtains

$$T + s\,dl - (T + dT)\cos d\alpha = 0 \qquad (8.1)$$

$$p\,dl - (T + dT)\sin d\alpha = 0 \qquad (8.2)$$

where T is the tension in the track per unit width, p is the normal pressure exerted by the terrain, and s is the shear stress acting on the track element. Other parameters are illustrated in Figure 8.3. If the vehicle is operating at low slips and the shear stress s is not significant, its effects on normal pressure distribution may be neglected. In this section the prediction of the normal pressure distribution under a track in the absence of shear stresses is discussed. At high slips, however, the effects of the shear stress will be significant and should be included in the prediction of normal pressure distribution. This is discussed later in Section 8.3.

If the shear stress is neglected, and $d\alpha$ is small, which implies $\cos d\alpha = 1$ and $\sin d\alpha = d\alpha$, Eqn (8.1) may be rewritten as

$$dT = 0 \text{ and } T = \text{Constant}$$

and Eqn. (8.2) becomes

$$T = p\frac{dl}{d\alpha} = pR \qquad (8.3)$$

where R is the radius of curvature of the track and is expressed by

$$R = \left[1 + \left(\frac{dz}{dx}\right)^2\right]^{\frac{3}{2}} \bigg/ \frac{d^2z}{dx^2} \tag{8.4}$$

Substituting Eqn (8.4) into Eqn (8.3), one obtains

$$T\frac{d^2z}{dx^2} - p\left[1 + \left(\frac{dz}{dx}\right)^2\right]^{\frac{3}{2}} = 0 \tag{8.5}$$

This is the basic equation that governs the shape of the track segment between two adjacent roadwheels in contact with the terrain only, such as CF shown in Figure 8.2. Dependent upon whether the terrain is being loaded, unloaded or reloaded, the pressure–sinkage relationship will take different forms as described in Chapter 4.

When the terrain under the track is subject to continuously increasing load, such as that under segment EF shown in Figure 8.2 (equivalent to loading along the path OA or AC shown in Figures 4.11, 4.22 and 4.30), the pressure–sinkage relationship may be described by Eqn (4.1) or (4.40) or (4.60), dependent on the type of terrain.

In the following analysis of track–terrain interaction, the mineral terrain described in Section 4.1 is used as an example to illustrate the procedures involved. Over the mineral terrain, the pressure–sinkage relationship under continuously increasing load is described by Eqn (4.1) and therefore Eqn. (8.5) can be rewritten as

$$T\frac{d^2z}{dx^2} - kz^n\left[1 + \left(\frac{dz}{dx}\right)^2\right]^{\frac{3}{2}} = 0 \tag{8.6}$$

Setting $dz/dx = u$, the above equation becomes a first-order differential equation:

$$T\frac{u\,du}{\left(1 + u^2\right)^{\frac{3}{2}}} = kz^n dz \tag{8.7}$$

Integrating the above equation, one obtains

$$\frac{-T}{\sqrt{1 + (dz/dx)^2}} = \frac{kz^{n+1}}{(n+1)} + C_1 \tag{8.8}$$

where C_1 is an integration constant. Equation (8.8) may be rewritten as

$$dx = dz \bigg/ \sqrt{\left[\left(\frac{-T}{kz^{n+1}/(n+1) + C_1}\right)^2 - 1\right]} \qquad (8.9)$$

Integrating the above equation, one obtains

$$x = \int dz \bigg/ \sqrt{\left[\left(\frac{-T}{kz^{n+1}/(n+1) + C_1}\right)^2 - 1\right]} + C_2 \qquad (8.10)$$

where C_2 is an integration constant.

Equation (8.10) defines the shape of track segment EF, shown in Figure 8.2, over a mineral terrain. For other types of terrain a similar approach can be followed to obtain the equation that governs the shape of the track segment.

After a roadwheel has passed, the pressure exerted on the terrain along track segment CD, shown in Figure 8.2, is gradually reduced. As described in Section 4.1, during unloading the pressure–sinkage relationship is expressed by Eqn (4.23) and the shape of track segment CD is, therefore, governed by

$$T\frac{d^2 z}{dx^2} - k_u(z - z_a)\left[1 + \left(\frac{dz}{dx}\right)^2\right]^{\frac{3}{2}} = 0 \qquad (8.11)$$

where $z_a = z_u - kz_u^n/k_u$ and z_u is the sinkage of the preceding roadwheel, such as that at point B in Figure 8.2.

Equation (8.11) can be integrated twice to obtain the following equation governing the shape of track segment CD:

$$x = \int dz \bigg/ \sqrt{\left[\left(\frac{-T}{k_u z(z/2 - z_a) + C_1}\right)^2 - 1\right]} + C_2 \qquad (8.12)$$

Under segment DE of the track shown in Figure 8.2, the terrain is being reloaded during which the pressure–sinkage relationship is described by Eqn (4.23). Therefore, Eqn (8.12) also defines the shape of track segment DE. As the track is modelled as a flexible belt, the transition from BC to CD and from DF to FG, shown in Figure 8.2, should be smooth.

This means that track segment CD should be tangent to roadwheel i at C. Similarly, track segment DF should also be tangent to roadwheel $i + 1$ at F.

At point C, defined by coordinates x_{ri} and z_{ri}, the condition that should be met to ensure smooth transition may be expressed by

$$\frac{dz}{dx}\bigg|_{x=x_{ri}} = -\tan\varphi_{ri} \quad (i = 1, 2, \ldots, N-1) \tag{8.13}$$

where N is the number of roadwheels on a track and φ_{ri} is the contact angle of the wheel i with the track on the right from the vertical, as shown in Figure 8.2(b).

Making use of Eqn (8.12), one may rewrite the above equation as follows:

$$\left[\frac{-T}{k_u z_{ri}(z_{ri}/2 - z_{ai}) + C_{1i}}\right]^2 - 1 = \tan^2\varphi_{ri} \quad (i = 1, 2, \ldots, N-1) \tag{8.14}$$

or

$$k_u z_{ri}\left(\frac{z_{ri}}{2} - z_{ai}\right) + C_{1i} = -T\cos\varphi_{ri} \quad (i = 1, 2, \ldots, N-1) \tag{8.15}$$

where $z_{ai} = z_{ui} - kz_{ui}^n/k_u$ and z_{ui} is the sinkage of point B.

From Eqn (8.15), C_{1i}, which is the integration constant for roadwheel i, can be defined in terms of φ_{ri}:

$$C_{1i} = -T\cos\varphi_{ri} - k_u z_{ri}\left(\frac{z_{ri}}{2} - z_{ai}\right) \tag{8.16}$$

It should be noted that $z_{ri} = z_{ui} - R(1 - \cos\varphi_{ri})$, where R is the radius of the roadwheel.

Point F in Figure 8.2, where track segment DF should be tangent to roadwheel $i + 1$, may be below or above the level of point B of roadwheel i (i.e. z_{ui}). If point F is level with or above point B (i.e. $z_{li+1} \leq z_{ui}$), then under the entire segment DF the terrain is being reloaded (corresponding to the reloading path BA in Figures 4.11, 4.22 and 4.30). Accordingly, the slope of segment DF can be derived by differentiating Eqn (8.12) with the integration constant C_{1i} defined by Eqn (8.16):

$$\frac{dz}{dx} = \sqrt{\left(\frac{-T}{k_u(z - z_{ri})(z - 2z_{ai} + z_{ri})/2 - T\cos\varphi_{ri}}\right)^2 - 1} \tag{8.17}$$

The slope of the tangent to roadwheel $i+1$ can be defined by

$$\frac{dz}{dx} = \frac{(x_{ci+1} - x)}{(z - z_{ci+1})} = \frac{\sqrt{R^2 - (z - z_{ci+1})^2}}{(z - z_{ci+1})} \qquad (8.18)$$

where x_{ci+1} and z_{ci+1} are the coordinates of the centre of roadwheel $i+1$.

At point F, the slope of track segment DF should match that of the tangent to roadwheel $i+1$. As a consequence the sinkage z_{li+1} at point F is defined by setting Eqn (8.17) equal to Eqn (8.18) and $z = z_{li+1}$.

$$z_{li+1} = z_{ai} - \frac{T}{Rk_u} + \sqrt{(z_{ai} - z_{ri})^2 + \frac{2T}{Rk_u}\left[\frac{T}{2Rk_u} - z_{ai} + R\cos\varphi_{ri} + z_{ci+1}\right]} \qquad (8.19)$$

Equation (8.19) defines z_{li+1} as a function of φ_{ri}, T and other parameters. For a given set of parameters, if the value of z_{li+1} is found to be greater than the sinkage z_{ui} of roadwheel i, then point F will be below point B.

When point F is below point B (i.e. $z_{li+1} > z_{ui}$), then under segment DE the terrain is being reloaded (corresponding to the reloading path BA in Figures 4.11, 4.22 and 4.30), whereas under segment EF the pressure applied to the terrain exceeds the maximum value applied to the preceding roadwheel at B and as a result additional sinkage occurs (corresponding to the loading path AC in Figures 4.11, 4.22 and 4.30). Consequently, the shape of track segment DE is governed by Eqn (8.12), whereas the shape of segment EF is governed by Eqn (8.10). Since the track is modelled as a flexible belt, the transition from segment DE to segment EF must also be smooth. This means that both segments should have the same slope at E. The slope of DE at E (i.e. $z = z_{ui}$) can be determined from Eqn (8.17) by setting $z = z_{ui}$, and is expressed as

$$\left.\frac{dz}{dx}\right|_{z=z_{ui}} = \sqrt{\left(\frac{-T}{k_u(z_{ui} - z_{ri})(z_{ui} - 2z_{ai} + z_{ri})/2 - T\cos\varphi_{ri}}\right)^2 - 1} = S_E \qquad (8.20)$$

The slope of track segment EF is governed by Eqn (8.9). The integration constant C_1 in Eqn (8.9) can be obtained by setting the slope of track segment EF at E equal to that defined by Eqn (8.20):

$$C_1 = \frac{-\left[T + \left(\sqrt{S_E^2 + 1}\right)kz_{ui}^{n+1}/(n+1)\right]}{\sqrt{S_E^2 + 1}} \qquad (8.21)$$

The general expression for the slope of segment EF is therefore given by

$$\frac{dz}{dx} = \sqrt{\left(\frac{-T}{k(z^{n+1} - z_{ui}^{n+1})/(n+1) - T/\sqrt{S_E^2 + 1}}\right)^2 - 1} \qquad (8.22)$$

Track segment EF should be tangent to roadwheel $i + 1$ at F. By setting Eqn (8.18) equal to Eqn (8.22), the sinkage z_{li+1} at F can be defined by

$$\frac{kR}{n+1}\left(z_{li+1}^{n+1} - z_{ui}^{n+1}\right) - T\left[\frac{R}{\sqrt{S_E^2 + 1}} - (z_{li+1} - z_{ci+1})\right] = 0 \qquad (8.23)$$

This indicates that when point F is below point B (i.e. $z_{li+1} > z_{ui}$) the sinkage z_{li+1} at point F can be determined by solving Eqn (8.23) for a given set of parameters including φ_{ri}, T and z_{ui}.

It should be mentioned that Eqns (8.19) and (8.23) only define the sinkage z_{li+1} at which the slope of track segment DF will match that of a corresponding point at the same sinkage on the circumference of roadwheel $i + 1$. For point F on track segment DF actually matching the corresponding point on the circumference of roadwheel $i + 1$, additional requirements should be met.

Considering the horizontal projection of the track segment between the centres of two adjacent roadwheels, i and $i + 1$ shown in Figure 8.2, one finds that one of the following two conditions should be satisfied, dependent upon the sinkage of point F:

(a) When point F is above point B (i.e. $z_{li+1} < z_{ui}$):

$$R(\sin\varphi_{ri} + \sin\varphi_{li+1}) + \int_{z_{mi}}^{z_{ri}} dz \bigg/ \sqrt{\left[\frac{-T}{k_u(z - z_{ri})(z - 2z_{ai} + z_{ri})/2 - T\cos\varphi_{ri}}\right]^2 - 1}$$

$$+ \int_{z_{mi}}^{z_{li+1}} dz \bigg/ \sqrt{\left[\frac{-T}{k_u(z - z_{ri})(z - 2z_{ai} + z_{ri})/2 - T\cos\varphi_{ri}}\right]^2 - 1} \qquad (8.24)$$

$$= l_0 \cos\delta$$

where l_0 is the distance between the centres of two adjacent roadwheels, δ is the inclination of the vehicle frame with respect to the horizontal, and z_{mi} is the ordinate of point D which is the minimum sinkage of the track between two adjacent roadwheels. At point D the slope of the track is zero. Therefore, by setting dz/dx in Eqn (8.17) to zero, z_{mi} can be expressed by

$$z_{mi} = z_{ai} + \sqrt{(z_{ai} - z_{ri})^2 - \frac{2T}{k_u}(1 - \cos\varphi_{ri})} \qquad (8.25)$$

(b) When point F is below point B (i.e. $z_{l+1} > z_{ui}$):

$$R(\sin\varphi_{ri} + \sin\varphi_{li+1}) + \int_{z_{mi}}^{z_{ri}} dz / \sqrt{\left[\frac{-T}{k_u(z-z_{ri})(z-2z_{ai}+z_{ri})/2 - T\cos\varphi_{ri}}\right]^2 - 1}$$

$$+ \int_{z_{mi}}^{z_{ui}} dz / \sqrt{\left[\frac{-T}{k_u(z-z_{ri})(z-2z_{ai}+z_{ri})/2 - T\cos\varphi_{ri}}\right]^2 - 1} \qquad (8.26)$$

$$+ \int_{z_{ui}}^{z_{li+1}} dz / \sqrt{\left[\frac{-T}{k/(n+1)(z^{n+1}-z_{ui}^{n+1}) - T/\sqrt{(S_E^2+1)}}\right]^2 - 1}$$

$$= l_0 \cos\delta$$

It should be noted that in Eqns (8.24) and (8.26), there are four basic unknowns for a track system with rigidly connected roadwheels; φ_{ri} for roadwheel i, vehicle inclination angle δ, track tension T, and sinkage of the first roadwheel, z_{u1}. All other parameters, such as z_{ai}, z_{li+1}, z_{mi} and z_{ri} can be expressed in terms of the four basic unknowns. It should be mentioned that the sinkage z_{ui} of the lowest point of roadwheel i (i.e. point B shown in Figure 8.2) can be expressed in terms of z_{u1} and δ:

$$z_{ui} = z_{u1} + (i-1)l_0 \sin\delta (i = 1, 2, \ldots, N) \qquad (8.27)$$

For each track segment between two adjacent roadwheels, an equation of the form of Eqn (8.24) or Eqn (8.26), which contains δ, T, z_{u1} and φ_{ri}, can be derived. Therefore, for a track–roadwheel system with N roadwheels, $N-1$ equations containing δ, T, z_{u1} and φ_{ri} ($i = 1, 2, \ldots, N-1$) can be established. To completely define the deflected shape of $N-1$ track segments between adjacent roadwheels, three additional independent equations containing δ, z_{u1} and T should be established. These additional equations can be derived by considering the equilibrium of the normal forces and moments acting on the track–roadwheel system as a whole, and the overall track length. In the evaluation of the overall track length, the effect of track tension on elongation is taken into consideration. The solution to this set of $N+2$ equations will yield the values of δ, T, z_{u1} and φ_{ri} ($i = 1, 2, \ldots, N-1$). Thus, the shape of the deflected track from the front to the rear roadwheels in contact with the terrain is completely defined. Making use of the known pressure–sinkage relationship and response to repetitive loading of the terrain, the normal pressure distribution under a track–roadwheel system with roadwheels rigidly connected to the vehicle frame can then be determined, under steady-state forward motion and with shear stress on the track–terrain interface being neglected.

For a track system with independent roadwheel suspensions, the sinkages of the individual roadwheels are independent of each other. Thus, for a system with N roadwheels, there are $2N+2$

unknowns, which include $N - 1$ track contact angles φ_{ri} ($i = 1, 2, ..., N - 1$), N roadwheel sinkages z_{ui} ($i = 1, 2, ..., N$), the track tension T and the vertical displacement and inclination of the vehicle frame. Similar to the track–roadwheel system with N roadwheels rigidly connected to the vehicle frame, $N - 1$ equations can be established for the track segments between roadwheels. Also N equations for the equilibrium of normal forces acting on the roadwheels can be derived. In addition, three equations can be formulated by considering the equilibrium of normal forces and moments acting on the track–roadwheel system as a whole and the overall track length. In the evaluation of the overall track length, the effect of tension on elongation of the track and that of the deflection of the springs of the independent suspensions are taken into consideration. This means that in this case the decrease in the sag of the top run of the track, the deflections of the suspension springs, and the elongation of the track under tension compensate for the increase in the length of the track between the front and rear roadwheels in contact with the terrain. The $N + 2$ independent equations can then be solved for the $N + 2$ unknowns mentioned above. Thus, the deflected shape of the track between the front and rear roadwheels and the sinkages of the independently suspended roadwheels can be defined. As a result, the normal pressure distribution under a track–roadwheel system with independent suspensions can be determined, under steady-state forward motion and with shear stress on the track–terrain interface being neglected.

8.2 Prediction of Shear Stress Distribution under a Flexible Track

The tractive performance of a tracked vehicle is closely related to its normal pressure and shear stress distributions on the track–terrain interface. In Section 8.1 the prediction of normal pressure distribution, with the effects of the shear stress being neglected, is discussed. In this section the shearing action of a flexible track is analysed and a method for predicting the shear stress distribution is presented. Based on these, the effects of shear stresses on the normal pressure distribution are discussed in the next section.

To predict the shear stress distribution under a track, the shear stress–shear displacement relationship, the shear strength and the response to repetitive shear loading of the terrain, as discussed in Chapter 5, must be known. Over a given terrain, the shear stress at a given point on the track–terrain interface is a function of the shear displacement, measured from the point where shearing (or reshearing) begins, and the normal pressure at that point. The shear displacement developed under a flexible track, shown in Figure 8.4, may be determined from the analysis of the slip velocity V_j, similar to that for a rigid track or a rigid wheel (Bekker, 1956; Wong and Reece, 1967a; Wong et al., 1984). The slip velocity V_j of a point P on a flexible track relative to the terrain surface is the tangential component of the absolute velocity V_a shown in Figure 8.4. The magnitude of the slip velocity V_j is expressed by

$$\begin{aligned} V_j &= V_t - V\cos\alpha \\ &= r\omega - r\omega(1 - i)\cos\alpha \\ &= r\omega[1 - (1 - i)\cos\alpha] \end{aligned} \quad (8.28)$$

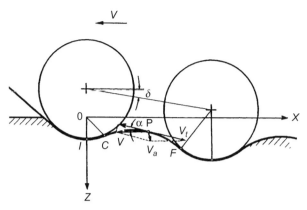

Figure 8.4: Slip velocity of a point on a flexible track in contact with deformable terrain

where r and ω are the pitch radius and angular speed of the sprocket, respectively, i is the slip of the track, α is the angle between the tangent to the track at point P and the horizontal, V_t is the theoretical forward speed of the vehicle (i.e. $V_t = r\omega$), and V is the actual forward speed of the vehicle. The shear displacement j along the track–terrain interface is given by

$$\begin{aligned}
j &= \int_0^t r\omega\big[1 - (1-i)\cos\alpha\big]dt \\
&= \int_0^l r\omega\big[1 - (1-i)\cos\alpha\big]\frac{dl}{r\omega} \\
&= \int_0^l \left[1 - (1-i)\frac{dx}{dl}\right]dl \\
&= l - (1-i)x
\end{aligned} \qquad (8.29)$$

where l is the distance along the track between point P and the point where shearing (or reshearing) begins, and x is the corresponding horizontal distance between point P and the initial shearing (or reshearing) point.

If the shear stress–displacement relationship of the terrain is described by Eqn (5.2), then the shear stress distribution may be expressed by

$$s(x) = \big[c + p(x)\tan\phi\big]\left[1 - \exp\left(-\frac{(l-(1-i)x)}{K}\right)\right] \qquad (8.30)$$

where $p(x)$ is the normal pressure on the track, which is a function of x.

For terrain with a shear stress–displacement relationship described by Eqn (5.3) or (5.5) a similar approach may be followed to derive an appropriate expression for the shear stress distribution.

In using Eqn (8.30) to predict the shear stress distribution under a track, attention should be paid to the following:

(a) As mentioned previously, an element of the terrain under a moving track is subject to the repetitive normal load applied by the consecutive roadwheels. The behaviour of the terrain under repetitive shear loading should, therefore, be taken into account in predicting shear stress distribution under a track. Based on the behaviour of the terrain under repetitive shear loading described in Section 5.3, for an idealized case shown in Figure 5.8, the shear displacement should be calculated either from the initial shearing point for the first track segment, such as point A, or from the points where reshearing begins for the subsequent track segments, such as points A_1, A_2 and A_3. This will lead to the development of shear stress under various segments of the track as illustrated in Figure 5.8(b).

(b) As pointed out previously, the normal pressure under a track is rarely uniformly distributed. Consequently, shearing takes place under varying normal load beneath a track segment. However, the shear stress–shear displacement relationships of the terrain are usually measured under constant normal pressures. To predict the shear stress under varying normal pressure, a special procedure must be followed. This procedure is illustrated in Figure 8.5 for an idealized case with discrete steps in normal pressure. Although in practice the normal pressure under the track usually varies continuously, its variation may be represented by small steps. The shear stress–shear displacement relationships of a terrain under different normal pressures of p_1, p_2 and p_3 are shown in Figure 8.5(a). If the normal pressure on a track segment increases by equal steps from 0 to p_3 and then decreases by the same steps from p_3 to 0, as shown in Figure 8.5(b), then during the loading part of the cycle, from 0 to p_3, the variation of shear stress with shear–displacement under the track segment will follow the path $OS_1 S_2 S_3$. It will be

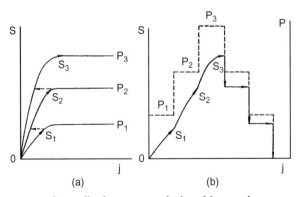

Figure 8.5: (a) Shear stress–shear displacement relationships under constant normal pressure; (b) shear stress–shear displacement relationships under varying normal pressure

noted that the path OS$_1$ in Figure 8.5(b) is identical to the path OS$_1$ in Figure 8.5(a) and that paths S$_1$S$_2$ and S$_2$S$_3$ in Figure 8.5(b) correspond to those shown in Figure 8.5(a). This is based on the observation of the shearing behaviour of terrain described previously in Section 5.3, which indicates that the maximum shear stress under a given normal pressure is reached only after a certain amount of shear displacement has taken place. During the unloading part of the cycle, however, the shear stress decreases instantaneously with the decrease of normal pressure for a terrain with predominantly frictional characteristics, as shown in Figure 8.5(b).

(c) Over terrains exhibiting both frictional and cohesive properties, the shear stress developed on the track–terrain interface is composed of two components. One is due to friction and depends on the normal pressure and shear displacement calculated from the initial shearing point (or reshearing point) as described above. The other is due to cohesion (or adhesion) and does not depend on normal pressure. It is dependent on the shear displacement measured from the initial shearing point of the track.

Based on the above analysis of the shear displacement under a flexible track and the development of shear stress under varying normal pressure, the shear stress distribution under a track can be predicted.

8.3 Effects of Shear Stresses on the Normal Pressure Distribution

In Section 8.1, an analytical method for predicting the normal pressure distribution under a flexible track by neglecting the effects of shear stresses is presented. As a result, the tension is assumed to be constant throughout the track. At high slips, however, the shear stresses on the track–terrain interface become significant, and their presence changes the tension of the track from one segment to another. Since the shape of the track segment between adjacent roadwheels is dependent upon track tension, the normal pressure distribution is affected by the presence of shear stresses.

As described in the preceding section, the shear stress on the track–terrain interface varies along the contact length. Thus, the track tension changes from one segment to another. To simplify the analysis, in determining the deflected shape of track segments between adjacent roadwheels, such as CF shown in Figure 8.4, an average track tension is used. This average track tension is taken as the mean of the tensions at points C and F, which are determined by the shear stress distribution on the track–terrain interface. It can be shown that the shear stresses acting on track segments (such as CF) between adjacent roadwheels are usually relatively low and hence cause only a small change in track tension between C and F. This is primarily due to the fact that the normal pressure on the track segment (such as CF in Figure 8.2) between adjacent roadwheels is usually much lower than that immediately under the roadwheel (such as AC in Figure 8.2). Therefore, using an average track tension in determining the deflected shape of track segments between adjacent roadwheels is reasonable. Following this approach, an iterative procedure is developed for predicting the normal pressure and shear stress distributions under a flexible track at a given slip. The basic steps in this procedure are outlined below:

(a) Following the procedure described in Section 8.1, first calculate the normal pressure distribution by neglecting the presence of shear stresses;

(b) Use the method developed in the preceding section to calculate the shear stress distribution under the track for a given slip, using the normal pressure distribution determined at the preceding step;

(c) Based on the shear stress distribution, calculate the tractive effort, which can be obtained by integrating the shear stresses over the contact area, and determine the difference between the tensions at either end of each track segment between adjacent roadwheels, such as that between C and F shown in Figure 8.4; calculate the average tension in each track segment between adjacent roadwheels;

(d) Use the average tensions in the track segments to initiate an iterative process to recalculate the normal pressure distribution. Check whether the conditions that define the eqilibrium of forces and moments acting on the track–roadwheel system and the overall track length are satisfied. If the conditions are not met, select a set of new values for the variables, including the tension in the track segment behind the rear roadwheel. When a tension change is necessary, increase or decrease all tensions by a constant amount. If the tension between the sprocket and the front roadwheel decreases to zero, slack may be introduced.

(e) Repeat the procedure described above until the errors in the equilibrium equations for forces and moments and in the calculations of the overall track length are less than the preassigned values.

In the equation for moment equilibrium, the moment due to drawbar pull, which causes load transfer from the front to the rear, has been taken into consideration (see Figure 8.2(a)). Furthermore, the role that the shear stresses play in supporting the vertical load applied to the track has also been taken into account in the calculations.

It should be mentioned that the presence of shear stresses not only affects the normal pressure distribution, but also may induce additional sinkage, usually referred to as slip–sinkage. For the types of terrain described in Chapter 5, it was found that slip–sinkage is not significant, and is therefore not included in the present analysis. However, if slip–sinkage is found to be significant, for the terrain in question, it can be incorporated into the present analytical framework.

8.4 Prediction of Motion Resistance and Drawbar Pull

When the normal pressure and shear stress distributions under a tracked vehicle at a given slip have been determined, the tractive performance of the vehicle can then be predicted. The tractive performance of an off-road vehicle is usually characterized by its motion resistance, tractive effort, and drawbar pull (the difference between tractive effort and motion resistance) as functions of slip.

The external motion resistance R_t of the track can be determined from the horizontal component of the normal pressure acting on the track in contact with the ground. For a vehicle with two tracks, R_t can be described by

$$R_t = 2b \int_0^{L_t} p \sin \alpha \, dl \tag{8.31}$$

where b is the contact width of the track, L_t is the length of track in contact with the terrain, p is normal pressure, and α is the angle of the track element with respect to the horizontal.

If the track sinkage is greater than the ground clearance of the vehicle, the belly (hull) will be in contact with the terrain, giving rise to an additional drag, known as belly drag R_{be}. It can be determined from the horizontal components of the normal and shear stresses acting on the belly–terrain interface and is described by

$$R_{be} = b_b \left[\int_0^{L_b} p_b \sin \alpha_b \, dl + \int_0^{L_b} s_b \cos \alpha_b \, dl \right] \tag{8.32}$$

where b_b is the contact width of the belly, L_b is the contact length of the belly, α_b is the angle of the belly with respect to the horizontal, and p_b and s_b are the normal and shear stresses on the belly–terrain interface, respectively.

The tractive effort F of the vehicle can be calculated from the horizontal component of the shear stress acting on the track in contact with the terrain. For a vehicle with two tracks, F is given by

$$F = 2b \int_0^{L_t} s \cos \alpha \, dl \tag{8.33}$$

where s is the shear stress on the track–terrain interface.

Since both the normal pressure p and shear stress s are functions of track slip, the track motion resistance R_t, belly drag R_{be} (if any) and tractive effort F vary with slip.

For a track with rubber pads, part of the total tractive effort is generated by the rubber–terrain shearing. To predict the tractive effort developed by the rubber pads, the portion of the vehicle weight supported by the rubber pads should be estimated and the characteristics of rubber–terrain shearing should be taken into account.

It should be mentioned that the tractive effort F calculated by Eqn (8.33) is due to the shearing action of the track across the grouser tips (or due to rubber–terrain shearing for a track with rubber pads). For a track with long grousers, additional thrust will be developed due to the shearing action on the vertical surfaces on either side of the track. According to Reece (1964), the maximum shear force per unit track length $S_{v\,max}$ from the two sides of a track is given by

$$S_{v\,max} = 4ch\sin^2\left(\frac{\pi}{4}+\frac{\phi}{2}\right) + 2hz_m \gamma \tan\left(\frac{\pi}{4}+\frac{\phi}{2}\right)\cos\left(\frac{\pi}{2}-\phi\right) \quad (8.34)$$

where c, γ and ϕ are the cohesion, weight density, and angle of shearing resistance of the terrain, respectively, h is the grouser height, and z_m is the mean sinkage of the grouser.

The relationship between the shear force developed on the two sides of a track and shear displacement may be assumed to be similar to that between the shear stress and shear displacement described in Chapter 5. Therefore, the distribution of the shear force on the two vertical sides along the length of the track can be predicted in the same way as that described in Section 8.2, and the thrust F_v from the two sides of a track can be expressed by

$$F_v = \int_0^{L_t} S_v \cos\alpha\, dl \quad (8.35)$$

where S_v is the shear force per unit length of the track.

The drawbar pull F_d of the vehicle can be considered as the difference between the tractive effort and the total motion resistance (including the belly drag, if any), and can be expressed by

$$\begin{aligned} F_d &= F + 2F_v - R_t - R_{be} \\ &= 2b\int_0^{L_t} s\cos\alpha\, dl + 2\int_0^{L_t} S_v\cos\alpha\, dl - 2b\int_0^{L_t} p\sin\alpha\, dl \\ &\quad - b_b\left[\int_0^{L_b} p_b \sin\alpha_b\, dl + \int_0^{L_b} s_b \cos\alpha_b\, dl\right] \end{aligned} \quad (8.36)$$

From this equation, the relationship between drawbar pull F_d and track slip i can be determined.

The analyses and procedures described above are implemented in the computer-aided method NTVPM. In its development, particular attention has been paid to its user-friendliness. For instance, the dialogue box format for inputting all vehicle and terrain data is adopted for the convenience of the user. It runs on Microsoft Windows operating systems. Figure 8.6 shows the control centre, as displayed on the computer monitor screen, for the operation of the latest version of NTVPM.

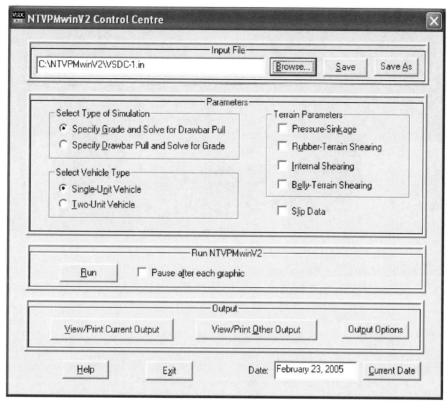

Figure 8.6: Control centre for the operation of the computer-aided method NTVPM, as displayed on the monitor screen

8.5 Experimental Substantiation

To validate the basic features of the computer-aided method NTVPM, a series of field tests was conducted. The first set of tests was performed using an armoured personnel carrier M113A1 and a two-unit, articulated tracked vehicle Bv202 on a sandy terrain (referred to as LETE sand), two types of muskeg (referred to as Petawawa muskeg A and B) and two types of snow-covered terrain (referred to as Petawawa snow A and B) (Wong et al., 1984). The instrumented M11A1 and Bv202 used in the tests are shown in Figures 8.7 and 8.8, respectively. The basic vehicle parameters of the M113A1 and the Bv202 used in field tests are given in Tables 8.1 and 8.2, respectively. The parameters characterizing the pressure–sinkage relations and the response to repetitive loading of the terrains on which vehicle tests were conducted are summarized in Tables 8.3 to 8.5. The shear strength parameters of these terrains are given in Table 8.6.

To measure the normal pressure on the track–terrain interface of the M113A1, four Kulite IPT-750 flush stainless steel diaphragm pressure transducers were used. These transducers

Figure 8.7: Instrumented M11A1 used in field tests

Figure 8.8: Instrumented Bv202 used in field tests

employ semiconductor strain gauge elements, bonded directly to the inside surface of the diaphragm. The diameter of the diaphragm in contact with the terrain is 1.9 cm (¾ in.). The transducers were mounted on a track link of the M113A1: two on the rubber pad and the other two on the metal part of the link, as indicated in Figure 8.9. The signals from the

Table 8.1: Basic parameters of M113A1 used in field tests

Total weight, kN	88.71
CG x-coordinate* (ahead of the midpoint of the track), cm	12.7
CG y-coordinate (above ground), cm	99
Number of roadwheels on one track	5
Roadwheel radius, cm	30.5
Distance between roadwheel centres (average), cm	66.7
Initial track tension, kN	10
Number of supporting rollers	0
Drawbar hitch x-coordinate (behind the midpoint of the track), cm	229
Drawbar hitch y-coordinate (above ground), cm	68
Track type	Metal links with rubber pads
Track weight per unit length, kN/m	0.67
Track width, cm	38

*Dimensions indicated in this table are for the vehicle stationary on level, hard surface.

Table 8.2: Basic parameters of Bv202 used in field tests

Parameters	Front unit	Rear unit
Total weight, kN	20.73	11.48
CG x-coordinate* (ahead of the midpoint of the track), cm	15	29.4
CG y-coordinate (above ground), cm	99.5	85.5
Number of roadwheels on one track	5	5
Roadwheel radius, cm	21	21
Distance between roadwheel centres (average), cm	45	45
Initial track tension, kN	2.69	2.69
Number of supporting rollers	1	1
x-coordinate for the connecting link pivot, cm	105 (behind the midpoint of the track)	169.5 (ahead of the midpoint of the track)
y-coordinate for the connecting link pivot (above ground), cm	52	41
Connecting link length, cm	50	
Drawbar hitch x-coordinate, cm		116 (behind the midpoint of the track)
Drawbar hitch y-coordinate (above ground), cm		47
Track type	Reinforced rubber belt	Reinforced rubber belt
Track weight per unit length, kN/m	0.255	0.255
Track width, cm	50	50

*Dimensions indicated in this table are for the vehicle stationary on level, hard surface.

Table 8.3: Values of the pressure–sinkage and repetitive loading parameters for a sandy terrain (LETE sand)

k_c (kN/m^{n+1})	k_ϕ (kN/m^{n+2})	n	k_0 (kN/m^3)	A_u (kN/m^4)
102	5301	0.793	0	503 000

Table 8.4: Values of the pressure–sinkage and repetitive loading parameters for two snow covers

Snow Type	Petawawa Snow A		Petawawa Snow B	
Section of the pressure sinkage curve	Before failure of the crust	After failure of the crust	Before failure of the crust	After failure of the crust
k_{p1} (kPa)	3.2	52.7	16.3	10.8
k_{p2} (kPa/m)	234	−48	0	0
k_{z1} (cm)	0.9	14.2	24.8	41.0
k_{z2} (cm^2)	39.7	67.3	0	0
k_0 (kN/m^3) A_u (kN/m^4)	0 109 600		0 25 923	

Table 8.5: Values of the pressure–sinkage and repetitive loading parameters for two types of Muskeg

Muskeg type	Petawawa Muskeg A	Petawawa Muskeg B
k (kN/m^3)	29	762
m_m (kN/m^3)	51	97
k_0 (kN/m^3)	123	147
A_u (kN/m^4)	23 540	29 700

Table 8.6: Shear strength parameters of various types of terrain

Terrain type	Type of shearing	Cohesion (adhesion) (kPa)	Angle of shearing resistance (degrees)	K (cm)	K_r	K_w (cm)
LETE sand	Internal	1.3	31.1	1.2	–	–
	Rubber–sand	0.7	27.5	1.0	–	–
Petawawa Snow A & B	Internal	0.4	24.0	–	0.655	2.2
	Rubber–snow	0.12	16.4	0.4	–	–
Petawawa Muskeg A	Peat (internal)	2.8	39.4	3.1	–	–
Petawawa Muskeg B	Peat (internal)	2.6	39.2	3.1	–	–

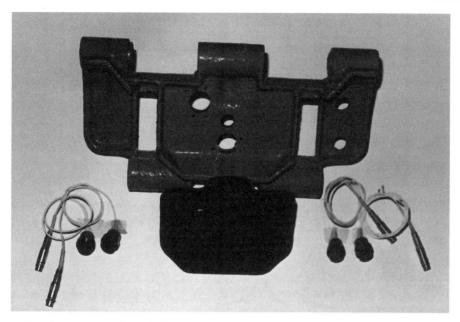

Figure 8.9: Instrumented track link of M11A1 and transducers for measuring the normal pressure on the track–terrain interface

transducers were transmitted through an adapter and a cable to a multi-channel signal conditioner installed inside the test vehicle. The transducer signals, after being amplified, were recorded on an oscillograph.

In addition to the normal pressure on the track–terrain interface, a number of other performance parameters of the test vehicle were monitored during tests. To measure the dynamic sinkage of the front and rear roadwheels with respect to the terrain surface, as well as the trim angle of the vehicle, a special measuring system was developed. In this system, two linear displacement transducers were employed to monitor the displacements of the vehicle body at locations immediately above the centres of the front and rear roadwheels with respect to the terrain surface. Over snow, skis were used as the sensing elements, as shown in Figures 8.7 and 8.8 for the M11A1 and the Bv202, respectively. Over sandy terrain or muskeg, bicycle wheels, instead of skis, were used as the sensing elements. Two additional displacement transducers were used to monitor the displacements of centres of the front and rear roadwheels relative to the vehicle body at locations immediately above them. Based on these measurements, the dynamic sinkages of the front and rear roadwheels of the track system with respect to the terrain surface can be derived.

The distance that the vehicle travelled during tests was measured using a fifth wheel running in the rut made by the track. It was found that the fifth wheel performed its intended function satisfactorily. The revolutions of the sprockets of the tracks of the test vehicles were measured

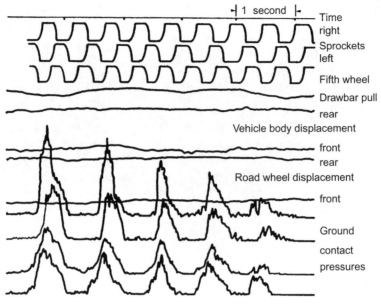

Figure 8.10: Records of the parameters monitored during field tests

using photoelectric counters and recorded on a galvanometer type oscillograph. Based on the signals from the fifth wheel and those from the sprocket counters, the slips of the tracks can be derived. A time signal was also recorded. Thus, based on the fifth wheel signals and the corresponding time signals, vehicle speed can be determined. The drawbar load was applied to the test vehicle by a towed vehicle through braking. The drawbar pull was monitored by a strain-gauge type load cell installed between the test vehicle and the towed vehicle. During tests, a specific level of braking was applied to the towed vehicle to maintain a desired drawbar load as steady as possible. Figure 8.10 shows a typical record of the performance parameters monitored during a test, which include ground contact pressures at four locations of the track link, distance measured by the fifth wheel, left and right sprocket revolutions, front and rear roadwheel displacements relative to the vehicle body, vehicle body displacements immediately above the centres of the front and rear roadwheels with respect to the terrain surface, time and drawbar pull. The data recorded by the oscillograph were later digitized and processed.

The instrumentation for monitoring the performance parameters of the two-unit, articulated tracked vehicle Bv202 was similar to that for the M11A1. For the Bv202, two pressure transducers were installed on the contact surface of a rubber grouser and two on the rubber belt, as shown in Figure 8.11.

Another set of field tests was conducted using another two-unit, articulated tracked vehicle Bv206, which is similar in form to the Bv202 mentioned previously, on a snow-covered

Figure 8.11: Pressure transducers mounted on the rubber belt track of Bv202

terrain referred to as Fernie snow (Wong, 1992a). Drawbar performance of the vehicle was measured under two snow conditions: one in its virgin state (referred to as undisturbed) and the other after being compacted by the passage of the tracks of the test vehicle (referred to as preconditioned). The basic parameters of the Bv206 used in field tests are summarized in Table 8.7. During performance testing, the drawbar load was varied continuously at a slow rate, so as to obtain a continuous plot of drawbar pull versus slip. The pressure–sinkage parameters for the snow under the two different conditions (undisturbed and preconditioned) are summarized in Table 8.8, whereas the shear strength parameters are given in Table 8.9.

As examples, Figures 8.12–8.15 show the correlations between the measured normal pressure distributions under the rubber pads of the M113A1 and the predicted ones by NTVPM on LETE sand, Petawawa Muskeg B, and Petawawa Snow A and B, respectively. Comparisons of the measured and predicted drawbar pull–slip relationships of the M11A1 on LETE sand, Petawawa Muskeg B and Petawawa Snow A are shown in Figures 8.16–8.18, respectively. The correlations between the measured and predicted drawbar pull–slip relationships of the two-unit, articulated tracked vehicle Bv202 on LETE sand, Petawawa Muskeg A and Petawawa Snow B are shown in Figures 8.19–8.21, respectively. Comparisons of the measured and predicted drawbar pull–slip relationships of the two-unit, articulated tracked vehicles Bv206 on the undisturbed and preconditioned Fernie snow are shown in Figures 8.22 and 8.23, respectively (Wong, 1992a). The response to repetitive loading of the Fernie snow was not available. In the predictions of its tractive performance, estimated values for the repetitive loading parameters k_o and A_u, based on data of similar snow, were used. It should

Table 8.7: Basic parameters of Bv206 used in field tests

Parameters	Front unit	Rear unit
Total weight, kN	28.06	17.95
CG x-coordinate* (ahead of the midpoint of the track), cm	0	29.8
CG y-coordinate (above ground), cm	103.8	103.5
Number of roadwheels on one track	5	5
Roadwheel radius, cm	19	19
Distance between roadwheel centres (average), cm	49	49
Initial track tension, kN	3.53	4.69
Number of supporting rollers	1	1
x-coordinate for the connecting link pivot, cm	139 (behind the midpoint of the track)	163.9 (ahead of the midpoint of the track)
y-coordinate for the connecting link pivot (above ground), cm	54.8	54.5
Connecting link length, cm	76	
Drawbar hitch x-coordinate, cm		130.1 (behind the midpoint of the track)
Drawbar hitch y-coordinate (above ground), cm		59.5
Track type	Reinforced rubber belt	Reinforced rubber belt
Track weight per unit length, kN/m	0.33	0.33
Track width, cm	62	62

*Dimensions indicated in this table are for the vehicle stationary on level, hard surface.

Table 8.8: Pressure–sinkage parameters of undisturbed and preconditioned Fernie snow

Snow condition	n	k (kPa/m^n)
Undisturbed	0.71	192.89
Preconditioned	0.31	319.66

Note: The pressure–sinkage relation takes the form of $p = kz^n$, where p is pressure, k and n are the pressure–sinkage parameters, and z is sinkage.

Table 8.9: Shear strength parameters of undisturbed and preconditioned Fernie snow

Snow condition	Type of shearing	c, cohesion (adhesion) (kPa)	ϕ, angle of shearing resistance (degrees)	K, shear deformation parameter (cm)
Undisturbed	Internal	1.96	25	5.5
	Rubber	0.21	11.4	0.4
Preconditioned	Internal	0.32	29.4	2.5
	Rubber	0.02	12	0.4

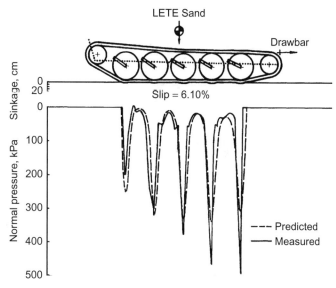

Figure 8.12: Measured and predicted normal pressure distribution by NTVPM under the track pad of the M11A1 on LETE sand

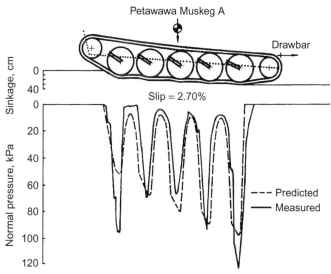

Figure 8.13: Measured and predicted normal pressure distribution by NTVPM under the track pad of the M11A1 on Petawawa Muskeg A

be mentioned that the predicted drawbar performances on the undisturbed and preconditioned Fernie snow shown in the figures are based on the average terrain values shown in Tables 8.8 and 8.9. The measured drawbar pulls are expressed in terms of their mean values $\pm$ standard deviations.

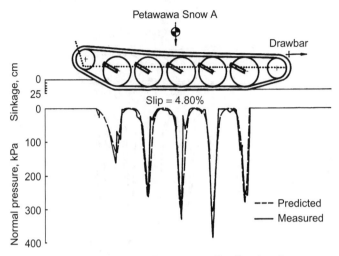

Figure 8.14: Measured and predicted normal pressure distribution by NTVPM under the track pad of the M11A1 on Petawawa Snow A

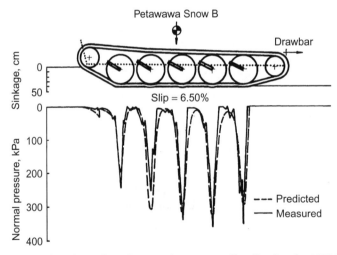

Figure 8.15: Measured and predicted normal pressure distribution by NTVPM under the track pad of the M11A1 on Petawawa Snow B

In addition to the field tests for validating the basic features of NTVPM described above, a study of the correlation between the tractive performance of a specially designed test vehicle in deep snow predicted by NTVPM and field test data was conducted in Sweden (Bodin, 2002).

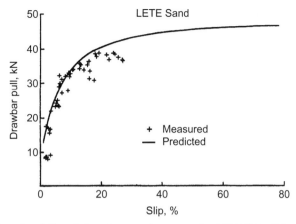

Figure 8.16: Measured and predicted drawbar performance by NTVPM of the M11A1 on LETE sand

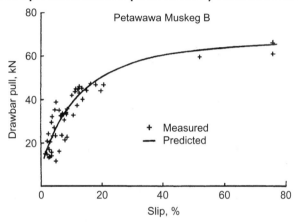

Figure 8.17: Measured and predicted drawbar performance by NTVPM of the M11A1 on Petawawa Muskeg B

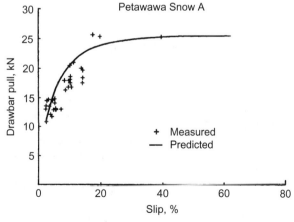

Figure 8.18: Measured and predicted drawbar performance by NTVPM of the M11A1 on Petawawa Snow A

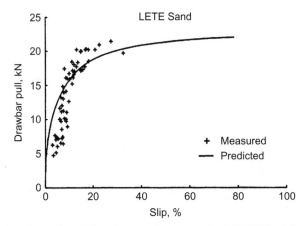

Figure 8.19: Measured and predicted drawbar performance by NTVPM of the Bv202 on LETE sand

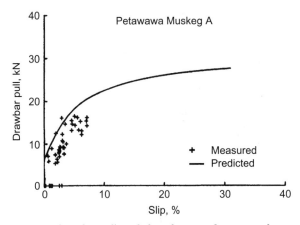

Figure 8.20: Measured and predicted drawbar performance by NTVPM of the Bv202 on Petawawa Muskeg A

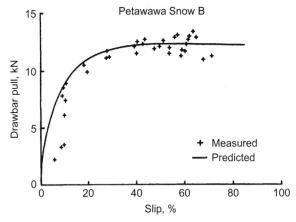

Figure 8.21: Measured and predicted drawbar performance by NTVPM of the Bv202 on Petawawa Snow B

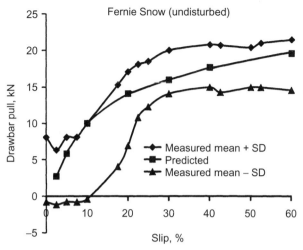

Figure 8.22: Measured and predicted drawbar performance by NTVPM of the Bv206 on undisturbed Fernie snow (Reprinted by permission of the Council of the Institution of Mechanical Engineers from Wong, 1992a)

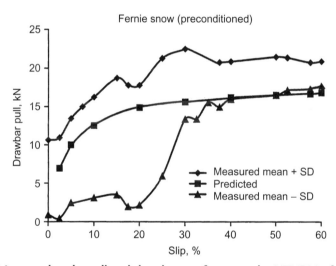

Figure 8.23: Measured and predicted drawbar performance by NTVPM of the Bv206 on preconditioned Fernie snow (Reprinted by permission of the Council of the Institution of Mechanical Engineers from Wong, 1992a)

8.6 Comparisons of Predictions by NTVPM with Test Data

1. It is seen from Figures 8.12–8.15 that the normal pressure distributions under the rubber pad of the track of the M11A1 predicted by NTVPM correlate reasonably well with the measured ones on LETE sand, Petawawa Muskeg A, and Petawawa Snow A and B, respectively, considering the variability of terrain conditions in the field. This confirms

the capability of NTVPM in realistically predicting the normal pressure distributions under a vehicle with flexible tracks over different types of terrain. It also provides evidence to support the assumption that the track with relatively short track pitch may be idealized as a flexible and extensible belt.

It is noted that the measured maximum normal pressure under the track pad on LETE sand is approximately 500 kPa, which is 11.36 times the nominal ground pressure of 44 kPa, obtained from dividing the vehicle weight by the product of track width and length on the ground of the two tracks. On the Petawawa Muskeg A, the measured maximum normal pressure is approximately 120 kPa, which is 2.73 times the nominal ground pressure. On Petawawa Snow A and B, the measured maximum normal pressures are 382 and 355 kPa, which are 8.68 and 8.07 times the nominal ground pressure, respectively. This indicates that the actual normal pressure distribution on the track–terrain interface varies significantly with terrain conditions and that using the nominal ground pressure as a performance indicator for tracked vehicles is not meaningful. On firm ground, such as LETE sand, the normal pressures are concentrated under the roadwheels and track segments between adjacent roadwheels support little load. Consequently, high peak pressures under the roadwheels and low pressures on track segments between adjacent roadwheels are observed. On soft terrain, such as muskeg, the track segments between roadwheels supports load and as a result, the peak pressures under the roadwheels are reduced. Data presented in Figures 8.12–8.15 also show that the actual value of MMP, based on the mean value of the measured maximum pressures under all roadwheel stations, varies with terrain conditions. This provides further evidence that the empirical methods for predicting the values of MMP presented in Chapter 7, which completely ignore the effect of terrain conditions, are unrealistic.

The reasonably close agreement between the measured normal pressure distributions and predicted ones is due to the facts that NTVPM takes into account the response of the terrain to repetitive normal and shear loadings described in Chapter 4 and Chapter 5, respectively, and that it is based on a detailed analysis of track–terrain interaction as presented in Sections 8.1–8.4. As pointed out previously, an element of the terrain under the track is subject to the repetitive loading of consecutive roadwheels and track segments. The terrain, after being compacted by the preceding roadwheel, may become harder (or stiffer). Consequently, this would promote the concentration of normal pressure immediately under the roadwheels and significant pressure peaks are generally observed, even on a terrain like muskeg which is relatively soft in its undisturbed (virgin) state.

Furthermore, the behaviour of the terrain during the loading–unloading–reloading cycle described in Chapter 4 also explains why it is possible that the pressure at a point on the track segment between two adjacent roadwheels can be as low as zero, while the sinkage at that point, measured from the original terrain surface, is not zero.

2. It is observed from Figures 8.16–8.18 and from Figures 8.19–8.21 that there is a reasonably close agreement between the measured and predicted drawbar performance of the M113A1 and the Bv202, respectively, over LETE sand, Petawawa muskeg and Petawawa snow, considering the variability of terrain conditions in the field. It can also be seen from Figures 8.22 and 8.23 that there is reasonable correlation between the measured and predicted drawbar performance of the Bv206 over the undisturbed and preconditioned Fernie snow. The predicted drawbar performance based on the average values of terrain parameters generally lies within the boundaries of the measured mean values ± standard deviations.

The reasonably close agreement between the measured and predicted drawbar performance is primarily due to the fact that the NTVPM takes into account the response of the terrain to repetitive shearing and the shearing characteristics of the terrain under varying (or cyclic) normal pressure. The behaviour of the terrain under repetitive shearing and under varying normal pressure has a significant effect on the development of shear stress under the track, hence its tractive effort, as discussed in Section 5.3.

In summary, the reasonably close agreement between the measured and predicted normal pressure distributions under the track and that between the measured and predicted drawbar performances over a range of terrains indicate that the basic features of the computer-aided method NTVPM, have been experimentally substantiated. Since NTVPM takes into account all major design parameters of the vehicle and all relevant terrain characteristics, it is particularly suited for the engineer in the evaluation of vehicle design concepts and in the optimization of vehicle design parameters prior to prototyping. It is also a useful tool for the procurement manager in comparing the tractive performance of vehicle candidates prior to field testing. The practical applications of NTVPM to the design and performance evaluation of vehicles with flexible tracks, from the traction perspective, are presented in the following chapter.

CHAPTER 9

Applications of the Computer-Aided Method NTVPM to Parametric Analysis of Vehicles with Flexible Tracks

As noted in the previous chapter, the computer-aided method NTVPM takes into account all major vehicle design parameters and all pertinent terrain characteristics. Thus, it is particularly suited for detailed parametric analysis of the performance and design of vehicles with flexible tracks. NTVPM can be applied to evaluating competing designs, to optimizing vehicle design parameters, to assessing the effects on performance of terrain conditions, or to evaluating vehicle candidates for a given mission and environment from the procurement perspective (Wong, 1992a, 1994b, 1995, 1997, 1999, 2007; Bodin, 2002; Wong and Preston-Thomas, 1988; Wong and Huang, 2005, 2006a, b, 2008).

To demonstrate its applications to detailed parametric analysis, NTVPM is employed to evaluate the effects of track system configuration, initial track tension, suspension setting, longitudinal location of the centre of gravity, vehicle weight, track width, and the location of the drive sprocket on the performances of a baseline vehicle, and its two variants with different track–roadwheel configurations over two types of terrain. The applications of NTVPM to evaluate the effects of the configuration of the articulation joint on the performance of a two-unit articulated vehicle are presented. Analysis and evaluation of detracking risks using a special module of NTVPM is discussed. An example of employing NTVPM to assist an off-road vehicle manufacturer in the development of a new high-mobility version of an infantry fighting vehicle is also described.

The baseline tracked vehicle of the single-unit type used in the study is similar to an armoured personnel carrier widely used in many parts of the world. Its basic design parameters are given in Table 9.1. Its roadwheel–suspension parameters are presented in Table 9.2. It can be seen from the tables that for the baseline vehicle, its track pitch is relatively short of 15 cm, the ratio of roadwheel diameter to track pitch is 4.07 and the ratio of roadwheel spacing to track pitch is 4.43. Accordingly, the short-pitch track of the baseline vehicle may well be idealized as a flexible track. Consequently, NTVPM can be employed to evaluate its performance.

The parametric analysis of the design and performance of the baseline vehicle was carried out over a snow-covered terrain and a clayey soil. The snow-covered terrain, referred to as

Table 9.1: Basic parameters of Vehicle A (baseline vehicle)

Sprung weight, kN	\multicolumn{2}{c}{100.57}
Unsprung weight, kN	\multicolumn{2}{c}{10}
Sprung weight CG x-coordinate*, cm	\multicolumn{2}{c}{198}
Sprung weight CG y-coordinate*, cm	\multicolumn{2}{c}{48.1}
Sprocket centre x-coordinate*, cm	\multicolumn{2}{c}{0}
Sprocket centre y-coordinate*, cm	\multicolumn{2}{c}{0}
Sprocket pitch radius, cm	\multicolumn{2}{c}{21.4}
Roadwheel diameter, cm	\multicolumn{2}{c}{61}
Average distance between roadwheels, cm	\multicolumn{2}{c}{66.4}
Idler centre x-coordinate*, cm	\multicolumn{2}{c}{402.3}
Idler centre y-coordinate*, cm	\multicolumn{2}{c}{−15.1}
Idler pitch radius, cm	\multicolumn{2}{c}{21.9}
Drawbar hitch x-coordinate*, cm	\multicolumn{2}{c}{427.5}
Drawbar hitch y-coordinate*, cm	\multicolumn{2}{c}{12.7}
Track type	\multicolumn{2}{c}{Segmented metal track with rubber pads}
Track weight per unit length, kN/m	\multicolumn{2}{c}{0.67}
Track width, cm	\multicolumn{2}{c}{38}
Track pitch, cm	\multicolumn{2}{c}{15}
Ground clearance on level, hard ground, cm	\multicolumn{2}{c}{41.7}
Belly (hull) width, cm	\multicolumn{2}{c}{170}
Coordinates defining the longitudinal shape of the belly, cm	x-coordinates*	y-coordinates*
	−21.4	−34.3
	−4.1	14.3
	417	14.3

*Coordinate origin is at the centre of the sprocket which is located at the front. Positive x- and y-coordinates are to the rear and down, respectively.

Hope Valley snow, represents a snow cover with average depth of 128 cm found in a mountainous region in the USA (Harrison, 1975). The clayey soil represents a fine-grained, mineral soil with high moisture content, found near Bangkok, Thailand (Bekker, 1969).

The Bekker equation, Eqn (4.1), was used to characterize the pressure–sinkage relationships of the Hope Valley snow and the clayey soil, while Eqns (4.23) and (4.24) were employed to describe their responses to repetitive normal loading. The values of the pressure–sinkage parameters of the Hope Valley snow and of the clayey soil were taken from Harrison (1975) and Bekker (1969), respectively. Certain terrain parameters required as input to NTVPM are, however, not available from the references. For these parameters, estimated values based on those of similar terrains were used. The values of the pressure–sinkage and of the repetitive loading parameters of these two types of terrain are summarized in Table 9.3.

Table 9.2: Roadwheel – suspension system parameters of Vehicle A, Vehicle A (6W) and Vehicle A (8W)

Vehicle configuration	Wheel radius (cm)	Torsion arm pivots		Torsion bar stiffness (kN-m/deg)	Torsion arm angle at free position** (degrees)	Torsion arm length (cm)	Trailing or leading arm
		x-coordinates* (cm)	y-coordinates* (cm)				
Vehicle A (baseline vehicle)	30.5	39.69	8.73	0.1668	43	31.75	T***
	30.5	106.36	8.73	0.1668	43	31.75	T
	30.5	173.04	8.73	0.1668	43	31.75	T
	30.5	239.71	8.73	0.1668	43	31.75	T
	30.5	306.39	8.73	0.1668	43	31.75	T
Vehicle A (6W)	23.5	39.69	8.73	0.1668	50.25	37.5	T
	23.5	93.03	8.73	0.1668	50	37.5	T
	23.5	146.37	8.73	0.1668	50	37.5	T
	23.5	199.71	8.73	0.1668	50	37.5	T
	23.5	253.05	8.73	0.1668	48.5	37.5	T
	23.5	306.38	8.73	0.1668	48.5	37.5	T
Vehicle A (8W)	30.5	39.69	8.73	0.1043	44	31.75	T
	30.5	77.79	8.73	0.1043	43.5	31.75	T
	30.5	115.89	8.73	0.1043	43	31.75	T
	30.5	153.99	8.73	0.1043	43	31.75	T
	30.5	192.09	8.73	0.1043	43	31.75	T
	30.5	230.19	8.73	0.1043	43	31.75	T
	30.5	268.29	8.73	0.1043	43	31.75	T
	30.5	306.39	8.73	0.1043	43	31.75	T

*Coordinate origin is at the centre of the sprocket. Positive x- and y-coordinates are to the rear and down, respectively.
**Torsion arm angle under no load condition.
***T – Trailing arm.

Table 9.4 summarizes the values of the shear strength parameters of the two types of terrain examined in this study. Equation (5.2) was used to characterize the internal and rubber–terrain shear stress–displacement relationships of both types of terrain. The shear strength of the terrains was described using the Mohr-Coulomb relation, Eqn (5.16).

When the terrain is highly compressible, such as the Hope Valley snow, the track sinkage may be greater than the vehicle ground clearance and the vehicle belly (hull) may be in contact with the terrain surface. In this case, the belly–terrain shearing parameters are required as input to NTVPM. The estimated values of the parameters characterizing the belly–terrain shearing on the Hope Valley snow are also given in Table 9.4.

It should be noted that while in this study Eqn (4.1) was selected to represent the pressure–sinkage relationship and Eqn (5.2) was chosen to represent the shear stress–shear displacement relationship of the two types of terrain examined, NTVPM can accommodate other

Table 9.3: Pressure–sinkage and repetitive loading parameters for the Hope Valley snow and the clayey soil used in the study

Terrain parameters	Hope Valley snow	Clayey soil
k_c (kN/m^{n+1})	6.16	20.68
k_ϕ (kN/m^{n+2})	149.35	814.30
n	1.53	1.0
k_o (kN/m^3)	0	0
A_u (kN/m^4)	40 000	78 820

Table 9.4: Shear strength parameters for the Hope Valley snow and the clayey soil used in the study

Terrain type	Type of shearing	Cohesion (adhesion) c (kPa)	Angle of shearing resistance ϕ (degrees)	Shear deformation parameter K (cm)
Hope Valley snow	Internal	0.76	23.2	4.24
	Rubber-terrain	0.12	16.4	0.39
	Belly-terrain	0	10	–
Clayey soil	Internal	3.45	11	2.54
	Rubber-terrain	1.85	9.7	2.12
	Belly-terrain	–	–	–

types of equation for characterizing terrain behaviour. For instance, Eqn (4.42) or (4.60) may be selected to represent the pressure–sinkage relationship of muskeg or snow with ice layers, respectively. Also, Eqn (5.3) or (5.5) may be chosen to represent the shear stress–shear displacement relationship of the terrain.

Vehicle performance parameters predicted by NTVPM include the normal and shear stress distributions under the track, the mean maximum pressure (MMP) based on the predicted normal pressure distribution on the track–terrain interface, the sinkage and track motion resistance distributions along the track contact length, belly load and belly drag (if the vehicle belly is in contact with the terrain surface), track motion resistance, thrust (tractive effort), drawbar pull, and tractive efficiency as functions of track slip. These performance parameters provide detailed information for design and performance evaluation of vehicles with flexible tracks under various operating environments.

9.1 Effects of Initial Track Tension and Track System Configuration

To examine the effects of track system configuration on vehicle mobility, the performances of three configurations with different roadwheel arrangements were simulated using NTVPM.

These include the track system of the baseline vehicle with five roadwheels on each track, referred to as Vehicle A, the track system of a six-roadwheel vehicle, referred to as Vehicle A (6 W), and the track system of an eight-overlapping roadwheel vehicle, referred to as Vehicle A (8 W). They are schematically shown in Figure 9.1. To compare the effects of track system configuration on performance on a common basis, the basic design parameters of Vehicle A (6 W) and of Vehicle A (8 W), such as their sprung weight, unsprung weight, track contact length and width, location of the centre of gravity, etc., are taken to be the same as those of Vehicle A shown in Table 9.1. Because of the difference in the number of roadwheels, the diameter of the roadwheels of Vehicle A (6 W) is smaller than that for Vehicle A, in order to maintain the same track contact length. Also in Vehicle A (6 W), there is a roller to support the top run of the track. The diameter of the roadwheels of Vehicle A (8 W) is the same as that of Vehicle A. The roadwheel–suspension parameters of Vehicle A (6 W) and vehicle A (8 W) are different from those of Vehicle A, as shown in Table 9.2. They are chosen in such a way so as to keep the overall suspension characteristics of the three configurations as close as possible. For instance, the overall vertical stiffnesses of the suspensions of the three configurations are similar and vary in a limited range between 452 and 479 kN/m. This ensures that their fundamental natural frequencies for bounce (heave) are close. The fundamental natural frequencies for bounce for Vehicle A, Vehicle A (6 W), and Vehicle A (8 W) are 1.54, 1.49, and 1.54 Hz, respectively. Under static load on level, hard ground, the ground clearances measured at the front end of the vehicle belly of Vehicle A, Vehicle A (6 W), and Vehicle A

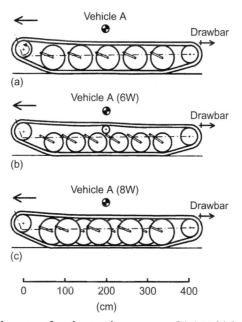

Figure 9.1: Schematic diagrams for the track systems of (a) Vehicle A, (b) Vehicle A (6 W) and (c) Vehicle A (8 W)

(8 W) are 41.7, 41.7, and 41.8 cm, respectively, which are essentially identical. The inclination angles with respect to the horizontal of the bellies (referred to as the trim angles) of Vehicle A, Vehicle A (6 W), and Vehicle A (8 W) under static load on level, hard ground are $-0.660°$, $-0.672°$, and $-0.666°$, respectively, which are approximately the same. The negative sign indicates that the vehicle belly takes a nose-down attitude. It should be noted that when the track sinkage is greater than vehicle ground clearance and the vehicle belly is in contact with the terrain surface, the trim angle of the vehicle belly has noticeable effects on vehicle performance, which is examined in detail later.

The overlapping roadwheel system of Vehicle A (8 W), was widely used in many types of combat vehicle produced in Germany during World War II (Ogorkiewicz, 1991). In this system, the centres of the two rows of roadwheels on a track are shifted relative to each other in the longitudinal direction. This provides a more uniform ground pressure distribution along the length of the track, while retaining the desirable characteristics of large diameter roadwheels with sufficiently large suspension travels. The more uniform pressure distribution reduces track sinkage and track motion resistance, resulting in improved vehicle mobility over soft terrain. The effects of the overlapping roadwheel system on performance are discussed in detail later. It should be mentioned that while this system may improve the uniformity of ground pressure distribution, it would increase design complexity and manufacturing costs. Furthermore, on terrain with high cohesion (or adhesion), the spacing between the overlapping roadwheels may be more susceptible to clogging with terrain materials. However, this problem may be alleviated by proper design of the roadwheel. The overlapping roadwheel arrangement represents an innovative concept from the traction perspective.

The initial track tension is the tension in the track system when the vehicle is stationary on level, hard ground. Usually, it is deduced from the sag of the top run of the track, based on the mechanics of catenary. If the initial track tension is low, the track will be loose. The track segments between roadwheels cannot support much load and the vehicle weight is primarily supported by the track segments (links) immediately under the roadwheels. In this case, the tracked vehicle in essence behaves like a multi-wheel vehicle. On the other hand, if the initial track tension is high, the track will be tight and track segments between roadwheels can support substantial load. This reduces the peak normal pressure under the track, hence track sinkage and track motion resistance. On soft terrain with high compressibility, such as deep snow, where the track sinkage exceeds the vehicle ground clearance, the vehicle belly comes into contact with the terrain surface. In this case, increasing the initial track tension has two major effects. Firstly, it reduces the sinkages of both the track and the belly, hence lowering the track motion resistance and the belly drag, caused by vehicle belly–terrain interaction. Secondly, it reduces the load supported by the belly, hence increasing the load applied on the track. This enables the track to develop higher thrust on most terrains with a significant frictional component in their shear strength. This indicates that on highly compressible terrain, where the belly contacts the terrain surface, increasing the initial track tension has particularly significant effects on improving vehicle performance (Wong, 1995, 1997, 1999, 2007; Wong and Preston-Thomas, 1988; Wong and Huang, 2005, 2006a, 2008). This is discussed in detail later in the chapter.

A. On Hope Valley snow

As noted previously, the Hope Valley snow has an average depth of 128 cm. The values of its pressure–sinkage parameters, such as k_c and k_ϕ, are relatively low, as shown in Table 9.3. This indicates that the Hope Valley snow is highly compressible and that it poses a challenge to vehicle mobility.

On highly compressible terrain, track sinkage may exceed vehicle ground clearance, and the vehicle belly may be, partially or fully, in contact with the terrain surface. Under these circumstances, the total motion resistance of the vehicle consists of two major components. One is the track motion resistance which is related to track sinkage. The other is the motion resistance caused by vehicle belly–terrain interaction, referred to as the belly drag. Belly drag is induced by the shear force when the belly is sliding on the terrain surface and by the bulldozing effect when the belly is inclined with respect to the direction of motion of the vehicle. Therefore, belly drag is related to the load supported by the belly which is a function of belly sinkage and to the trim angle of the vehicle belly. As noted previously, the trim angle is the angle between the belly and the horizontal. The effect of belly trim angle on belly drag is illustrated in Figure 9.2. If the belly takes a nose-down attitude, for which the belly trim angle is designated as negative noted previously, then most part of the belly (except the front part) will not be in contact with the terrain surface. Belly drag is primarily caused by the bulldozing effect of the front part of the belly. On the other hand, if the belly takes a nose-up attitude, for which the belly trim angle is designated positive, then belly drag is caused by both the

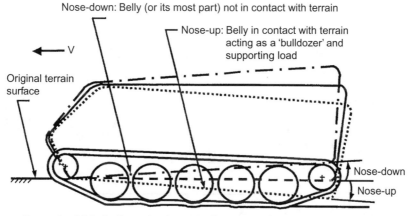

Figure 9.2: Effect of vehicle belly attitude on belly–terrain interaction, when track sinkage is equal to or greater than vehicle ground clearance (Reprinted by permission of *Int. J. Heavy Vehicle Systems* from Wong and Huang, 2005)

bulldozing effect of the belly and the shear force caused by the sliding of the belly on the terrain surface. This indicates that if the belly is in contact with the terrain surface and takes a nose-up attitude, it will have significant adverse effects on vehicle performance by introducing a substantial belly drag.

Figures 9.3, 9.4, and 9.5 show the normal pressure distributions and sinkages, on the Hope Valley snow, under the tracks of Vehicle A, Vehicle A (6 W), and Vehicle A (8 W), respectively, at initial track tension to vehicle weight ratios of 2.5, 10, 20, and 40% and at 20% slip. They were predicted using NTVPM (Wong, 2007). The ratio of the initial track tension to vehicle weight is referred to as the initial track tension coefficient. For the widely used armoured personnel carrier mentioned previously, on which Vehicle A is based, an initial track tension coefficient of approximately 10% is recommended for normal use by its manufacturer. In this study, following the usual practice, performance parameters of different track system configurations are compared at a slip of 20%.

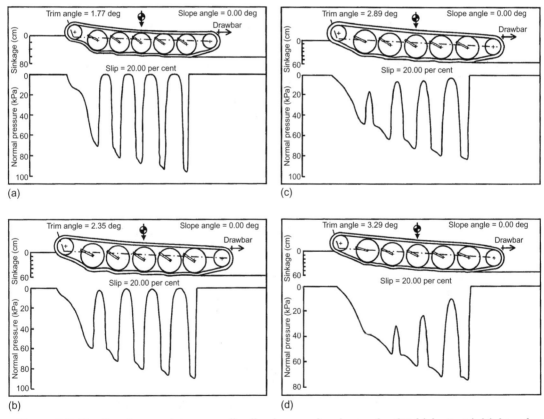

Figure 9.3: Predicted normal pressure distributions under the track of Vehicle A at initial track tension coefficients of (a) 2.5%, (b) 10%, (c) 20% and (d) 40% at 20% slip on the Hope Valley snow (Reprinted by permission of *ISTVS* from Wong, 2007)

As shown in Figures 9.3, 9.4 and 9.5, the vehicle bellies for the three configurations are either partially or fully in contact with the snow surface. From Figure 9.3 and Table 9.5, it is seen that for Vehicle A with the value of the initial track tension coefficient less than 10%, the track sinkages under both the front and rear roadwheels are greater than the ground clearance, hence its belly is in full contact with the snow surface. On the other hand, from Figure 9.5 and Table 9.5, for Vehicle A (8 W) with the value of the initial track tension higher than 10%, the track sinkage under the front roadwheel is less than the ground clearance, while that under the rear roadwheel is greater than the ground clearance, hence the belly is in partial contact with the snow surface. With the vehicle belly in contact with the terrain, it supports part of the vehicle weight, while the remainder is applied to the track. The normal pressure distributions under the tracks shown in the figures are those that the load applied to the tracks is reduced by that supported by the vehicle belly.

It is shown that for all three configurations examined, as the initial track tension coefficient increases, the pressure on the track segments between roadwheels increases, which indicates

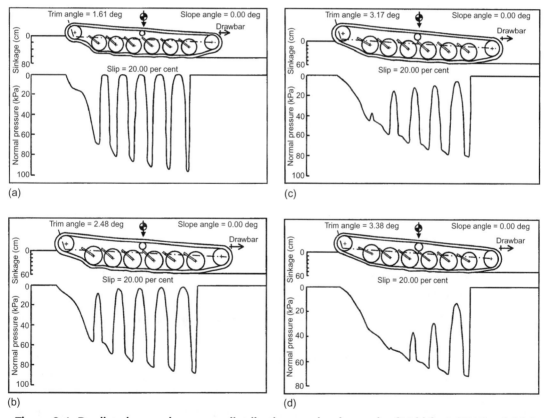

Figure 9.4: Predicted normal pressure distributions under the track of Vehicle A (6 W) at initial track tension coefficients of (a) 2.5%, (b) 10%, (c) 20% and (d) 40% at 20% slip on the Hope Valley snow (Reprinted by permission of *ISTVS* from Wong, 2007)

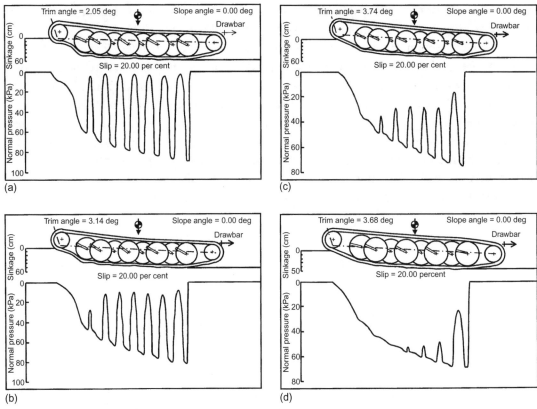

Figure 9.5: Predicted normal pressure distributions under the track of Vehicle A (8 W) at initial track tension coefficients of (a) 2.5%, (b) 10%, (c) 20% and (d) 40% at 20% slip on the Hope Valley snow (Reprinted by permission of *ISTVS* from Wong, 2007)

that they take up more load. This reduces the peak pressures under the track, leading to improvement in performance. With the same value of the initial track tension coefficient, as the number of roadwheels increases, the uniformity of pressure distribution under the track improves. The uniformity of pressure distribution may be expressed in terms of the variation of the peak pressure under roadwheels. This may be illustrated, for instance, by comparing the pressure distributions under the tracks in Figure 9.3(d), Figure 9.4(d) and Figure 9.5(d).

Table 9.5 shows that for a given track system configuration, both the front and rear roadwheel sinkages decrease with the increase of the initial track tension coefficient. For Vehicle A, increasing the value of the initial track tension coefficient from 2.5 to 40%, the sinkage of the front roadwheel decreases from 50.8 to 33.7 cm, representing a reduction of 33.7%. Varying the initial track tension coefficient in the same range, the rear roadwheel sinkage decreases from 62.2 to 52.6 cm, representing a reduction of 15.4%. For Vehicle A (6 W) and Vehicle A (8 W), the variations of the front and rear roadwheel sinkages with the initial track tension coefficient show similar trend to that for Vehicle A.

Table 9.5: Effects of track system configuration and initial track tension on performance parameters at 20% slip on Hope Valley snow

Vehicle configuration	Initial track tension coeff. %	Belly load coeff. %	Belly drag coeff. %	Track motion resistance coeff. %	Total motion resistance coeff. %	Thrust coeff. %	Drawbar pull coeff. %	Tractive efficiency %	Front roadwheel sinkage (cm)	Rear roadwheel sinkage (cm)	Belly trim angle (degrees)
Vehicle A (baseline vehicle)	2.5	31.61	6.61	13.85	20.46	18.24	−2.22	–	50.8	62.2	1.77
	10	24.02	5.26	12.42	17.68	20.37	2.69	10.58	45.6	59.7	2.35
	20	18.37	4.20	10.87	15.07	23.43	8.36	28.53	40.1	56.7	2.89
	30	15.20	3.59	9.70	13.29	26.13	12.84	39.32	35.8	54.3	3.27
	40	13.81	3.26	8.95	12.21	27.44	15.23	44.41	33.7	52.6	3.29
Vehicle A (6W)	2.5	32.35	6.69	13.78	20.47	17.82	−2.65	–	51.1	62.1	1.61
	10	21.83	4.83	11.97	16.80	20.51	3.71	14.45	43.9	58.8	2.48
	20	16.11	3.77	10.31	14.08	24.28	10.20	33.61	38.0	55.6	3.17
	30	13.68	3.25	9.24	12.49	26.56	14.07	42.37	34.8	53.3	3.36
	40	12.48	2.97	8.51	11.48	28.28	16.80	47.52	32.7	51.6	3.38
Vehicle A (8W)	2.5	21.26	4.54	11.95	16.49	19.83	3.34	13.47	46.0	58.7	2.05
	10	14.97	3.49	10.44	13.93	22.65	8.72	30.78	39.1	55.8	3.14
	20	11.50	2.81	8.95	11.76	26.78	15.02	44.86	33.7	52.6	3.74
	30	10.38	2.55	8.21	10.76	29.10	18.34	50.41	31.2	50.9	3.81
	40	10.22	2.49	7.81	10.29	30.18	19.89	52.72	29.9	49.9	3.68

At the same initial track tension coefficient, the front and rear roadwheel sinkages decrease with the increase of the number of roadwheels. For instance, at the initial track tension coefficient of 20%, the front roadwheel sinkage of Vehicle A with five roadwheels is 40.1 cm and that of Vehicle A (8 W) with eight roadwheels is 33.7 cm, representing a reduction of 16%. At the initial track tension coefficient of 20%, the rear roadwheel sinkage of Vehicle A is 56.7 cm and that of Vehicle A (8 W) is 52.6 cm, representing a reduction of 7.2%.

Comparing the front roadwheel sinkage of 50.8 cm for Vehicle A at the initial track tension coefficient of 2.5% with that of 29.9 cm for Vehicle A (8 W) at the initial track tension coefficient of 40%, the reduction is 41.1%. Comparing the rear roadwheel sinkage of 62.2 cm for Vehicle A at the initial track tension coefficient of 2.5% with that of 49.9 cm for Vehicle A (8 W) at the initial track tension coefficient of 40%, the reduction is 19.8%. All of these indicate that both the track system configuration and initial track tension coefficient have significant effects on the front and rear roadwheel sinkages.

Table 9.5 shows that the track system configuration and initial track tension coefficient also affect the belly trim angle. In general, the belly trim angle increases with the increase of the initial track tension coefficient. This is primarily due to the increase of drawbar pull with the increase of the initial track tension coefficient. Higher drawbar pull causes greater load transfer from the front to the rear. For the same initial track tension coefficient, increasing the number of roadwheels generally causes an increase in the belly trim angle. Again, this is primarily due to the increase of the drawbar pull with the increase of the number of roadwheels, which causes a greater load transfer from the front to the rear.

On the Hope Valley snow, with the vehicle belly contacting the snow surface in part or in full, it supports part of the vehicle weight and reduces the normal load applied to the track. Figure 9.6 shows the variations of the belly load coefficient (i.e. the ratio of the load supported by the belly to the vehicle weight) with the initial track tension coefficient for the three configurations. It is seen that the belly load coefficient decreases with the increase of the initial track tension coefficient or the number of roadwheels. Table 9.5 gives the values of the belly load coefficient at the initial track tension coefficients of 2.5, 10, 20 and 40% for the three configurations at 20% slip. It shows that for Vehicle A, increasing the initial track tension coefficient from 2.5 to 40%, the belly load coefficient decreases from 31.61 to 13.81%, representing a reduction of 56.3%. For Vehicle A (6 W), the variation of belly load coefficient with the initial track tension coefficient shows a similar trend to that for Vehicle A. For Vehicle A (8 W), increasing the initial track tension coefficient over the same range, the belly load coefficient decreases from 21.26 to 10.22%, representing a reduction of 51.9%. In comparison with the belly load coefficient of 31.61% for Vehicle A at the initial track tension coefficient of 2.5%, the belly load coefficient of 10.22% for Vehicle A (8 W) at the initial track tension coefficient of 40% represents a reduction of 67.7%.

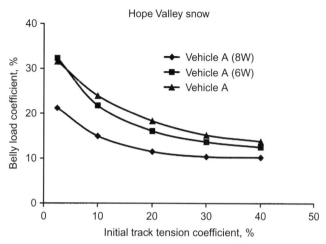

Figure 9.6: Variations of the belly load coefficient with the initial track tension coefficient for the three vehicle configurations at 20% slip on the Hope Valley snow

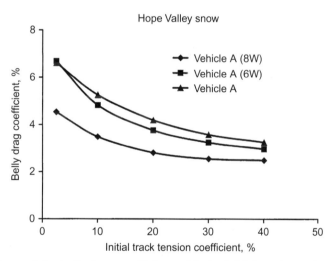

Figure 9.7: Variations of the belly drag coefficient with the initial track tension coefficient for the three vehicle configurations at 20% slip on the Hope Valley snow

Figure 9.7 shows the variations of the belly drag coefficient (i.e. the ratio of the belly drag to the vehicle weight) with the initial track tension coefficient for the three configurations. It is seen that the belly drag coefficient decreases with the increase of the initial track tension coefficient or the number of roadwheels. Table 9.5 gives the values of the belly drag coefficient at the initial track tension coefficients of 2.5, 10, 20 and 40% for the three configurations at 20% slip. It shows that for Vehicle A, increasing the initial track tension coefficient from 2.5 to 40%, the belly drag coefficient decreases from 6.61 to 3.26%, representing a reduction of 50.7%. For Vehicle A (6 W), the variation of belly drag coefficient with the initial track tension coefficient shows a similar trend to that for Vehicle A. For Vehicle A (8 W), increasing the initial track tension coefficient over the same range, the belly drag coefficient decreases from 4.54 to 2.49%, representing a reduction of 45.2%. In comparison with the belly drag coefficient of 6.61% for Vehicle A at the initial track tension coefficient of 2.5%, the belly drag coefficient of 2.49% for Vehicle A (8 W) at the initial track tension coefficient of 40% represents a reduction of 62.3%. It should be noted that the belly drag is related to both belly sinkage and the belly trim angle. While increasing the initial track tension coefficient generally results in an increase in the belly trim angle for a given vehicle configuration, as shown in Table 9.5, the effects of increasing the initial track tension coefficient on the reduction of both the front and rear roadwheel sinkages, hence belly sinkage, outweighs the increase of the belly trim angle. As a result, for a given track system configuration, the belly drag coefficient decreases with the increase of the initial track tension coefficient.

As noted earlier, the track motion resistance due to track–terrain interaction is related to track sinkage. Figure 9.8 shows the variations of the track motion resistance coefficient (i.e. the ratio of the track motion resistance due to track–terrain interaction to the vehicle weight) with the initial track tension coefficient for the three configurations. It is seen that the track motion resistance coefficient decreases with the increase of the initial track tension coefficient or the number of roadwheels. Table 9.5 gives the values of the track motion resistance coefficient at the initial track tension coefficients of 2.5, 10, 20 and 40% for the three configurations at 20% slip. It shows that for Vehicle A, increasing the initial track tension coefficient from 2.5 to 40%, the track motion resistance coefficient decreases from 13.85 to 8.95%, representing a

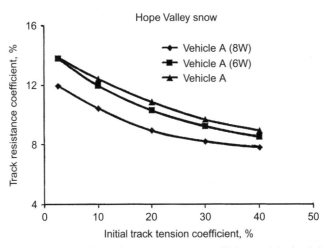

Figure 9.8: Variations of the track motion resistance coefficient with the initial track tension coefficient for the three vehicle configurations at 20% slip on the Hope Valley snow

reduction of 35.4%. For Vehicle A (6W), the variation of track motion resistance coefficient with the initial track tension coefficient shows similar trend to that for Vehicle A. For Vehicle A (8W), increasing the initial track tension coefficient over the same range, the track motion resistance coefficient decreases from 11.95 to 7.81%, representing a reduction of 34.6%. In comparison with the track motion resistance coefficient of 13.85% for Vehicle A at the initial track tension coefficient of 2.5%, the track motion resistance coefficient of 7.81% for Vehicle A (8W) at the initial track tension coefficient of 40% represents a reduction of 43.6%.

Figure 9.9 shows the variations of the total motion resistance coefficient (i.e. the sum of the track motion resistance coefficient and the belly drag coefficient) with the initial track tension coefficient for the three configurations. It is seen that the total motion resistance coefficient decreases with the increase of the initial track tension coefficient or the number of road-wheels. Table 9.5 gives the values of the total motion resistance coefficient at the initial track tension coefficients of 2.5, 10, 20 and 40% for the three configurations at 20% slip. It shows that for Vehicle A, increasing the initial track tension coefficient from 2.5 to 40%, the total motion resistance coefficient decreases from 20.46 to 12.21%, representing a reduction of 40.3%. For Vehicle A (6W), the variation of total motion resistance coefficient with the initial track tension coefficient shows a similar trend to that for Vehicle A. For Vehicle A (8W), increasing the initial track tension coefficient over the same range, the total motion resistance coefficient decreases from 16.49 to 10.29%, representing a reduction of 37.6%. In comparison with the total motion resistance coefficient of 20.46% for Vehicle A at the initial track tension coefficient of 2.5%, the total motion resistance coefficient of 10.29% for Vehicle A (8W) at the initial track tension coefficient of 40% represents a reduction of 49.7%.

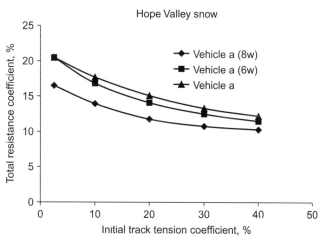

Figure 9.9: Variations of the total motion resistance coefficient with the initial track tension coefficient for the three vehicle configurations at 20% slip on the Hope Valley snow

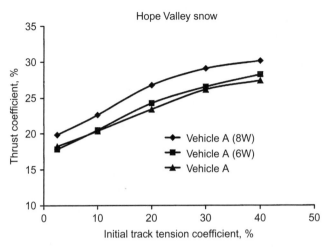

Figure 9.10: Variations of the thrust coefficient with the initial track tension coefficient for the three vehicle configurations at 20% slip on the Hope Valley snow

As mentioned previously, with the increase of the initial track tension, the load supported by the belly decreases and a higher proportion of the vehicle weight is applied to the track. Since the Hope Valley snow has a significant frictional component in its shear strength, it leads to the increase of thrust with the increase of the initial track tension. Furthermore, with a tighter track, track segments between roadwheels flatten out. Consequently, the horizontal component of the shear stress on the track–terrain interface becomes larger. This would also enhance the development of higher thrust by the track. Figure 9.10 shows the variations of the thrust coefficient (i.e. the ratio of the thrust (tractive effort) developed by the track to the vehicle weight) with the initial track tension coefficient for the three configurations. It is seen that the thrust coefficient increases with the increase of the initial track tension coefficient or the number of roadwheels. Table 9.5 gives the values of the thrust coefficient at the initial track tension coefficients of 2.5, 10, 20 and 40% for the three configurations at 20% slip. It shows that for Vehicle A, increasing the initial track tension coefficient from 2.5 to 40%, the thrust coefficient increases from 18.24 to 27.44%, representing an increase of 50.4%. For Vehicle A (6 W), the variation of thrust coefficient with the initial track tension coefficient shows a similar trend to that for Vehicle A. For Vehicle A (8 W), increasing the initial track tension coefficient over the same range, the thrust coefficient increases from 19.83 to 30.18%, representing an increase of 52.2%. In comparison with the thrust coefficient of 18.24% for Vehicle A at the initial track tension coefficient of 2.5%, the thrust coefficient of 30.18% for Vehicle A (8 W) at the initial track tension coefficient of 40% represents an increase of 65.5%.

As discussed in Section 6.2, the drawbar pull coefficient (i.e. the ratio of the drawbar pull to the vehicle weight, or the difference between the thrust coefficient and the total motion resistance coefficient on a level ground) at 20% slip is a widely accepted parameter for characterizing

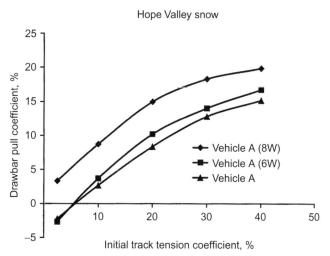

Figure 9.11: Variations of the drawbar pull coefficient with the initial track tension coefficient for the three vehicle configurations at 20% slip on the Hope Valley snow

the performance of off-road vehicles. It represents not only the capability of the vehicle to pull (or push) implements or working machinery, but also its ability to negotiate grades or to accelerate. For instance, if the drawbar pull coefficient obtained on a level ground is 20%, then it indicates that the vehicle will be able to climb a maximum grade of 20% or it will be able to accelerate at a rate of $0.2\,g$ (or 20% of the acceleration due to gravity). Figure 9.11 shows the variations of the drawbar pull coefficient with the initial track tension coefficient for the three configurations. It is seen that the drawbar pull coefficient increases with the increase of the initial track tension coefficient or the number of roadwheels. It should be pointed out that for Vehicle A and Vehicle A (6 W), if the initial track tension coefficient is less than 5%, then the two vehicles at 20% slip cannot propel themselves. In other words, the drawbar pull coefficient is negative. However, for Vehicle A (8 W), even if the initial track tension coefficient is as low as 2.5%, the vehicle is still mobile and is able to develop positive drawbar pull. Table 9.5 gives the values of the drawbar pull coefficient at the initial track tension coefficients of 2.5, 10, 20 and 40% for the three configurations at 20% slip. It shows that for Vehicle A, increasing the initial track tension coefficient from 10 to 40%, the drawbar pull coefficient increases from 2.69 to 15.23%, representing an increase of 466.2%. For Vehicle A (6 W), the variation of drawbar pull coefficient with the initial track tension coefficient shows a similar trend to that for Vehicle A. For Vehicle A (8 W), increasing the initial track tension coefficient over the same range, the drawbar pull coefficient increases from 8.72 to 19.89%, representing an increase of 128.1%. In comparison with the drawbar pull coefficient of 2.69% for Vehicle A at the initial track tension coefficient of 10%, the drawbar pull coefficient of 19.89% for Vehicle A (8 W) at the initial track tension coefficient of 40% represents an increase of 639.4%.

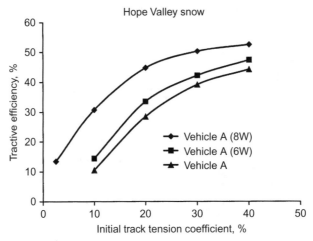

Figure 9.12: Variations of the tractive efficiency with the initial track tension coefficient for the three vehicle configurations at 20% slip on the Hope Valley snow

As noted in Section 6.2, the tractive (drawbar) efficiency (i.e. the ratio of drawbar power to the power input to the drive sprocket of a tracked vehicle, as defined by Eqn (6.15)) at 20% slip is another widely used parameter for characterizing the performance of off-road vehicles. It represents the efficiency in converting the power at the sprockets to the power at the drawbar pitch for performing productive work. Figure 9.12 shows the variations of the tractive efficiency with the initial track tension coefficient for the three configurations. It is seen that the tractive efficiency increases with the increase of the initial track tension coefficient or the number of roadwheels. It should be noted that for Vehicle A and Vehicle A (6W), the drawbar pull at 20% slip is negative when the initial track tension coefficient is below 5%. Accordingly, the tractive efficiency is not shown in the figure. Table 9.5 gives the values of the tractive efficiency at various initial track tension coefficients for the three configurations at 20% slip. It shows that for Vehicle A, increasing the initial track tension coefficient from 10 to 40%, the tractive efficiency increases from 10.58 to 44.41%, representing an increase of 319.8%. For Vehicle A (6W), the variation of tractive efficiency with the initial track tension coefficient shows a similar trend to that for Vehicle A. For Vehicle A (8W), increasing the initial track tension coefficient over the same range, the tractive efficiency increases from 30.78 to 52.72%, representing an increase of 71.3%. In comparison with the tractive efficiency of 10.58% for Vehicle A at the initial track tension coefficient of 10%, the tractive efficiency of 52.72% for Vehicle A (8W) at the initial track tension coefficient of 40% represents an increase of 398.3%.

B. *On clayey soil*

As shown in Table 9.3, the values of the pressure–sinkage parameters of the clayey soil, such as k_c and k_ϕ, are higher than those for the Hope Valley snow. This indicates that it is firmer and less compressible than the Hope Valley snow.

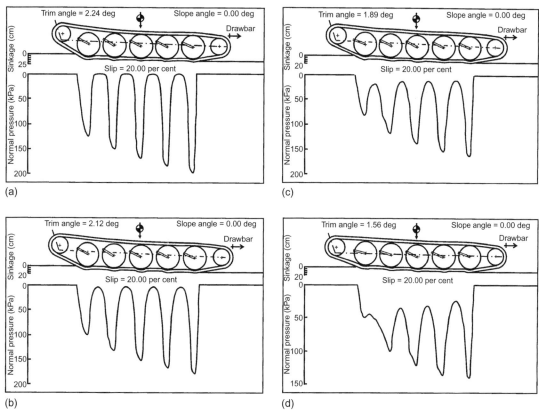

Figure 9.13: Predicted normal pressure distributions under the track of Vehicle A at initial track tension coefficients of (a) 2.5%, (b) 10%, (c) 20% and (d) 40% at 20% slip on the clayey soil (Reprinted by permission of *Int. J. Heavy Vehicle Systems* from Wong and Huang, 2008, copyright © Inderscience Enterprises Ltd)

Figures 9.13, 9.14 and 9.15 show the normal pressure distributions and sinkages, on the clayey soil, under the tracks of Vehicle A, Vehicle A (6W), and Vehicle A (8W), respectively, at initial track tension coefficients of 2.5, 10, 20 and 40% and at 20% slip. They were predicted using NTVPM (Wong and Huang, 2008). It can be seen that the track sinkages of all three configurations are lower than the respective ground clearances. Consequently, the vehicle bellies do not come into contact with the soil surface, as shown in the figures. Under these circumstances, the belly trim angle no longer has any direct effects on performance. However, similar to operating on the Hope Valley snow, for all three configurations, the pressure on the track segments between roadwheels increases, as the initial track tension coefficient increases. This reduces the peak pressures under the track, leading to lower track sinkage and track motion resistance, hence improved performance. It is also noted that as the number of roadwheels increases, the uniformity of pressure distributions under the tracks improves, particularly for the track system configuration of Vehicle A (8W) with eight overlapping roadwheels.

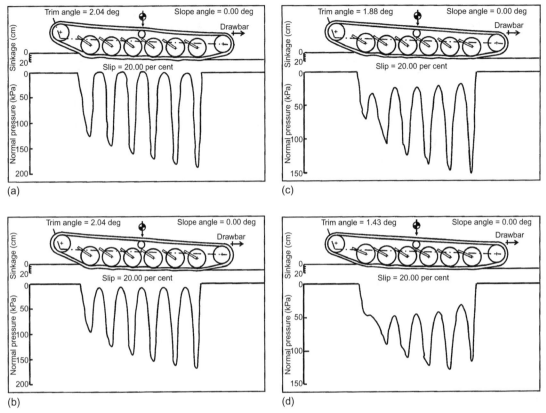

Figure 9.14: Predicted normal pressure distributions under the track of Vehicle A (6W) at initial track tension coefficients of (a) 2.5%, (b) 10%, (c) 20% and (d) 40% at 20% slip on the clayey soil (Reprinted by permission of *Int. J. Heavy Vehicle Systems* from Wong and Huang, 2008, copyright © Inderscience Enterprises Ltd)

As the track sinkage is less than the ground clearance, the vehicle weight is entirely supported by the track for all three configurations. The variations of the mean maximum pressure (MMP) with the initial track tension coefficient for the three configurations are shown in Figure 9.16. The values of the MMP shown in the figure are the mean values of the peak pressures under the tracks predicted by NTVPM. It is seen that the value of MMP decreases with the increase of the initial track tension coefficient or the number of roadwheels. Table 9.6 gives the values of the MMP at the initial track tension coefficients of 2.5, 10, 20 and 40% for the three configurations at 20% slip. It shows that for Vehicle A, increasing the initial track tension coefficient from 2.5 to 40%, the MMP decreases from 165.8 to 110.1 kPa, representing a reduction of 33.6%. For Vehicle A (6W), the variation of the MMP with the initial track tension coefficient shows similar trend to that for Vehicle A. For Vehicle A (8W), increasing the initial track tension coefficient over the same range, the MMP decreases from 125.3 to 89.1 kPa, representing a reduction of 28.9%. In comparison with the MMP of 165.8 kPa for Vehicle A

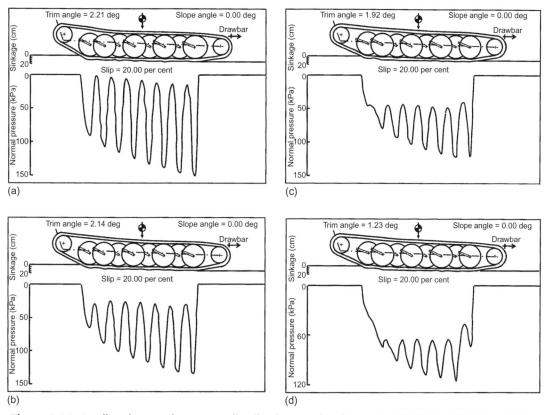

Figure 9.15: Predicted normal pressure distributions under the track of Vehicle A (8 W) at initial track tension coefficients of (a) 2.5%, (b) 10%, (c) 20% and (d) 40% at 20% slip on the clayey soil (Reprinted by permission of *Int. J. Heavy Vehicle Systems* from Wong and Huang, 2008, copyright © Inderscience Enterprises Ltd)

at the initial track tension coefficient of 2.5%, the MMP of 89.1 kPa for Vehicle A (8 W) at the initial track tension coefficient of 40% represents a reduction of 46.3%.

The track sinkage is related to the normal pressure exerted on the track, which in turn is indicated by the MMP. Table 9.6 gives the values of the front roadwheel sinkage at the initial track tension coefficients of 2.5, 10, 20 and 40% for the three configurations at 20% slip. It shows that for Vehicle A, increasing the initial track tension coefficient from 2.5 to 40%, the front roadwheel sinkage decreases from 12.7 to 5.1 cm, representing a reduction of 59.8%. For Vehicle A (6 W), the variation of front roadwheel sinkage with the initial track tension coefficient shows a similar trend to that for Vehicle A. For Vehicle A (8 W), increasing the initial track tension coefficient over the same range, the front roadwheel sinkage decreases from 9.1 to 3.3 cm, representing a reduction of 70.3%. In comparison with the front roadwheel sinkage of 12.7 cm for Vehicle A at the initial track tension coefficient of 2.5%, the front roadwheel sinkage of 3.3 cm for Vehicle A (8 W) at the initial track tension coefficient

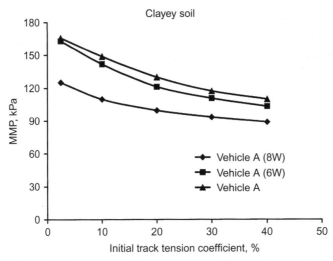

Figure 9.16: Variations of the mean maximum pressure (MMP) with the initial track tension coefficient for the three vehicle configurations at 20% slip on the clayey soil

of 40% represents a reduction of 74%. Table 9.6 also gives the values of the rear roadwheel sinkage at the initial track tension coefficients of 2.5, 10, 20 and 40% for the three configurations at 20% slip. It shows that for Vehicle A, increasing the initial track tension coefficient from 2.5 to 40%, the rear roadwheel sinkage decreases from 20.1 to 14.1 cm, representing a reduction of 30%. For Vehicle A (6 W), the variation of rear roadwheel sinkage with the initial track tension coefficient shows similar trend to that for Vehicle A. For Vehicle A (8 W), increasing the initial track tension coefficient over the same range, the rear roadwheel sinkage decreases from 15.1 to 11.2 cm, representing a reduction of 25.8%. In comparison with the rear roadwheel sinkage of 20.1 cm for Vehicle A at the initial track tension coefficient of 2.5%, the rear roadwheel sinkage of 11.2 cm for Vehicle A (8 W) at the initial track tension coefficient of 40% represents a reduction of 44.5%.

As noted earlier, the track motion resistance due to track–terrain interaction is related to track sinkage. Figure 9.17 shows the variations of the track motion resistance coefficient with the initial track tension coefficient for the three configurations. It is seen that the track motion resistance coefficient decreases with the increase of the initial track tension coefficient or the number of roadwheels. Table 9.6 gives the values of the track motion resistance coefficient at the initial track tension coefficients of 2.5, 10, 20 and 40% for the three configurations at 20% slip. It shows that for Vehicle A, increasing the initial track tension coefficient from 2.5 to 40%, the track motion resistance coefficient decreases from 10.09 to 4.46%, representing a reduction of 55.8%. For Vehicle A (6 W), the variation of track motion resistance coefficient with the initial track tension coefficient shows a similar trend to that for Vehicle A. For Vehicle A (8 W), increasing the initial track tension coefficient over the same range, the

Table 9.6: Effects of track system configuration and initial track tension on performance parameters at 20% slip on clayey soil

Vehicle configuration	Initial track tension coeff. %	Belly load coeff. %	Belly drag coeff. %	Track motion resistance coeff. %	Total motion resistance coeff. %	Thrust coeff. %	Drawbar pull coeff. %	Tractive efficiency %	Front roadwheel sinkage (cm)	Rear roadwheel sinkage (cm)	MMP (kPa)
Vehicle A (baseline vehicle)	2.5	0	0	10.09	10.09	17.16	7.07	32.95	12.7	20.1	165.8
	10	0	0	8.25	8.25	17.65	9.40	42.58	10.5	18.4	148.8
	20	0	0	6.32	6.32	18.71	12.39	52.97	8.1	16.3	130.2
	30	0	0	5.16	5.16	19.58	14.42	58.91	6.2	15.0	117.6
	40	0	0	4.46	4.46	20.16	15.70	62.29	5.1	14.1	110.1
Vehicle A (6W)	2.5	0	0	9.12	9.12	16.44	7.32	35.62	12.7	19.1	162.9
	10	0	0	7.10	7.10	17.18	10.08	46.93	9.8	17.1	141.9
	20	0	0	5.17	5.17	18.71	13.54	57.87	7.1	14.9	121.5
	30	0	0	4.22	4.22	19.66	15.44	62.81	5.5	13.7	110.9
	40	0	0	3.76	3.76	20.27	16.51	65.15	4.7	12.9	103.5
Vehicle A (8W)	2.5	0	0	5.31	5.31	16.51	11.20	54.26	9.1	15.1	125.3
	10	0	0	4.12	4.12	18.30	14.18	61.99	6.6	13.5	110.0
	20	0	0	3.31	3.31	19.73	16.42	66.59	4.7	12.4	99.8
	30	0	0	3.02	3.02	20.47	17.45	68.18	3.7	11.8	93.7
	40	0	0	2.83	2.83	20.99	18.16	69.23	3.3	11.2	89.1

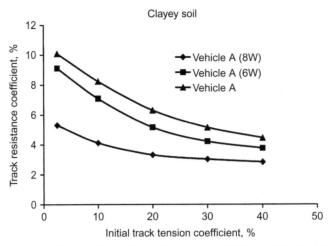

Figure 9.17: Variations of the track motion resistance coefficient with the initial track tension coefficient for the three vehicle configurations at 20% slip on the clayey soil

track motion resistance coefficient decreases from 5.31 to 2.83%, representing a reduction of 46.7%. In comparison with the track motion resistance coefficient of 10.09% for Vehicle A at the initial track tension coefficient of 2.5%, the track motion resistance coefficient of 2.83% for Vehicle A (8 W) at the initial track tension coefficient of 40% represents a reduction of 71.9%. As noted previously, on the clayey soil, the vehicle belly is not in contact with the terrain surface and there is no belly drag. As a result, the total motion resistance coefficient is equal to the track motion resistance coefficient.

Figure 9.18 shows the variations of the thrust coefficient with the initial track tension coefficient for the three configurations. It is seen that the thrust coefficient increases with the increase of the initial track tension coefficient or the number of roadwheels. Table 9.6 gives the values of the thrust coefficient at the initial track tension coefficients of 2.5, 10, 20 and 40% for the three configurations at 20% slip. It shows that for Vehicle A, increasing the initial track tension coefficient from 2.5 to 40%, the thrust coefficient increases from 17.16 to 20.16%, representing an increase of 17.5%. For Vehicle A (6 W), the variation of thrust coefficient with the initial track tension coefficient shows a similar trend to that for Vehicle A. For Vehicle A (8 W), increasing the initial track tension coefficient over the same range, the thrust coefficient increases from 16.51 to 20.99%, representing an increase of 27.1%. In comparison with the thrust coefficient of 17.16% for Vehicle A at the initial track tension coefficient of 2.5%, the thrust coefficient of 20.99% for Vehicle A (8 W) at the initial track tension coefficient of 40% represents an increase of 22.3%.

Figure 9.19 shows the variations of the drawbar pull coefficient, which is the difference between the thrust coefficient and the track motion resistance coefficient, with the initial

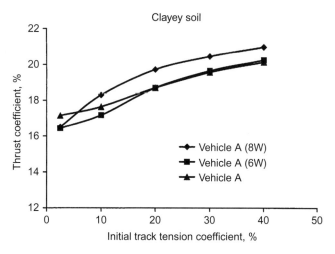

Figure 9.18: Variations of the thrust coefficient with the initial track tension coefficient for the three vehicle configurations at 20% slip on the clayey soil

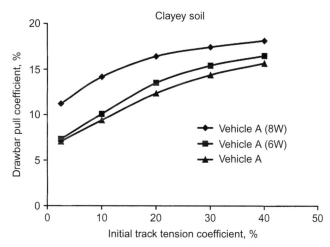

Figure 9.19: Variations of the drawbar pull coefficient with the initial track tension coefficient for the three vehicle configurations at 20% slip on the clayey soil

track tension coefficient for the three configurations. It is seen that the drawbar pull coefficient increases with the increase of the initial track tension coefficient or the number of roadwheels. Table 9.6 gives the values of the drawbar pull coefficient at the initial track tension coefficients of 2.5, 10, 20 and 40% for the three configurations at 20% slip. It shows that for Vehicle A, increasing the initial track tension coefficient from 2.5 to 40%, the drawbar pull coefficient increases from 7.07 to 15.70%, representing an increase of 122.1%. For Vehicle A (6 W), the variation of drawbar pull coefficient with the initial track tension coefficient shows a similar trend to that for Vehicle A. For Vehicle A (8 W), increasing the initial track tension

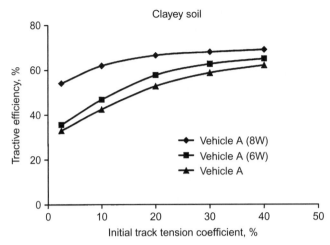

Figure 9.20: Variations of the tractive efficiency with the initial track tension coefficient for the three vehicle configurations at 20% slip on the clayey soil

coefficient over the same range, the drawbar pull coefficient increases from 11.2 to 18.16%, representing an increase of 62.1%. In comparison with the drawbar pull coefficient of 7.07% for Vehicle A at the initial track tension coefficient of 2.5%, the drawbar pull coefficient of 18.16% for Vehicle A (8 W) at the initial track tension coefficient of 40% represents an increase of 156.9%.

Figure 9.20 shows the variations of the tractive efficiency with the initial track tension coefficient for the three configurations. It is seen that the tractive efficiency increases with the increase of the initial track tension coefficient or the number of roadwheels. Table 9.6 gives the values of the tractive efficiency at various initial track tension coefficients for the three configurations at 20% slip. It shows that for Vehicle A, increasing the initial track tension coefficient from 2.5 to 40%, the tractive efficiency increases from 32.95 to 62.29%, representing an increase of 89%. For Vehicle A (6W), the variation of tractive efficiency with the initial track tension coefficient shows a similar trend to that for Vehicle A. For Vehicle A (8 W), increasing the initial track tension coefficient over the same range, the tractive efficiency increases from 54.26 to 69.23%, representing an increase of 27.6%. In comparison with the tractive efficiency of 32.95% for Vehicle A at the initial track tension coefficient of 2.5%, the tractive efficiency of 69.23% for Vehicle A (8 W) at the initial track tension coefficient of 40% represents an increase of 110.4%.

C. Summary

Based on the results described above, it can be seen that on both the Hope Valley snow and the clayey soil, the track system configuration has significant effects on vehicle performance.

Among the three configurations examined, the track system configuration of Vehicle (8 W) with eight overlapping roadwheels has the highest performance on both types of terrain. The performance of the configuration of Vehicle (6 W) with six roadwheels falls between those of the other two configurations. The prime reason for the configuration with eight overlapping roadwheels to have the highest performance is that it has the lowest peak pressures under the track among the three configurations. On terrains, such as the clayey soil, where the track sinkage is lower than the vehicle ground clearance, lowering the peak pressures under the track with larger number of roadwheels reduces the track sinkage and track motion resistance, leading to improved performance. On highly compressible terrain, such as the Hope Valley snow, where the vehicle belly is either partially or fully in contact the terrain surface, increasing the number of roadwheels reduces the sinkages of both the track and the belly, leading to lower track motion resistance and belly drag. This indicates that increasing the number of roadwheels has a more significant impact on vehicle performance on softer terrain than on firmer terrain.

This physical understanding of the effects of the number of roadwheels given above is well borne out by the results presented earlier. For instance, on the clayey soil, the drawbar pull coefficient at 20% slip for Vehicle A (8 W) is approximately 50.9% higher than that for Vehicle A at the initial track tension coefficient of 10% (see Table 9.6), whereas over the Hope Valley snow, the corresponding improvement is approximately 224.2% (see Table 9.5).

When the track sinkage is less than the vehicle ground clearance, as is the case for the three configurations operating on the clayey soil, the major effect of increasing the initial track tension is to increase the load supported by the track segments between roadwheels. This leads to the reduction in the peak normal pressures and the mean maximum pressure (MMP) on the track–terrain interface, hence lower track sinkage and track motion resistance. The increase of the initial track tension also enhances the development of vehicle thrust. This is because when the initial track tension increases, the track becomes tighter and the track segments between roadwheels flatten out. As a result, the horizontal component of the shear stress on the track–terrain interface becomes larger, leading to the increase of vehicle thrust.

On highly compressible terrain, such as the Hope Valley snow, where the vehicle belly is partially or fully in contact with the terrain surface, increasing the initial track tension has two major effects. First, it reduces the sinkages of both the track and the belly, hence lowering the track motion resistance and the belly drag. Second, it reduces the load supported by the belly, hence increasing the load applied to the track. This enables the track to develop higher thrust on terrain with a significant frictional component in its shear strength. This indicates that on highly compressible terrain where the vehicle belly contacts the terrain surface, increasing the initial track tension has even more significant effects on improving vehicle performance.

This physical understanding of the effects of the initial track tension on performance given above is well borne out by the results presented earlier. For instance, for Vehicle A on the clayey soil, increasing the initial track tension coefficient from 10 to 40%, the drawbar pull coefficient at 20% slip increases by approximately 67% (see Table 9.6), whereas over the Hope Valley snow, the corresponding improvement is approximately 466.2% (see Table 9.5).

It should be mentioned that the effects of initial track tension on performance is dependent on the design features of the vehicle. For instance, for agricultural or industrial tractors with friction drive between the sprocket and the rubber belt (band) track, to minimize slippage between the sprocket and the track while transmitting the driving torque (or force), the initial track tension coefficient under normal operating conditions is already set at a high level. Under these circumstances, further increasing the initial track tension coefficient would not necessarily have any significant impact on the improvement of vehicle performance, as its effects may have already reached a plateau.

In many cases, while increasing the initial track tension improves vehicle performance, particularly on highly compressible terrain, it may increase the wear and tear of the track system. As a practical measure to accommodate these conflicting factors, it is suggested that a central initial track tension regulating system controlled by the driver be developed (Wong, 1986a, b, 1992a, 1994b, 1995, 1997, 1999, 2007; Wong and Preston-Thomas, 1998; Wong and Huang, 2005, 2008). This remotely controlled regulating system enables the driver to conveniently increase the initial track tension, when traversing soft terrain is anticipated. It also enables the driver to reduce the initial track tension when the vehicle operates on firm terrain to minimize the wear and tear of the track system due to high initial track tension. The central initial track tension regulating system for improving tracked vehicle mobility is analogous to the central tyre inflation system for improving the mobility of off-road wheeled vehicles. It appears that in many cases retrofitting existing tracked vehicles with the central initial track tension regulating system may be one of the most cost-effective means to enhance their soft ground mobility, as it is simply an add-on device. In most cases, its installation requires only relatively minor modifications to existing vehicles.

For vehicles designed for operation over soft terrain for an extended period of time, a case may be made for the development of an automatic initial track tension regulating system controlled by the slip of the track, in order to further improve its effectiveness (Wong and Huang, 2006a, 2008; Wong, 2007).

For the cases studied above, if increasing the number of roadwheels from five to eight is coupled with the increase of the initial track tension from 10 to 40%, then on the Hope Valley snow the drawbar pull coefficient and tractive efficiency will increase from 2.69 to 19.89% and from 10.58 to 52.72%, respectively, representing an increase in the drawbar pull coefficient of 639.4% and in the tractive efficiency of 398.3% (see Table 9.5). On the clayey soil, if increasing the number of roadwheels from five to eight is coupled with the increase of the initial track tension from 10 to 40%, then the drawbar pull coefficient and tractive efficiency will increase from 9.4 to 18.16% and from 42.58 to 69.23%, respectively, representing an increase in the drawbar pull coefficient of 93.2% and in the tractive efficiency of 62.6%. This indicates that by a combination of increasing the initial track tension and the number of roadwheels, significant improvements in performance can be achieved, particularly on highly compressible terrain.

While both the track system configuration and the initial track tension have noticeable effects on performance on both the Hope Valley snow and the clayey soil examined in this study, their effects are more significant on highly compressible terrain, such as the Hope Valley snow, than on firmer terrain, such as the clayey soil.

9.2 Effects of Suspension Setting

On highly compressible terrain, such as the Hope Valley snow, when the belly comes into contact with the terrain surface, the attitude of the vehicle belly may have noticeable effects on performance. As discussed in the preceding section and illustrated in Figure 9.2, if the belly takes a nose-down attitude in operation, then most parts of the belly (except the front part) will not be in contact with the terrain surface. On the other hand, if the belly takes a nose-up attitude, then the belly comes into contact, in part or in full, with the terrain surface and supports part of the vehicle weight. This reduces the load on the track, hence vehicle thrust on terrain having a significant frictional component in its shear strength. Also, this introduces a substantial belly drag. This indicates that if the belly is in contact with the terrain surface and takes a nose-up attitude, it will have significant adverse effects on vehicle performance. Vehicle belly attitude is related to suspension setting. In this section, the effects on performance of suspension settings are examined.

The performances of Vehicle A with three different suspension settings were simulated using NTVPM, at an initial track tension coefficient of 10% on the Hope Valley snow and on the clayey soil. The three suspension settings differ in the torsion arm angles of roadwheels at the free positions (i.e. at the free position, the torsion bar is not subject to any torque, or the load applied at the end of the torsion arm is zero). The torsion arm angle is measured from the horizontal in the clockwise direction, as indicated in Figure 9.1. The torsion arm angles of the five roadwheels at the free positions for the three suspension settings, referred to as the Standard, S2 and S3, are shown in Table 9.7. The Standard setting is the original setting of Vehicle A shown in Table 9.2, in which the torsion arms of all roadwheels at free positions are set at the same angle of 43°. This results in a nose-down attitude for the vehicle belly with a negative trim angle of $-0.66°$, when the vehicle is stationary on level, hard ground. In suspension setting S2, the torsion arm angles at free positions are set in a descending order from 51.6° at the front roadwheel to 34.4° at the rear roadwheel, while maintaining an angle of 43° for the torsion arm at the middle (or the third) roadwheel. This results in a nose-up attitude for the belly with a positive trim angle of $+1.33°$, when the vehicle is stationary on

Table 9.7: Various suspension settings for Vehicle A: Standard, S2 and S3

Roadwheel	Torsion arm angle at free position* (degrees)		
	Standard suspension setting	Suspension setting S2**	Suspension setting S3**
1	43	51.6	34.4
2	43	47.3	38.7
3	43	43	43
4	43	38.7	47.3
5	43	34.4	51.6

*Torsion arm angle under no load condition.
**Other suspension parameters are the same as those of Vehicle A shown in Table 9.2.

level, hard ground. For the suspension setting S3, the torsion arm angles at free positions are set in an ascending order from 34.4° at the front roadwheel to 51.6° at the rear roadwheel, while maintaining an angle of 43° for the torsion arm at the middle roadwheel. This results in a nose-down attitude for the belly with a negative trim angle of $-2.67°$, when the vehicle is stationary on level, hard ground. It should be noted that when the vehicle operates on deformable terrain with drawbar load applied, the attitude of the vehicle belly and its trim angle may be quite different from that when the vehicle is stationary on level, hard ground. It should also be mentioned that the suspension settings described above were selected for the purpose of providing a sufficiently wide variety for the study of the effects of suspension setting on performance. Some of the settings selected for this study are not necessarily recommended for use in practice.

It should be noted that suspension settings affect not only the attitude of the vehicle belly, but also the load distribution on the roadwheels, hence the normal pressure distribution on the track–terrain interface. In many cases, this also affects vehicle performance.

A. *On Hope Valley snow*

Table 9.8 shows that both the front and rear roadwheel sinkages, hence track sinkages for the three suspension settings at 20% slip, are greater than the ground clearance on the Hope Valley snow. Consequently, the vehicle belly is fully in contact with the terrain. It also shows that the roadwheel sinkages, particularly the rear roadwheel sinkages, for the three settings vary in a narrow range. Figure 9.21 shows the variations of the belly trim angle with slip for the three suspension settings. It is seen that operating on the Hope Valley snow, the belly trim angles for the three suspension settings are all positive, indicating a nose-up attitude, although when the vehicle is stationary on level, hard ground, the belly trim angles for the Standard

Table 9.8: Effects of suspension setting on performance parameters of Vehicle A with initial track tension coefficient of 10% at 20% slip on Hope Valley snow

Terrain	Suspension setting	Belly load coeff. %	Belly drag coeff. %	Track motion resistance coeff. %	Total motion resistance coeff. %	Thrust coeff. %	Drawbar pull coeff. %	Tractive efficiency %	Front roadwheel sinkage (cm)	Rear roadwheel sinkage (cm)	Belly trim angle (degrees)
Hope Valley snow	Standard	24.02	5.26	12.42	17.68	20.37	2.69	10.56	45.6	59.7	2.35
	S2	29.42	6.67	12.36	19.03	18.53	−0.50	–	48.9	59.5	2.78
	S3	19.95	4.20	12.82	17.02	21.86	4.84	17.71	42.7	60.4	1.90

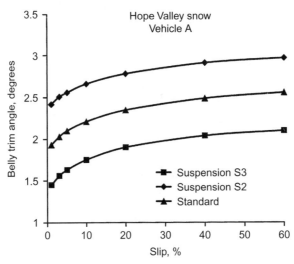

Figure 9.21: Variations of the belly trim angle with slip for the three suspension settings of Vehicle A on the Hope Valley snow

setting and setting S3 are negative, as noted previously. As shown in Table 9.8 and Figure 9.21, at 20% slip, the trim angle for the setting S3 is 1.9°, followed by 2.35° for the Standard setting and 2.78° for the setting S2.

Table 9.8 gives the values of the performance parameters of the vehicle with the three suspension settings at the initial track tension coefficient of 10% and at 20% slip. It can be seen that the belly load coefficient for the setting S2 is 29.42% whereas that for the setting S3 is 19.95%, representing a reduction of 32.2%. As noted previously, belly drag is related to the belly trim angle. Since the belly is in full contact with the snow and the trim angle for the setting S3 is the lowest, its belly drag coefficient is the lowest among the three settings. The belly drag coefficient for the setting S2 is 6.67%, whereas that for the setting S3 is 4.20%, representing a reduction of 37%.

The values of the track motion resistance coefficient for the three suspension settings vary in a narrow range. However, the setting S2 has a slightly lower value than those of the other two. This is attributable to the setting S2 having the highest value of belly load coefficient, which results in the lowest load applied to the track, hence the lowest track motion resistance coefficient. The total motion resistance coefficient, which is the sum of the belly drag coefficient and the track motion resistance coefficient, of the setting S2 is 19.03%, whereas that of the setting S3 is 17.02%, representing a reduction of 10.6%. The thrust coefficient for the setting S2 is 18.53%, whereas that for the setting S3 is 21.83%, representing an increase of 18%. This is attributable to a higher proportion of the vehicle weight being applied to the track for the setting S3 than that for the setting S2.

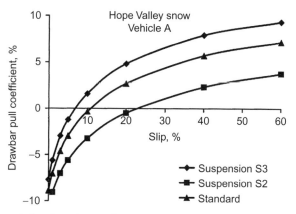

Figure 9.22: Variations of the drawbar pull coefficient with slip for the three suspension settings of Vehicle A on the Hope Valley snow

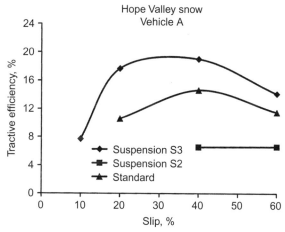

Figure 9.23: Variations of the tractive efficiency with slip for the three suspension settings of Vehicle A on the Hope Valley snow

As a result of higher thrust coefficient and lower total motion resistance coefficient for the setting S3, its drawbar performance is the highest among the three settings, as shown in Figure 9.22. It is interesting to note that with the suspension setting S2, the vehicle is unable to propel itself at slip less than approximately 23%. The drawbar pull coefficient at 20% slip for the setting S3 is 4.84%, whereas that for the setting S2 is -0.5% indicating that it is unable to propel itself, as shown in Table 9.8. For the Standard suspension setting, the drawbar pull coefficient at 20% slip is 2.69 %. This indicates that the drawbar pull coefficient at 20% slip for the setting S3 is 80% higher than that for the Standard setting at the same slip. Figure 9.23 shows the variations of tractive efficiency with slip for Vehicle A with the three suspension settings. It shows that the tractive efficiency for the setting S3 is highest among the three settings. Table 9.8 indicates that the tractive efficiency at 20% slip for the setting S3 is 17.71%, which is 67.5% higher than that for the Standard setting.

The results presented above indicate that when the vehicle belly is in full contact with the terrain surface, the suspension setting, hence the vehicle belly trim angle, has significant effects on vehicle mobility. Among the three suspension settings examined, the setting S3 provides the vehicle with the highest performance. This is primarily attributable to its belly trim angle being the lowest when operating on the Hope Valley snow. This leads to the lowest values of the belly load coefficient, belly drag coefficient, and the total motion resistance coefficient, as well as the highest thrust coefficient, drawbar pull coefficient, and tractive efficiency for the suspension setting S3, among the three settings studied.

B. On clayey soil

Operating on the clayey soil, the track sinkage for the three suspension settings is less than the vehicle ground clearance and the vehicle belly is not in contact with the terrain surface. Consequently, the belly trim angle no longer has any direct effect on vehicle performance. The major effect of suspension setting is on the load distribution among the roadwheels, hence the pressure distribution on the track–terrain interface. Figure 9.24 shows the variations of the mean maximum pressure (MMP) with slip for the three suspension settings on the clayey soil. The values of the MMP shown are based on the pressure distribution on the track–terrain interface predicted by NTVPM. As noted previously, the value of MMP is an indication of the uniformity of the pressure distribution on the track–terrain interface. It can be seen that the value of the MMP for suspension setting S2 is lowest among the three settings over the slip range shown, whereas that for the setting S3 is the highest. Table 9.9

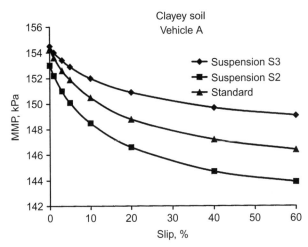

Figure 9.24: Variations of the mean maximum pressure (MMP) with slip for the three suspension settings of Vehicle A on the clayey soil

Table 9.9: Effects of suspension setting on performance parameters of Vehicle A with initial track tension coefficient of 10% at 20% slip on clayey soil

Terrain	Suspension setting	Belly load coeff. %	Belly drag coeff. %	Track motion resistance coeff. %	Total motion resistance coeff. %	Thrust coeff. %	Drawbar pull coeff. %	Tractive efficiency %	Front roadwheel sinkage (cm)	Rear roadwheel sinkage (cm)	MMP (kPA)
Clayey soil	Standard	0	0	8.25	8.25	17.65	9.40	42.60	10.5	18.4	148.8
	S2	0	0	7.98	7.98	17.88	9.90	44.29	10.2	18.1	146.6
	S3	0	0	8.42	8.42	17.50	9.08	41.50	10.8	18.5	150.9

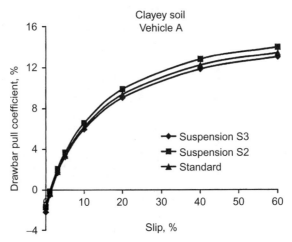

Figure 9.25: Variations of the drawbar pull coefficient with slip for the three suspension settings of Vehicle A on the clayey soil

gives the values of performance parameters of the vehicle with the three suspension settings with the initial track tension coefficient of 10% and at 20% slip on the clayey soil. It shows that their values vary in a narrow range, although the value of the track motion resistance coefficient for the setting S2 is slightly lower than those of the other two settings, and that the values of the thrust coefficient, drawbar pull coefficient, and tractive efficiency for the setting S2 are slightly higher than those of the other two settings. Figure 9.25 shows the variations of the drawbar pull coefficient with slip for the three suspension settings. It can be seen that over the range of slip shown, the drawbar pull coefficient for the setting S2 is slightly higher than that for the Standard setting, and in turn the drawbar pull coefficient for the Standard setting is slightly higher than that for the setting S3. At 20% slip, the drawbar pull coefficient of 9.9% for the setting S2 is 5.3% higher than that for the Standard setting and is 9% higher than that for the setting S3. Figure 9.26 shows the variations of the tractive efficiency with slip for the three suspension settings. It can be seen that over the range of slip shown, the tractive efficiency for the setting S2 is slightly higher than that for the Standard setting, and in turn the tractive efficiency for the Standard setting is slightly higher than that for the setting S3. At 20% slip, the tractive efficiency of 44.29% for the setting S2 is 4% higher than that for the Standard setting and is 6.7% higher than that for the setting S3.

C. Summary

The results presented above indicate that when the track sinkage is less than the vehicle ground clearance and the vehicle belly is not in contact with the terrain surface, the suspension setting affects the load distribution among the roadwheels, hence the pressure distribution on the track–terrain interface. In this case, the suspension setting has only a slight impact on performance. On the clayey soil, as shown in Table 9.9, among the three suspension settings examined, the

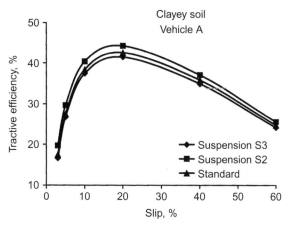

Figure 9.26: Variations of the tractive efficiency with slip for the three suspension settings of Vehicle A on the clayey soil

setting S2 provides the vehicle with slightly better performance than the Standard setting, and in turn the Standard setting provides slightly better performance than the setting S3.

On highly compressible terrain, such as the Hope Valley snow, where track sinkage exceeds vehicle ground clearance and the vehicle belly comes into contact with the terrain surface, the suspension setting affects not only the load distribution among the roadwheels, but also the belly trim angle. As the vehicle belly trim angle has significant effects on belly drag and vehicle traction, suspension setting has more significant effects on vehicle performance on softer terrain than on firmer terrain, as indicated in Tables 9.8 and 9.9.

It is interesting to point out that on the Hope Valley snow the suspension setting S3 provides the vehicle with much higher performance than the setting S2, whereas on the clayey soil, the suspension setting S2 provides slightly better performance than the setting S3. This indicates that the optimal suspension setting varies with terrain conditions. In view of this, it is suggested that for vehicles designed for operation over soft terrain for an extended period of time, a case may be made for the development of a suspension setting regulating system, so as to enable the driver to change suspension settings according to terrain conditions. For vehicles with hydropneumatic suspensions, it appears that a suspension setting regulating system controlled by the driver may be implemented with relative ease.

9.3 Effects of Longitudinal Location of the Centre of Gravity

The longitudinal location of the centre of gravity (CG) generally affects the load distribution among the roadwheels, hence roadwheel sinkages and belly trim angle.

To examine the effects of the longitudinal location of the CG, the performances of Vehicle A, Vehicle A (6 W) and Vehicle A (8 W) with five different longitudinal locations of the CG and

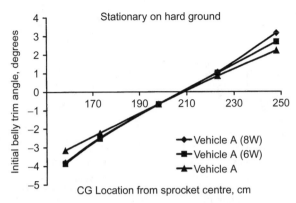

Figure 9.27: Variations of the belly trim angle with the longitudinal location of the centre of gravity for the three vehicle configurations at rest on level, hard ground

with the initial track tension coefficient of 10% were simulated using NTVPM on the Hope Valley snow and on the clayey soil. The five longitudinal CG locations are at distances of 158, 173, 198, 223 and 248 cm from the centre of the sprocket, which is located at the front of the vehicle, as shown in Figure 9.1(a). The five CG locations are equivalent to the CG at 44.5, 29.5 and 4.5 cm ahead and 20.5 and 45.5 cm behind the midpoint of the track contact length, respectively, when the vehicle is stationary on level, hard ground. It should be noted that the midpoint of the track contact length is located at a distance of 202.5 cm in the longitudinal direction from the centre of the sprocket. Figure 9.27 shows the variations of the initial belly trim angle with the longitudinal location of the CG for the three vehicle configurations with the initial track tension coefficient of 10% on level, hard ground. It shows that belly trim angles of the three vehicle configurations increase from negative (i.e. nose-down attitude) to positive (i.e. nose-up attitude) with the rearward shift of the CG.

A. On Hope Valley snow

Table 9.10 shows that on the Hope Valley snow, the vehicle bellies for the three configurations at 20% slip are either partially or fully in contact with the snow surface. For instance, for Vehicle A with the longitudinal location of the CG at a distance less than 198 cm from the centre of the sprocket, the track sinkages under both the front and rear roadwheels are greater than the ground clearance, hence its belly is in full contact with the snow surface. On the other hand, for Vehicle A (8 W) with the longitudinal location of the CG at a distance greater than 198 cm from the centre of the sprocket, the track sinkage under the front roadwheel is less than the ground clearance, while that under the rear roadwheel is greater than the ground clearance, hence the belly is in partial contact with the snow surface. With the vehicle belly in contact with the terrain, it supports part of the vehicle weight, while the remainder is applied to the track. Figure 9.28 shows the belly trim angles for the three configurations at various longitudinal locations of the CG on the Hope Valley snow. It shows that the belly trim angles

Table 9.10: Effects of longitudinal CG location on performance parameters of Vehicle A, Vehicle A (6W) and Vehicle A (8W), with initial track tension coefficient of 10% at 20% slip on Hope Valley snow

Vehicle configuration	Longitudinal CG location from sprocket centre (cm)	Belly load coeff. %	Belly drag coeff. %	Track motion resistance coeff. %	Total motion resistance coeff. %	Thrust coeff. %	Draw bar pull coeff. %	Tractive efficiency %	Front roadwheel sinkage (cm)	Rear roadwheel sinkage (cm)	Belly trim angle (degrees)
Vehicle A (baseline vehicle)	158	26.88	5.09	12.07	17.16	19.69	2.53	10.31	49.9	58.9	0.38
	173	25.38	5.01	12.12	17.13	20.01	2.88	11.52	48.4	59.0	1.04
	198	24.02	5.26	12.42	17.68	20.37	2.69	10.58	45.6	59.7	2.35
	223	24.61	6.30	12.71	19.03	20.48	1.47	5.76	40.3	60.3	4.35
	248	26.73	8.48	12.69	21.17	20.67	−0.50	–	29.3	60.4	7.59
Vehicle A (6W)	158	24.01	4.55	11.71	16.26	20.32	4.06	15.98	47.8	58.1	0.46
	173	22.77	4.51	11.70	16.21	20.36	4.15	16.30	46.6	58.3	1.12
	198	21.83	4.83	11.97	16.80	20.51	3.71	14.45	43.9	58.8	2.48
	223	22.69	5.94	12.12	18.06	20.57	2.51	9.74	38.2	59.2	4.67
	248	24.89	8.22	11.90	20.12	21.36	1.24	4.65	25.9	59.0	8.27
Vehicle A (8W)	158	15.97	3.02	10.48	13.50	22.55	9.05	32.10	44.5	55.7	0.59
	173	15.02	3.01	10.45	13.46	22.53	9.07	32.21	43.2	55.8	1.33
	198	14.97	3.49	10.44	13.93	22.65	8.72	30.78	39.1	55.8	3.14
	223	17.28	4.95	10.33	15.28	23.26	7.98	27.47	31.0	55.7	5.98
	248	21.59	7.59	10.37	17.96	24.35	6.39	21.00	19.8	56.1	9.37

for the three vehicle configurations on the Hope Valley snow all change to positive, indicating that the belly takes a nose-up attitude. It also shows that for a given vehicle configuration, the belly trim angle increases with the rearward shift of the CG. For the same CG location, the belly trim angle for Vehicle A (8 W) is slightly higher than that for Vehicle (6 W), and in turn the belly trim angle for Vehicle A (6 W) is slightly higher than that for Vehicle A.

Table 9.10 shows the variations of the vehicle belly load coefficient with the longitudinal location of the CG for the three vehicle configurations. For Vehicle A (8 W), the front and rear roadwheel sinkages are the lowest, and its belly load coefficient is, therefore, also the lowest for the same CG location, among the three vehicle configurations examined. It also shows that for a given vehicle configuration, shifting the CG location rearward, the belly load coefficient first decreases and reaches a minimum at an intermediate CG location. Further rearward shifting the CG results in an increase in the belly load coefficient. The variations of the belly drag coefficient with the longitudinal location of the CG for the three vehicle configurations are also shown in the table. The belly drag coefficient for Vehicle A (8 W) is the lowest for the same CG location, among the three vehicle configurations examined. It also shows that for a given vehicle configuration, shifting the CG location rearward, the belly drag coefficient first decreases slightly and reaches a minimum at an intermediate location. Further rearward shifting the CG results in an increase of the belly drag coefficient. Table 9.10 shows that when the longitudinal location of the CG is at a distance of 173 cm from the centre of the sprocket (i.e. the CG is 29.5 cm ahead of the midpoint of the track contact length), the values of belly drag coefficient for the three vehicle configurations reach their respective minimum. The minimum values of the belly drag coefficients for Vehicle A, Vehicle A (6 W) and Vehicle A (8 W) are 5.01, 4.51 and 3.01%, respectively. This indicates that the minimum belly drag coefficient for Vehicle A (8 W) is 39.9% lower than that for Vehicle A and is 33.3% lower than that for Vehicle A (6 W). As noted previously, the belly drag coefficient is a function of the belly sinkage, which is related to the front and rear roadwheel sinkages, and of the belly trim angle.

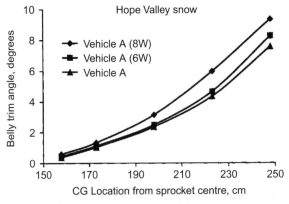

Figure 9.28: Variations of the belly trim angle with the longitudinal location of the centre of gravity for the three vehicle configurations at 20% slip on the Hope Valley snow

The effects of increasing the number of roadwheels on the reduction of both the front and rear roadwheel sinkages, hence the belly sinkage, outweigh that on the slight increase of the belly trim angle. As a result, for a given longitudinal location of the CG, the belly drag coefficient decreases with the increase of the number of roadwheels.

The variations of the track motion resistance coefficient with the longitudinal location of the CG for the three vehicle configurations are shown in Table 9.10. It shows that the track motion resistance coefficient varies in a narrow range with the rearward shift of the CG location for each of the three vehicle configurations. For the same CG location, the track motion resistance coefficient for Vehicle A (8 W) is the lowest, among the three vehicle configurations. The table shows the variations of the total motion resistance coefficient, which is the sum of the belly drag coefficient and the track motion resistance coefficient, with the longitudinal location of the CG for the three vehicle configurations. As shown in Table 9.10, when the longitudinal location of the CG is at a distance of 173 cm from the centre of the sprocket (i.e. the CG is 29.5 cm ahead of the midpoint of the track contact length), the values of the total motion resistance coefficient for the three vehicle configurations reach their respective minimum. The minimum values of the total motion resistance coefficients for Vehicle A, Vehicle A (6 W) and Vehicle A (8 W) are 17.13, 16.21 and 13.46%, respectively. This indicates that the minimum total motion resistance coefficient for Vehicle A (8 W) is 21.4% lower than that for Vehicle A and is 17% lower than that for Vehicle A (6 W).

The variations of the thrust coefficient with the longitudinal location of the CG for the three vehicle configurations are given in Table 9.10. It shows that the thrust coefficients increase slightly with the rearward shift of the CG location. The values of the thrust coefficients for Vehicle A and Vehicle A (6 W) for the same CG location are quite close, whereas that for Vehicle A (8 W) is noticeably higher. Figure 9.29 shows the variations of the drawbar pull coefficient, which is the difference between the thrust coefficient and the total motion

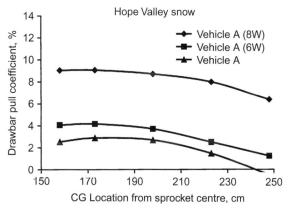

Figure 9.29: Variations of the drawbar pull coefficient with the longitudinal location of the centre of gravity for the three vehicle configurations at 20% slip on the Hope Valley snow

resistance coefficient, with the longitudinal location of the CG for the three vehicle configurations. Figure 9.29 and Table 9.10 show that when the longitudinal location of the CG is at a distance of 173 cm from the centre of the sprocket (i.e. the CG is 29.5 cm ahead of the midpoint of the track contact length), the values of the drawbar pull coefficient for the three vehicle configurations reach their respective maximum. The peak values of the drawbar pull coefficients for Vehicle A, Vehicle A (6W) and Vehicle A (8W) are 2.88, 4.15 and 9.07%, respectively. This indicates that the peak value of the drawbar pull coefficient for Vehicle A (8W) is 214.9% higher than that for Vehicle A and is 118.6% higher than that for Vehicle A (6W). Further rearward shifting of the CG location from a distance of 173 cm from the centre of the sprocket causes deterioration in performance. For Vehicle A, when the CG location is at a distance more than approximately 240 cm from the centre of the sprocket (i.e. the CG is more than 37.5 cm behind the midpoint of the track contact length), the drawbar pull coefficient at 20% slip is negative, which indicates that the vehicle is immobile. Figure 9.30 shows the variations of the tractive efficiency with the longitudinal location of the CG for the three vehicle configurations. Figure 9.30 and Table 9.10 show that when the longitudinal location of the CG is at a distance of 173 cm from the centre of the sprocket (i.e. the CG is 27.5 cm ahead of the midpoint of the track contact length), the values of the tractive efficiency for the three vehicle configurations reach their respective maximum. The peak values of the tractive efficiency for Vehicle A, Vehicle A (6W) and Vehicle A (8W) are 11.52, 16.30 and 32.21%, respectively. This indicates that the peak value of the tractive efficiency for Vehicle A (8W) is 179.6% higher than that for Vehicle A and is 97.6% higher than that for Vehicle A (6W).

B. *On clayey soil*

Operating on the clayey soil, the track sinkage for the three vehicle configurations with different longitudinal locations of the CG is less than the vehicle ground clearance and the vehicle belly is

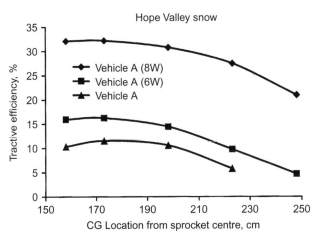

Figure 9.30: Variations of the tractive efficiency with the longitudinal location of the centre of gravity for the three vehicle configurations at 20% slip on the Hope Valley snow

not in contact with the terrain surface. Consequently, the belly trim angle no longer has any direct effect on vehicle performance. The major effect of the longitudinal location of the CG is on the load distribution among the roadwheels, hence the pressure distribution on the track–terrain interface. Figure 9.31 shows the variations of the mean maximum pressure (MMP) with the longitudinal location of the CG for the three vehicle configurations at 20% slip on the clayey soil. As noted previously, the value of MMP is an indication of the uniformity of the pressure distribution on the track–terrain interface. It is seen that at the same CG location, the value of the MMP for Vehicle A (8W) is lowest among the three configurations, whereas that for Vehicle A is the highest.

The track sinkage is related to the normal pressure exerted on the track, which is related to the MMP. Table 9.11 gives the values of the front roadwheel sinkage at various longitudinal locations of the CG for the three configurations. It shows that for Vehicle A, shifting the CG from 158 to 248 cm measured from the centre of the sprocket, the front roadwheel sinkage decreases from 16.6 to 1.4 cm. For Vehicle A (6W), the variation of front roadwheel sinkage with the rearward shifting of the CG location shows a similar trend to that for Vehicle A. For Vehicle A (8W), shifting the CG over the same range, the front roadwheel sinkage decreases from 11.9 to −1.5 cm, which indicates that the front roadwheel is off the ground. Comparing the front roadwheel sinkage of 10.5 cm for Vehicle A when the longitudinal location of the CG is at a distance of 198 cm from the centre of the sprocket, with that of 6.6 cm for Vehicle A (8W) at the same CG location, the reduction is 37.1%. Table 9.11 also gives the values of the rear roadwheel sinkage at various longitudinal locations of the CG for the three configurations. It shows that for Vehicle A, shifting the CG location from 158 to 248 cm measured from the centre of the sprocket, the rear roadwheel sinkage increases from 18.9 to 20.4 cm. For Vehicle A (6W), the variation of rear roadwheel sinkage with the longitudinal location of the CG shows

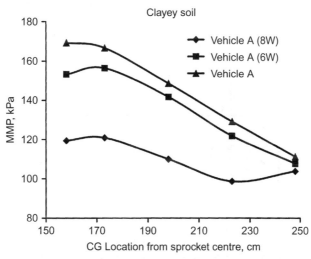

Figure 9.31: Variations of the mean maximum pressure (MMP) with the longitudinal location of the centre of gravity for the three vehicle configurations at 20% slip on the clayey soil

Table 9.11: Effects of longitudinal CG location on performance parameters of Vehicle A, Vehicle A (6W) and Vehicle A (8W) with initial track tension coefficient of 10% at 20% slip on clayey soil

Vehicle configuration	Longitudinal CG location from sprocket centre (cm)	Belly load coeff. %	Belly drag coeff. %	Track motion resistance coeff. %	Total motion resistance coeff. %	Thrust coeff. %	Drawbar pull coeff. %	Tractive efficiency %	Front roadwheel sinkage (cm)	Rear roadwheel sinkage (cm)	MMP (kPa)
Vehicle A (baseline vehicle)	158	0	0	9.50	9.50	17.56	8.06	36.75	16.6	18.9	169.3
	173	0	0	8.45	8.45	17.28	8.83	40.87	14.5	18.3	166.7
	198	0	0	8.25	8.25	17.65	9.40	42.58	10.5	18.4	148.8
	223	0	0	9.14	9.14	18.50	9.36	40.48	5.4	19.5	129.3
	248	0	0	9.78	9.78	19.51	9.73	39.90	1.4	20.4	111.3
Vehicle A (6W)	158	0	0	8.86	8.86	17.29	8.43	39.03	15.6	18.0	153.3
	173	0	0	7.74	7.74	16.87	9.13	43.29	13.9	17.3	156.5
	198	0	0	7.10	7.10	17.18	10.08	46.93	9.8	17.1	141.9
	223	0	0	7.95	7.95	18.27	10.32	45.18	4.5	18.4	121.9
	248	0	0	8.33	8.33	19.16	10.83	45.23	0.78	19.0	107.8
Vehicle A (8W)	158	0	0	5.05	5.05	17.39	12.34	56.79	11.9	13.7	119.4
	173	0	0	4.22	4.22	17.40	13.18	60.59	10.4	13.0	120.9
	198	0	0	4.12	4.12	18.30	14.18	61.99	6.6	13.6	110.0
	223	0	0	5.33	5.33	19.23	13.90	57.80	2.2	15.6	98.8
	248	0	0	5.76	5.76	19.73	13.97	56.65	−1.5	16.4	103.9

similar trend to that for Vehicle A. For Vehicle A (8 W), shifting the CG location over the same range, the rear roadwheel sinkage increases from 13.7 to 16.4 cm. Comparing the rear roadwheel sinkage of 18.4 cm for Vehicle A when the longitudinal location of the CG is at a distance of 198 cm from the centre of the sprocket, with that of 13.6 cm at the same CG location for Vehicle A (8 W), the reduction is 26.1%.

The track motion resistance is related to roadwheel sinkage. The variations of the track motion resistance coefficient with the longitudinal location of the CG for the three vehicle configurations are shown in Table 9.11. It shows that the track motion resistance coefficient varies in a narrow range with rearward shift of the CG location for each of the three vehicle configurations. The track motion resistance for Vehicle A (8 W) is the lowest for the same CG location, among the three vehicle configurations examined. As shown in the table, when the longitudinal location of the CG is at a distance of 198 cm from the centre of the sprocket (i.e. the CG is 4.5 cm ahead of the midpoint of the track contact length), the values of the track motion resistance coefficient for the three vehicle configurations reach their respective minimum. The minimum values of the track motion resistance coefficients for Vehicle A, Vehicle A (6 W) and Vehicle A (8 W) are 8.25, 7.10 and 4.12%, respectively. This indicates that the minimum track motion resistance coefficient for Vehicle A (8 W) is 50.1% lower than that for Vehicle A and is 42% lower than that for Vehicle A (6 W).

Table 9.11 shows the variations of the thrust coefficient with the longitudinal location of the CG for the three vehicle configurations. It shows that the values of the thrust coefficient increase slightly with rearward shift of the CG. The value of the thrust coefficient for Vehicle A (8 W) is the highest for the same CG location, among the three vehicle configurations examined, while that for Vehicle A is slightly higher than that for Vehicle A (6 W). Figure 9.32 shows the variations of the drawbar pull coefficient, which is the difference between the thrust coefficient and the track motion resistance coefficient, with the longitudinal location of the CG for the three vehicle configurations. It shows that the values of the drawbar pull coefficient for Vehicle A and Vehicle A (6 W) are quite close for the same CG location, while that for Vehicle A (8 W) is noticeably higher than those for the other two configurations. Figure 9.32 and Table 9.11 show that when the longitudinal location of the CG is at a distance of 198 cm from the centre of the sprocket (i.e. the CG is 4.5 cm ahead of the midpoint of the track contact length), the drawbar pull coefficient for Vehicle A (8 W) reaches a maximum, while for Vehicle A and Vehicle A (6 W), the peak values of the drawbar coefficient appear to occur at the CG location at a distance of 248 cm from the centre of the sprocket (i.e. the CG is 45.5 cm behind the midpoint of the track contact length). The peak values of the drawbar pull coefficient for Vehicle A, Vehicle A (6 W) and Vehicle A (8 W) are 9.73, 10.83 and 14.18%, respectively, over the range of the CG locations examined. This indicates that the peak value of the drawbar pull coefficient for Vehicle A (8 W) is 45.7% higher than that for Vehicle A and is 30.9% higher than that for Vehicle A (6 W). Figure 9.33 shows the variations of the tractive efficiency with the longitudinal location of the CG for the three vehicle configurations.

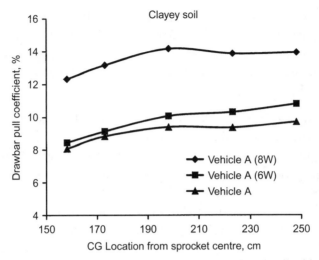

Figure 9.32: Variations of the drawbar pull coefficient with the longitudinal location of the centre of gravity for the three vehicle configurations at 20% slip on the clayey soil

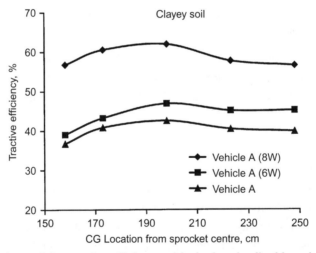

Figure 9.33: Variations of the tractive efficiency with the longitudinal location of the centre of gravity for the three vehicle configurations at 20% slip on the clayey soil

Figure 9.33 and Table 9.11 show that when the longitudinal location of the CG is at a distance of 198 cm from the centre of the sprocket (i.e. the CG is 4.5 cm ahead of the midpoint of the track contact length), the values of the tractive efficiency for the three vehicle configurations reach their respective maximum. The peak values of the tractive efficiency for Vehicle A, Vehicle A (6 W) and Vehicle A (8 W) are 42.58, 46.93 and 61.99%, respectively. This indicates that the peak value of the tractive efficiency for Vehicle A (8 W) is 45.6% higher than that for Vehicle A and is 32.1% higher than that for Vehicle A (6 W).

C. Summary

The results presented above indicate that when the track sinkage is less than the vehicle ground clearance and the vehicle belly is not in contact with the terrain surface, the longitudinal location of the CG affects the load distribution among the roadwheels, hence roadwheel sinkages. On the claycy soil, for Vehicle A (8 W), when the longitudinal location of the CG is at a distance of 198 cm from the centre of the sprocket (i.e. the CG is 4.5 cm ahead of the midpoint of the track contact length), the drawbar pull coefficient reaches a maximum, while for Vehicle A and Vehicle A (6 W), the peak values of the drawbar coefficient appear to occur at the CG location at a distance more than 198 cm from the centre of the sprocket. The values of the tractive efficiency reach their respective peaks for Vehicle A, Vehicle A (6 W) and Vehicle A (8 W), when the CG location is at a distance of 198 cm from the centre of the sprocket (i.e. the CG is 4.5 cm ahead of the midpoint of the track contact length). The peak value of the tractive efficiency for Vehicle A (8 W) is 45.6% higher than that for Vehicle A and is 32.1% higher than that for Vehicle A (6 W).

On highly compressible terrain, such as the Hope Valley snow, where the vehicle belly comes into contact with the terrain surface, the longitudinal location of the CG affects not only the load distribution among the roadwheels, but also vehicle belly–terrain interaction. On the Hope Valley snow, the values of the drawbar pull coefficient and tractive efficiency reach their respective peaks for Vehicle A, Vehicle A (6 W) and Vehicle A (8 W), when the CG location is at a distance of 173 cm from the centre of the sprocket (i.e. the CG is 29.5 cm ahead of the midpoint of the track contact length). The peak value of the drawbar pull coefficient for Vehicle A (8 W) is 214.9% higher than that for Vehicle A and is 118.6% higher than that for Vehicle A (6 W). The peak value of the tractive efficiency for Vehicle A (8 W) is 179.6% higher than that for Vehicle A and is 97.6% higher than that for Vehicle A (6 W).

In conclusion, the longitudinal location of the centre of gravity affects vehicle performance. The optimal longitudinal location of the CG for a tracked vehicle, where the drawbar pull coefficient or the tractive efficiency reaches its peak, primarily varies with the terrain on which it operates.

9.4 Effects of Vehicle Total Weight

It is generally recognized that the mobility of an off-road vehicle is sensitive to its weight, particularly on highly compressible terrain. To quantitatively examine the effects of the vehicle total weight (i.e. the sum of the sprung and unsprung weights of a vehicle), the performances of Vehicle A, Vehicle A (6 W) and Vehicle A (8 W) with five total weights of 90.57, 110.57, 130.57, 150.57 and 170.57 kN were simulated using NTVPM on the Hope Valley snow and on the clayey soil. To evaluate the effects of vehicle weight on the performances of the three vehicle configurations on a common basis, the initial track tension of 10 kN and the longitudinal location of the CG at a distance of 198 cm from the centre of the sprocket are kept the same for all three configurations. The roadwheel–suspension settings for the three configurations are given in Table 9.2.

A. On Hope Valley snow

Table 9.12 shows that on the Hope Valley snow, the vehicle bellies for the three configurations with different vehicle total weights at 20% slip are either partially or fully in contact with the snow surface. For instance, for Vehicle A with the five vehicle weights, the track sinkages under both the front and rear roadwheels are greater than the ground clearance, hence its belly is in full contact with the snow surface. On the other hand, for Vehicle A (8 W) with the weight less than 130.57 kN, the track sinkage under the front roadwheel is less than the ground clearance, while that under the rear roadwheel is greater than the ground clearance, hence the belly is in partial contact with the snow surface. With the vehicle belly in contact with the terrain, it supports part of the vehicle weight, while the remainder is applied to the track. Table 9.12 shows the belly trim angles for the three configurations at various weights on the Hope Valley snow. It shows that the belly trim angles for the three vehicle configurations at different weights vary in a fairly narrow range.

Table 9.12 shows the variations of the vehicle belly load coefficient with the vehicle total weight for the three vehicle configurations. It shows that for a given vehicle configuration, the belly load coefficient increases with the increase of the vehicle weight. For instance, for Vehicle A increasing the weight from 90.57 to 170.57 kN, the belly load coefficient increases from 16.96 to 37.36%, representing an increase of 120.3%. For Vehicle A (8 W), increasing the weight over the same range, the belly load coefficient increases from 9.07 to 28.98%, representing an increase of 219.5%. It should be pointed out that for the same weight the belly load coefficient for Vehicle A (8 W) is lower than those for Vehicle A and Vehicle A (6 W). For instance, at the weight of 170.57 kN, the belly load coefficients for Vehicle A, Vehicle A (6 W) and Vehicle A (8 W) are 37.36%, 34.75 and 28.98%, respectively. This indicates that the belly load coefficient for Vehicle A (8 W) is 22.4% lower than that for Vehicle A and is 16.6% lower than that for Vehicle A (6 W).

The variations of the belly drag coefficient with the vehicle total weight for the three vehicle configurations are given in Table 9.12. It shows that for a given vehicle configuration, the belly drag coefficient increases with the increase of the vehicle weight. For instance, for Vehicle A increasing the weight from 90.57 to 170.57 kN, the belly drag coefficient increases from 3.83 to 8.16%, representing an increase of 113.1%. For Vehicle A (8 W), increasing the weight over the same range, the belly drag coefficient increases from 2.26 to 6.44%, representing an increase of 185%. It should be pointed out that for the same weight the belly drag coefficient for Vehicle A (8 W) is lower than those for Vehicle A and Vehicle A (6 W). For instance, at the weight of 170.57 kN, the belly drag coefficients for Vehicle A, Vehicle A (6 W) and Vehicle A (8 W) are 8.16, 7.62 and 6.44%, respectively. This indicates that the belly drag coefficient for Vehicle A (8 W) is 21.1% lower than that for Vehicle A and is 15.5% lower than that for Vehicle A (6 W).

The variations of the track motion resistance coefficient with the vehicle total weight for the three vehicle configurations are given in Table 9.12. It shows that for a given vehicle

Table 9.12: Effects of vehicle weight on performance parameters of Vehicle A, Vehicle A (6W) and Vehicle A (8W), with initial track tension of 10kN at 20% slip on Hope Valley snow

Vehicle configuration	Vehicle total weight (kN)	Belly load coeff. %	Belly drag coeff. %	Track motion resistance coeff. %	Total motion resistance coeff. %	Thrust coeff. %	Drawbar pull coeff. %	Tractive efficiency %	Front roadwheel sinkage (cm)	Rear roadwheel sinkage (cm)	Belly trim angle (degrees)
Vehicle A (baseline vehicle)	90.57	16.96	3.83	13.55	17.38	22.23	4.86	17.48	42.4	57.1	2.71
	110.57	24.02	5.26	12.42	17.68	20.37	2.69	10.58	45.6	59.7	2.35
	130.57	29.54	6.42	11.40	17.82	19.03	1.21	5.04	47.3	61.5	2.25
	150.57	33.85	7.37	10.59	17.96	17.99	0.03	0.14	48.7	63.2	2.22
	170.57	37.36	8.16	9.95	18.11	17.17	−0.94	–	50.0	64.7	2.21
Vehicle A (6W)	90.57	15.16	3.48	12.99	16.47	22.11	5.64	20.42	40.7	56.2	2.92
	110.57	21.83	4.83	11.97	16.80	20.51	3.71	14.45	43.9	58.8	2.48
	130.57	27.18	5.95	10.99	16.94	19.34	2.40	9.91	45.6	60.7	2.34
	150.57	31.35	6.86	10.17	17.03	18.47	1.44	6.26	46.7	62.2	2.31
	170.57	34.73	7.62	9.50	17.12	17.79	0.67	2.99	47.8	63.6	2.29
Vehicle A (8W)	90.57	9.07	2.26	10.75	13.01	24.58	11.57	37.66	33.7	52.3	4.00
	110.57	14.97	3.49	10.44	13.93	22.65	8.72	30.78	39.1	55.8	3.14
	130.57	20.33	4.59	9.80	14.39	21.45	7.06	26.33	41.9	58.1	2.72
	150.57	25.02	5.58	9.18	14.75	20.49	5.74	22.40	43.6	59.8	2.56
	170.57	28.98	6.44	8.65	15.08	19.66	4.58	18.64	45.0	61.3	2.50

configuration, the track motion resistance coefficient generally decreases with the increase of the vehicle weight. This is because the belly load coefficient increases with the increase of the weight, resulting in a decrease in the proportion of the vehicle weight applied to the track, hence lowering the track motion resistance coefficient. For instance, for Vehicle A increasing the weight from 90.57 to 170.57 kN, the track motion resistance coefficient decreases from 13.55 to 9.95%, representing a decrease of 26.6%. For Vehicle A (8 W), increasing the weight over the same range, the track motion resistance coefficient decreases from 10.75 to 8.65%, representing a decrease of 19.5%. It should be pointed out that for the same weight the track motion resistance coefficient for Vehicle A (8 W) is lower than those for Vehicle A and Vehicle A (6 W). For instance, at the weight of 170.57 kN, the track motion resistance coefficients for Vehicle A, Vehicle A (6 W) and Vehicle A (8 W) are 9.95, 9.50 and 8.65%, respectively. This indicates that the track motion resistance coefficient for Vehicle A (8 W) is 13.1% lower than that for Vehicle A and is 8.9% lower than that for Vehicle A (6 W).

Table 9.12 shows the variations of the total motion resistance coefficient, which is the sum of the belly drag coefficient and the track motion resistance coefficient, with the vehicle total weight for the three vehicle configurations. As shown in the table, for a given vehicle configuration, the total motion resistance coefficient generally increases with the increase of the vehicle weight. This is because the increase of the belly drag coefficient with the increase of the weight is more than the decrease in the track motion resistance coefficient. For instance, for Vehicle A increasing the weight from 90.57 to 170.57 kN, the total motion resistance coefficient increases from 17.38 to 18.11%, representing an increase of 4.2%. For Vehicle A (8 W), increasing the weight over the same range, the total motion resistance coefficient increases from 13.01 to 15.08%, representing an increase of 15.9%. It should be pointed out that for the same weight the total motion resistance coefficient for Vehicle A (8 W) is lower than those for Vehicle A and Vehicle A (6 W). For instance, at the weight of 170.57 kN, the total motion resistance coefficients for Vehicle A, Vehicle A (6 W) and Vehicle A (8 W) are 18.11, 17.12 and 15.08%, respectively. This indicates that the total motion resistance coefficient for Vehicle A (8 W) is 16.7% lower than that for Vehicle A and is 11.9% lower than that for Vehicle A (6 W).

The variations of the thrust coefficient with the vehicle total weight for the three vehicle configurations are given in Table 9.12. As shown in the table, for a given vehicle configuration, the thrust coefficient decreases noticeably with the increase of the vehicle weight. This is because when the weight increases, the belly load coefficient increases considerably. As a result, the proportion of the vehicle weight applied to the track decreases substantially. Since the Hope Valley snow has a significant frictional component in its shear strength (see Table 9.4), the decrease in the proportion of the load applied to the track causes a considerable reduction in the thrust coefficient. For instance, for Vehicle A increasing the weight from 90.57 to 170.57 kN, the thrust coefficient decreases from 22.23 to 17.17%, representing a decrease of 22.8%. For Vehicle A (8 W), increasing the weight over the same range, the thrust coefficient decreases from 24.58 to 19.66%, representing a decrease of 20%. It should be pointed out that for the same weight the thrust coefficient for Vehicle A (8 W) is considerably higher than those for Vehicle A and Vehicle A (6 W), and that those of vehicle A and Vehicle A (6 W) are quite close. For instance, at the weight of 170.57 kN, the thrust coefficients for Vehicle A, Vehicle A (6 W) and Vehicle A (8 W) are 17.17, 17.79 and 19.66%, respectively. This indicates that the thrust coefficient for Vchicle A (8 W) is 14.5% higher than that for Vehicle A and is 10.5% higher than that for Vehicle A (6 W).

Figure 9.34 shows the variations of the drawbar pull coefficient, which is the difference between the thrust coefficient and the total motion resistance coefficient, with the vehicle total weight for the three vehicle configurations. Figure 9.34 and Table 9.12 show that for a given vehicle configuration, the drawbar pull coefficient decreases noticeably with the increase of the vehicle weight. This is because when the weight increases, the thrust coefficient decreases considerably, while the total motion resistance coefficient increases. For instance, for Vehicle A increasing the weight from 90.57 to 170.57 kN, the drawbar pull coefficient decreases from 4.86 to -0.94%, which indicates the vehicle is unable to propel itself with the weight of 170.57 kN. For Vehicle A (8 W), increasing the weight over the same range, the drawbar pull coefficient decreases from 11.57 to 4.58%, representing a decrease of 60.4%. It should be pointed out that for the same weight the drawbar pull coefficient for Vehicle A (8 W) is considerably higher than those for Vehicle A and Vehicle A (6 W). For instance, at the weight of 90.57 kN, the values of the drawbar pull coefficients for Vehicle A, Vehicle A (6 W) and Vehicle A (8 W) are 4.86, 5.64 and 11.57%, respectively. This indicates that the value of the drawbar pull coefficient for Vehicle A (8 W) is 138.1% higher than that for Vehicle A and is 105.1% higher than that for Vehicle A (6 W). At the weight of 170.57 kN, the drawbar pull coefficients for Vehicle A, Vehicle A (6 W) and Vehicle A (8 W) are -0.94, 0.67 and 4.58%, respectively. This indicates that Vehicle A is immobile with the weight of 170.57 kN, while

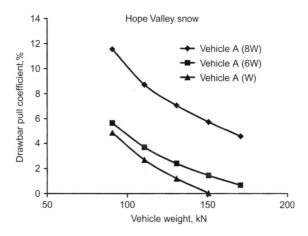

Figure 9.34: Variations of the drawbar pull coefficient with the vehicle weight for the three vehicle configurations at 20% slip on the Hope Valley snow

the drawbar pull coefficient for Vehicle A (8 W) is 583.6% higher than that for Vehicle A (6 W). This shows that the increase of the weight generally causes considerable deterioration in vehicle performance on highly compressible terrain.

Figure 9.35 shows the variations of the tractive efficiency with the vehicle total weight for the three vehicle configurations. Figure 9.35 and Table 9.12 show that for a given vehicle configuration, the tractive efficiency decreases noticeably with the increase of the vehicle weight. This is similar to the variations of the drawbar pull coefficient with the vehicle weight discussed above. For instance, for Vehicle A increasing the weight from 90.57 to 150.57 kN, the tractive efficiency decreases from 17.48 to 0.14%, representing a decrease of 99.2%. For Vehicle A (8 W), increasing the weight over the same range, the tractive efficiency decreases from 37.66 to 22.40%, representing a decrease of 40.5%. It should be pointed out that for the same weight the tractive efficiency for Vehicle A (8 W) is considerably higher than those for Vehicle A and Vehicle A (6 W). For instance, at the weight of 90.57 kN, the values of the tractive efficiency for Vehicle A, Vehicle A (6 W) and Vehicle A (8 W) are 17.48, 20.42 and 37.66%, respectively. This indicates that the value of the tractive efficiency for Vehicle A (8 W) is 115.4% higher than that for Vehicle A and is 84.4% higher than that for Vehicle A (6 W). At the weight of 150.57 kN, the values of the tractive efficiency for Vehicle A, Vehicle A (6 W) and Vehicle A (8 W) are 0.14, 6.26, and 22.4%, respectively. This indicates that the value of the tractive efficiency for Vehicle A (8 W) is 159-fold higher than that of Vehicle A and is 257.8% higher than that for Vehicle A (6 W).

B. On clayey soil

Operating on the clayey soil, the track sinkages for the three vehicle configurations with various weights are less than the vehicle ground clearance and the vehicle belly is not in contact

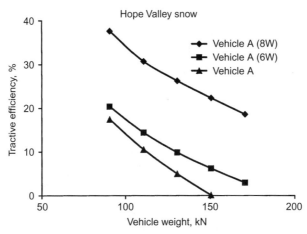

Figure 9.35: Variations of the tractive efficiency with the vehicle weight for the three vehicle configurations at 20% slip on the Hope Valley snow

with the terrain surface. Consequently, the belly trim angle no longer has any direct effect on vehicle performance. The major effect of the vehicle total weight is on the pressure exerted on the track–terrain interface. Figure 9.36 shows the variations of the mean maximum pressure (MMP) with the vehicle weight for the three configurations at 20% slip on the clayey soil. The values of the MMP shown are derived from the pressure distributions on the track–terrain interface predicted by NTVPM. As noted previously, the value of MMP is an indication of the uniformity of the pressure distribution on the track–terrain interface. It is seen that at the same weight, the value of the MMP for Vehicle A (8 W) is lowest among the three configurations, whereas that for Vehicle A is the highest. It also shows that the values of the MMP for Vehicle A and Vehicle A (6 W) are very close and that for Vehicle A (8 W) is considerably lower.

The track sinkage is related to the normal pressure exerted on the track, which is related to the MMP. Table 9.13 gives the values of the front roadwheel sinkage at various vehicle total weights for the three configurations. It shows that for Vehicle A, increasing the weight from 90.57 to 170.57 kN, the front roadwheel sinkage increases from 8.6 to 15.4 cm. For Vehicle A (6 W), the variation of front roadwheel sinkage with the weight shows a similar trend to that for Vehicle A. For Vehicle A (8 W), increasing the weight over the same range, the front roadwheel sinkage increases from 5.1 to 10.8 cm. Comparing the front roadwheel sinkage of 15.4 cm for Vehicle A with that of 10.8 cm for Vehicle A (8 W) at the same weight of 170.57 kN, it is seen that the reduction is 29.9%. Table 9.13 also gives the values of the rear roadwheel sinkage at various weights for the three configurations. It shows that for Vehicle A, increasing the weight from 90.57 to 170.57 kN, the rear roadwheel sinkage increases from 15.8 to 25 cm. For Vehicle A (6 W), the variation of rear roadwheel sinkage with the vehicle weight shows a similar trend to that for Vehicle A. For Vehicle A (8 W), increasing the weight over the same range, the rear roadwheel sinkage increases from 11.7 to 18.7 cm. Comparing

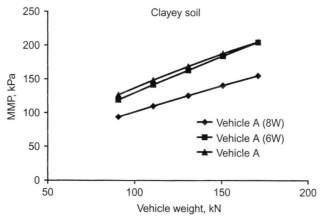

Figure 9.36: Variations of the mean maximum pressure (MMP) with the vehicle weight for the three vehicle configurations at 20% slip on the clayey soil

the rear roadwheel sinkage of 25 cm for Vehicle A with that of 18.7 cm for Vehicle A (8 W) at the same weight of 170.57 kN, it is seen that the reduction is 25.2%.

The track motion resistance is related to roadwheel sinkage. Table 9.13 shows the variations of the track motion resistance coefficient with the vehicle total weight for the three vehicle configurations. It shows that for a given vehicle configuration, the track motion resistance coefficient generally increases with the increase of the vehicle weight. For instance, for Vehicle A increasing the weight from 90.57 to 170.57 kN, the track motion resistance coefficient increases from 7.15 to 10.58%, representing an increase of 48%. For Vehicle A (8 W), increasing the weight over the same range, the track motion resistance coefficient increases from 3.53 to 5.60%, representing an increase of 58.6%. It should be pointed out that for the same weight the track motion resistance coefficient for Vehicle A (8 W) is lower than those for Vehicle A and Vehicle A (6 W). For instance, at the weight of 170.57 kN, the track motion resistance coefficients for Vehicle A, Vehicle A (6 W) and Vehicle A (8 W) are 10.58, 9.70 and 5.6%, respectively. This indicates that the track motion resistance coefficient for Vehicle A (8 W) is 47.1% lower than that for Vehicle A and is 42.3% lower than that for Vehicle A (6 W).

Table 9.13 shows the variations of the thrust coefficient with the vehicle total weight for the three vehicle configurations. As shown in the table, for a given vehicle configuration, the thrust coefficient decreases noticeably with the increase of the vehicle weight. It should be noted that the thrust developed by the shearing action of the track on the terrain is derived from the frictional and cohesive components of the shear strength of the terrain, as discussed previously. For the clayey soil, as shown in Table 9.4, it has a significant cohesive component in its shear strength. The part of the thrust derived from the frictional component of the shear strength of the terrain increases proportionally with the increase of vehicle weight. However, the part of the thrust derived from the cohesive component of the shear strength is the product of the cohesion of the terrain and the track contact area and is independent of the load applied to the track. As a result, the thrust does not increase proportionally with the increase of the vehicle weight, and the thrust coefficient decreases with the increase of vehicle weight. For instance, for Vehicle A increasing the weight from 90.57 to 170.57 kN, the thrust coefficient decreases from 18.79 to 16.13%, representing a decrease of 14.2%. For Vehicle A (8 W), increasing the weight over the same range, the thrust coefficient decreases from 19.65 to 16.29%, representing a decrease of 17.1%. It should be pointed out that for the same weight the thrust coefficient for Vehicle A (8 W) is higher than those for Vehicle A and Vehicle A (6 W), and that the thrust coefficient of Vehicle A is higher than that for Vehicle A (6 W). For instance, at the weight of 170.57 kN, the thrust coefficients for Vehicle A, Vehicle A (6 W) and Vehicle A (8 W) are 16.13%, 15.55 and 16.29%, respectively. This indicates that the thrust coefficient for Vehicle A (8 W) is approximately 1% higher than that for Vehicle A and is 4.8% higher than that for Vehicle A (6 W).

Figure 9.37 shows the variations of the drawbar pull coefficient with the vehicle total weight for the three vehicle configurations. Figure 9.37 and Table 9.13 show that for a given vehicle configuration, the drawbar pull coefficient decreases noticeably with the increase of the

Table 9.13: Effects of vehicle weight on performance parameters of Vehicle A, Vehicle A (6W) and Vehicle A (8W) with initial track tension of 10kN at 20% slip on clayey soil

Vehicle configuration	Vehicle total weight (kN)	Belly load coeff. %	Belly drag coeff. %	Track motion resistance coeff. %	Total motion resistance coeff. %	Thrust coeff. %	Drawbar pull coeff. %	Tractive efficiency %	Front roadwheel sinkage (cm)	Rear roadwheel sinkage (cm)	MMP (kPa)
Vehicle A (baseline vehicle)	90.57	0	0	7.15	7.15	18.79	11.64	49.57	8.6	15.8	126.9
	110.57	0	0	8.25	8.25	17.65	9.40	42.58	10.5	18.4	148.8
	130.57	0	0	9.19	9.19	16.92	7.73	36.56	12.3	20.7	169.3
	150.57	0	0	9.97	9.97	16.45	6.48	31.52	13.9	23.0	188.2
	170.57	0	0	10.58	10.58	16.13	5.55	27.51	15.4	25.0	205.6
Vehicle A (6W)	90.57	0	0	6.02	6.02	18.48	12.46	53.95	7.7	14.6	119.4
	110.57	0	0	7.10	7.10	17.18	10.08	46.93	9.8	17.1	141.9
	130.57	0	0	8.07	8.07	16.34	8.27	40.49	11.8	19.5	163.5
	150.57	0	0	8.94	8.94	15.82	6.88	34.80	13.8	21.3	184.8
	170.57	0	0	9.70	9.70	15.55	5.85	30.13	15.8	23.9	205.2
Vehicle A (8W)	90.57	0	0	3.53	3.53	19.65	16.12	65.63	5.1	11.7	93.8
	110.57	0	0	4.12	4.12	18.30	14.18	61.99	6.6	13.6	110.0
	130.57	0	0	4.66	4.66	17.38	12.72	58.55	8.1	15.3	125.8
	150.57	0	0	5.16	5.16	16.74	11.58	55.35	9.5	17.1	141.1
	170.57	0	0	5.60	5.60	16.29	10.69	52.51	10.8	18.7	155.8

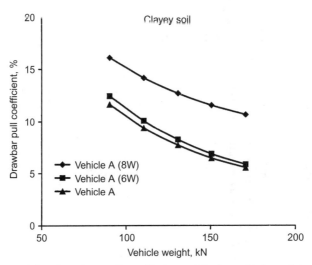

Figure 9.37: Variations of the drawbar pull coefficient with the vehicle weight for the three vehicle configurations at 20% slip on the clayey soil

vehicle weight. This is because when the weight increases, the thrust coefficient decreases, while the track motion resistance coefficient increases. For instance, for Vehicle A increasing the weight from 90.57 to 170.57 kN, the drawbar pull coefficient decreases from 11.64 to 5.55%, representing a decrease of 52.3%. For Vehicle A (8 W), increasing the weight over the same range, the drawbar pull coefficient decreases from 16.12 to 10.69%, representing a decrease of 33.7%. It should be pointed out that for the same weight the drawbar pull coefficient for Vehicle A (8 W) is considerably higher than those for Vehicle A and Vehicle A (6 W). For instance, at the weight of 90.57 kN, the values of the drawbar pull coefficients for Vehicle A, Vehicle A (6 W) and Vehicle A (8 W) are 11.64, 12.46 and 16.12%, respectively. This indicates that the value of the drawbar pull coefficient for Vehicle A (8 W) is 38.5% higher than that for Vehicle A and is 29.4% higher than that for Vehicle A (6 W). At the weight of 170.57 kN, the drawbar pull coefficients for Vehicle A, Vehicle A (6 W) and Vehicle A (8 W) are 5.55, 5.85 and 10.69%, respectively. This indicates that the drawbar pull coefficient for Vehicle A (8 W) is 92.6% higher than that for Vehicle A, and is 82.7% higher than that for Vehicle A (6 W).

Figure 9.38 shows the variations of the tractive efficiency with the vehicle total weight for the three vehicle configurations. Figure 9.38 and Table 9.13 show that for a given vehicle configuration, the tractive efficiency decreases noticeably with the increase of the vehicle weight. This is similar to the variations of the drawbar pull coefficient with the vehicle weight discussed above. For instance, for Vehicle A increasing the weight from 90.57 to 170.57 kN, the tractive efficiency decreases from 49.57 to 27.51%, representing a decrease of 44.5%. For Vehicle A (8 W), increasing the weight over the same range, the tractive efficiency decreases from 65.63 to 52.51%, representing a decrease of 20%. It should be pointed out that for the same weight the tractive efficiency for Vehicle A (8 W) is considerably higher than those for Vehicle A and

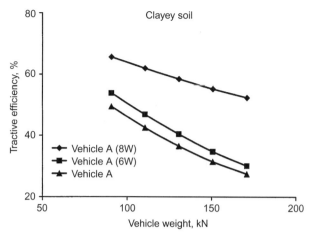

Figure 9.38: Variations of the tractive efficiency with the vehicle weight for the three vehicle configurations at 20% slip on the clayey soil

Vehicle A (6W), while those for Vehicle A and Vehicle A (6W) are quite close. For instance, at the weight of 90.57 kN, the values of the tractive efficiency for Vehicle A, Vehicle A (6W) and Vehicle A (8W) are 49.57, 53.95 and 65.63%, respectively. This indicates that the value of the tractive efficiency for Vehicle A (8W) is 32.4% higher than that for Vehicle A and is 21.6% higher than that for Vehicle A (6W). At the weight of 170.57 kN, the values of the tractive efficiency for Vehicle A, Vehicle A (6W), and Vehicle A (8W) are 27.51, 30.13 and 52.51%, respectively. This indicates that the value of the tractive efficiency for Vehicle A (8W) is 90.9% higher than that of Vehicle A and is 74.3% higher than that for Vehicle A (6W).

C. Summary

The results presented above indicate that when the track sinkage is less than the vehicle ground clearance and the vehicle belly is not in contact with the terrain surface, the vehicle weight affects the pressure exerted on the track–terrain interface, hence the track sinkage and track motion resistance. With the increase of the vehicle weight, track sinkage and track motion resistance coefficient increase. As discussed previously, vehicle thrust is developed by the shearing action of the track and is derived from the frictional and cohesive components of the shear strength of the terrain. For the clayey soil, it has a significant cohesive component in its shear strength. The part of the thrust derived from the frictional component of its shear strength increases proportionally with the increase of vehicle weight. However, the part of the thrust derived from the cohesive component of the shear strength is the product of the cohesion of the terrain and the track contact area and is independent of the load applied to the track. As a result, the thrust coefficient decreases with the increase of vehicle weight. Because of the increase of track motion resistance coefficient and the decrease of the thrust coefficient, the drawbar pull coefficient and the tractive efficiency decrease with the increase of the vehicle weight. It should be pointed out that the effects of vehicle weight on performance are related to vehicle configuration. For the same vehicle weight, the performance of Vehicle A (8 W) with eight overlapping roadwheels is the highest among the three configurations examined. For instance, at the weight of 90.57 kN, the values of the drawbar pull coefficients for Vehicle A with five roadwheels, Vehicle A (6 W) with six roadwheels and Vehicle A (8 W) with eight overlapping roadwheels are 11.64, 12.46 and 16.12%, respectively. This indicates that the value of the drawbar pull coefficient for Vehicle A (8 W) is 38.5% higher than that for Vehicle A and is 29.4% higher than that for Vehicle A (6 W). At the weight of 170.57 kN, the drawbar pull coefficients for Vehicle A, Vehicle A (6 W) and Vehicle A (8 W) are 5.55, 5.85 and 10.69%, respectively. This indicates that the drawbar pull coefficient for Vehicle A (8 W) is 92.6% higher than that for Vehicle A, and is 82.7% higher than that for Vehicle A (6 W).

On highly compressible terrain, such as the Hope Valley snow, where the vehicle belly comes into contact with the terrain surface, the vehicle weight affects not only the load exerted on the track–terrain interface, but also vehicle belly–terrain interaction. On the Hope Valley snow, the belly load coefficient and the belly drag coefficient increase with the increase of vehicle weight, while the load applied to the track and the track motion resistance coefficient decrease with the increase of vehicle weight. However, the increase in the belly drag coefficient is more than the decrease in the track motion resistance. As a result, the total motion resistance coefficient increases with the increase of vehicle weight. On the other hand, as the proportion of the load applied to the track decreases with the increase of vehicle weight, the thrust coefficient decreases accordingly. As a result, the drawbar coefficient and tractive efficiency decrease with the increase of vehicle weight for the three vehicle configurations. It should be pointed out that the effects of vehicle weight on performance are related to vehicle configuration. Among the three vehicle configurations examined, the performance of Vehicle A (8 W) is the highest for the same vehicle weight. For instance, at the weight of 90.57 kN, the values of the drawbar pull coefficients for Vehicle A, Vehicle A (6 W) and Vehicle A (8 W) are 4.86, 5.64 and 11.57%, respectively. This indicates that the value of the drawbar pull coefficient for Vehicle A (8 W) is 138.1% higher than that for Vehicle A and is 105.1% higher than that for Vehicle A (6 W). At the weight of 170.57 kN, the drawbar pull coefficients for Vehicle A, Vehicle A (6 W) and Vehicle A (8 W) are -0.94, 0.67 and 4.58%, respectively. This indicates that Vehicle A is immobile with the weight of 170.57 kN, while the drawbar pull coefficient for Vehicle A (8 W) is 583.6% higher than that for Vehicle A (6 W). This shows that the increase of the weight generally causes considerable deterioration in vehicle performance on highly compressible terrain.

9.5 Effects of Track Width

Increasing the track width increases the total track contact area. This generally results in the decrease in the normal pressure exerted on the track–terrain interface. On terrain with cohesion, the increase in track contact area also leads to the increase of vehicle thrust. This is because the component of vehicle thrust derived from the cohesion of the terrain is proportional to the total track contact area, as noted previously. To quantitatively examine the effects of track width, the performances of Vehicle A, Vehicle A (6 W) and Vehicle A (8 W) with five different track widths, 32, 38 (standard track width for Vehicle A), 44, 50 and 56 cm, were simulated using NTVPM on the Hope Valley snow and on the clayey soil. To evaluate the effects of track width on the performances of the three vehicle configurations on a common basis, the initial track tension coefficient of 10%, the longitudinal location of the CG at a distance of 198 cm from the centre of the sprocket, and the vehicle total weight of 110.57 kN are kept the same for all three configurations. It should be mentioned that in practice, increasing the track width would cause an increase of the unsprung weight. In the study presented here, this effect is, however, not taken into consideration.

A. On Hope Valley snow

Table 9.14 shows that on the Hope Valley snow, the vehicle bellies for the three configurations with different track widths at 20% slip are either partially or fully in contact with the snow surface. For instance, for Vehicle A with track width less than 50 cm, the track sinkages under both the front and rear roadwheels are greater than the ground clearance, hence its belly is in full contact with the snow surface. On the other hand, for Vehicle A (8 W) with track width greater than 32 cm, the track sinkage under the front roadwheel is less than the ground clearance, while that under the rear roadwheel is greater than the ground clearance, hence the belly is in partial contact with the snow surface. With the vehicle belly in contact with the terrain, it supports part of the vehicle weight, while the remainder is applied to the track. Table 9.14 shows the belly trim angles for the three configurations with various track widths on the Hope Valley snow. It shows that the belly trim angles for the three vehicle configurations generally increase with the increase of track width. For the same track width the belly trim angle for Vehicle A (8 W) is higher than those for the other two configurations. This is primarily due to a greater load transfer from the front to the rear for Vehicle A (8 W), as a result of its higher drawbar pull.

Table 9.14 shows the variations of the vehicle belly load coefficient with the track width for the three vehicle configurations. It shows that for a given vehicle configuration, the belly load coefficient decreases with the increase of the track width. For instance, for Vehicle A increasing the track width from 32 to 56 cm, the belly load coefficient decreases from 28.83 to 14.24%, representing a decrease of 50.6%. For Vehicle A (8 W), increasing the track width over the same range, the belly load coefficient decreases from 19.10 to 7.66%, representing a decrease of 59.9%. It should be pointed out that for the same track width the belly load coefficient for Vehicle A (8 W) is lower than those for Vehicle A and Vehicle A (6 W). For instance, at the track width of 56 cm, the belly load coefficients for Vehicle A, Vehicle A (6 W) and Vehicle A (8 W) are 14.24%, 12.76 and 7.66%, respectively. This indicates that the belly load coefficient for Vehicle A (8 W) is 46.2% lower than that for Vehicle A and is 40% lower than that for Vehicle A (6 W).

Table 9.14: Effects of track width on performance parameters of Vehicle A, Vehicle A (6W) and Vehicle A (8W), with initial track tension coefficient of 10% at 20% slip on Hope Valley snow

Vehicle configuration	Track width (cm)	Belly load coeff. %	Belly drag coeff. %	Track motion resistance coeff. %	Total motion resistance coeff. %	Thrust coeff. %	Drawbar pull coeff. %	Tractive efficiency %	Front roadwheel sinkage (cm)	Rear roadwheel sinkage (cm)	Belly trim angle (degrees)
Vehicle A (baseline vehicle)	32	28.83	6.21	11.42	17.63	19.22	1.58	6.59	47.6	61.2	2.16
	38	24.02	5.26	12.42	17.68	20.37	2.69	10.58	45.6	59.7	2.35
	44	20.08	4.47	13.24	17.71	21.35	3.64	13.63	43.6	58.2	2.55
	50	16.87	3.82	13.91	17.73	22.16	4.43	15.99	41.8	56.8	2.77
	56	14.24	3.28	14.44	17.72	22.83	5.11	17.90	40.0	55.4	2.99
Vehicle A (6W)	32	26.47	5.76	11.05	16.81	19.50	2.69	11.01	46.0	60.4	2.28
	38	21.83	4.83	11.97	16.80	20.51	3.71	14.45	43.9	58.8	2.48
	44	18.12	4.08	12.72	16.80	21.31	4.51	16.96	41.9	57.3	2.69
	50	15.14	3.47	13.31	16.78	21.97	5.19	18.88	40.0	55.8	2.91
	56	12.76	2.97	13.80	16.77	22.52	5.75	20.42	38.3	54.5	3.10
Vehicle A (8W)	32	19.10	4.33	9.85	14.18	21.91	7.73	28.23	41.9	57.8	2.76
	38	14.97	3.49	10.44	13.93	22.65	8.72	30.78	39.1	55.8	3.14
	44	11.94	2.86	10.87	13.73	22.20	9.47	32.66	36.6	54.0	3.48
	50	9.46	2.33	11.07	13.40	23.81	10.41	34.99	34.0	52.0	3.83
	56	7.66	1.92	11.26	13.18	24.19	11.01	36.41	31.9	50.4	4.09

The variations of the belly drag coefficient with the track width for the three vehicle configurations are given in Table 9.14. It shows that for a given vehicle configuration, the belly drag coefficient decreases with the increase of the track width. For instance, for Vehicle A increasing the track width from 32 to 56 cm, the belly drag coefficient decreases from 6.21 to 3.28%, representing a decrease of 47.2%. For Vehicle A (8 W), increasing the track width over the same range, the belly drag coefficient decreases from 4.33 to 1.92%, representing a decrease of 55.7%. It should be pointed out that for the same track width the belly drag coefficient for Vehicle A (8 W) is lower than those for Vehicle A and Vehicle A (6 W). For instance, at the track width of 56 cm, the belly drag coefficients for Vehicle A, Vehicle A (6 W) and Vehicle A (8 W) are 3.28, 2.97 and 1.92%, respectively. This indicates that the belly drag coefficient for Vehicle A (8 W) is 41.5% lower than that for Vehicle A and is 33.4% lower than that for Vehicle A (6 W).

Table 9.14 shows the variations of the track motion resistance coefficient with the track width for the three vehicle configurations. It shows that for a given vehicle configuration, the track motion resistance coefficient increases with the increase of the track width. It should be noted that the increase in the track width reduces the belly load coefficient, as shown in Table 9.14, and increases the proportion of load applied to the track. However, with the increase of track width, the track contact area increases more than the increase of the load on the track. As a result the normal pressure on the track decreases. Consequently, both the front and rear roadwheel sinkages decrease with the increase of the track width. As shown in Table 9.14, for Vehicle A, increasing the track width from 32 to 56 cm, representing an increase in track width of 75%, the front roadwheel sinkage decreases from 47.6 to 40 cm, representing a decrease of 16%, while the rear roadwheel sinkage decreases from 61.2 to 55.4 cm, representing a reduction of 9.5%. While increasing the track width reduces the track sinkage, the reduction in sinkage is not proportional to the increase of track width, as noted above. On the other hand, the track contact area is proportional to the track width, and as a result the total work done in compressing the terrain by the track, hence the track motion resistance coefficient increases with the increase of the track width. For instance, for Vehicle A increasing the track width from 32 to 56 cm, the track motion resistance coefficient increases from 11.42 to 14.44%, representing an increase of 26.4%. For Vehicle A (8 W), increasing the track width over the same range, the track motion resistance coefficient increases from 9.85 to 11.26%, representing an increase of 14.3%. It should be pointed out that for the same track width the track motion resistance coefficient for Vehicle A (8 W) is lower than those for Vehicle A and Vehicle A (6 W). For instance, at the track width of 56 cm, the track motion resistance coefficients for Vehicle A, Vehicle A (6 W) and Vehicle A (8 W) are 14.44, 13.80 and 11.26%, respectively. This indicates that the track motion resistance coefficient for Vehicle A (8 W) is 22% lower than that for Vehicle A and is 18.4% lower than that for Vehicle A (6 W).

The variations of the total motion resistance coefficient, which is the sum of the belly drag coefficient and the track motion resistance coefficient, with the track width for the three vehicle configurations are given in Table 9.14. As shown in the table, for Vehicle A and Vehicle A (6 W), the total motion resistance coefficient varies slightly with the track width. For Vehicle A (8 W), the total motion resistance coefficient decreases slightly with the increase of the track width. This is because the decrease of the belly drag coefficient with the increase of the track width more or less compensates for the increase in the track motion resistance coefficient. For instance, for Vehicle A increasing the track width from 32 to 56 cm, the total motion resistance coefficient increases from 17.63 to 17.72%, representing an increase of 0.5%. For Vehicle A (8 W), increasing the track width over the same range, the total motion resistance coefficient decreases from 14.18 to 13.18%, representing a decrease of 7.1%. It should be pointed out that for the same track width the total motion resistance coefficient for Vehicle A (8 W) is lower than those for Vehicle A and Vehicle A (6 W). For instance, at the track width of 56 cm, the total motion resistance coefficients for Vehicle A, Vehicle A (6 W) and Vehicle A (8 W) are 17.72, 16.77 and 13.18%, respectively. This indicates that the total motion resistance coefficient for Vehicle A (8 W) is 25.6% lower than that for Vehicle A and is 21.4% lower than that for Vehicle A (6 W).

Table 9.14 shows the variations of the thrust coefficient with the track width for the three vehicle configurations. As shown in the table, for a given vehicle configuration, the thrust coefficient increases with the increase of the track width. This is because when the track width increases, the belly load coefficient decreases. As a result, the proportion of the vehicle weight applied to the track increases. Since the Hope Valley snow has a significant frictional component in its shear strength (see Table 9.4), the increase in the proportion of the load applied to the track causes an increase in the thrust coefficient. In addition, as shown in Table 9.4, the Hope Valley snow also has some cohesion. The portion of the vehicle thrust derived from the cohesion of the terrain is proportional to the track contact area, which increases with the track width. This also contributes to the increase of the thrust coefficient with the increase of the track width. For instance, for Vehicle A increasing the track width from 32 to 56 cm, the thrust coefficient increases from 19.22 to 22.83%, representing an increase of 18.8%. For Vehicle A (8 W), increasing the track width over the same range, the thrust coefficient increases from 21.91 to 24.19%, representing an increase of 10.4%. It should be pointed out that for the same track width the thrust coefficient for Vehicle A (8 W) is higher than those for Vehicle A and Vehicle A (6 W), and that those of Vehicle A and Vehicle A (6 W) are quite close. For instance, at the track width of 56 cm, the thrust coefficients for Vehicle A, Vehicle A (6 W) and Vehicle A (8 W) are 22.83, 22.52 and 24.19%, respectively. This indicates that the thrust coefficient for Vehicle A (8 W) is 6% higher than that for Vehicle A and is 7.4% higher than that for Vehicle A (6 W).

Figure 9.39 shows the variations of the drawbar pull coefficient, which is the difference between the thrust coefficient and the total motion resistance coefficient, with the track width for the three vehicle configurations. Figure 9.39 and Table 9.14 show that for a given vehicle configuration, the drawbar pull coefficient increases with the increase of the track width. This

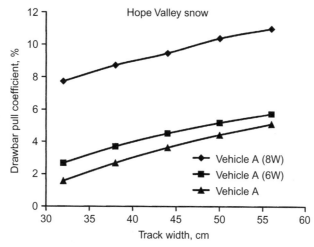

Figure 9.39: Variations of the drawbar pull coefficient with the track width for the three vehicle configurations at 20% slip on the Hope Valley snow

is because when the track width increases, the thrust coefficient increases, while the total motion resistance coefficient either remains more or less the same, in the case of Vehicle A and Vehicle A (6 W), or decreases slightly, in the case of Vehicle A (8 W). For instance, for Vehicle A increasing the track width from 32 to 56 cm, the drawbar pull coefficient increases from 1.58 to 5.11%, representing an increase of 223.4%. For Vehicle A (8 W), increasing the track width over the same range, the drawbar pull coefficient increases from 7.73 to 11.01%, representing an increase of 42.4%. It should be pointed out that for the same track width the drawbar pull coefficient for Vehicle A (8 W) is considerably higher than those for Vehicle A and Vehicle A (6 W). For instance, at the track width of 32 cm, the drawbar pull coefficients for Vehicle A, Vehicle A (6 W) and Vehicle A (8 W) are 1.58, 2.69 and 7.73%, respectively. This indicates that the drawbar pull coefficient for Vehicle A (8 W) is 389.2% higher than that for Vehicle A and is 187.4% higher than that for Vehicle A (6 W). At the track width of 56 cm, the values of the drawbar pull coefficients for Vehicle A, Vehicle A (6 W) and Vehicle A (8 W) are 5.11, 5.75 and 11.01%, respectively. This indicates that the value of the drawbar pull coefficient for Vehicle A (8 W) is 115.5% higher than that for Vehicle A and is 91.5% higher than that for Vehicle A (6 W). This shows that the increase of the track width generally improves vehicle performance on highly compressible terrain.

Figure 9.40 shows the variations of the tractive efficiency with the track width for the three vehicle configurations. Figure 9.40 and Table 9.14 show that for a given vehicle configuration, the tractive efficiency increases with the increase of the track width. This is similar to the variations of the drawbar pull coefficient with the track width discussed above. For instance, for Vehicle A increasing the track width from 32 to 56 cm, the tractive efficiency increases from 6.59 to 17.90%, representing an increase of 171.6%. For Vehicle A (8 W), increasing the track width over the same range, the tractive efficiency increases from 28.23 to 36.41%,

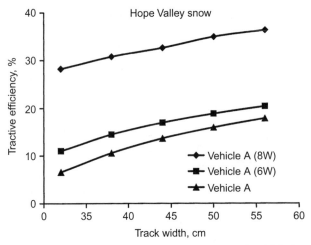

Figure 9.40: Variations of the tractive efficiency with the track width for the three vehicle configurations at 20% slip on the Hope Valley snow

representing an increase of 29%. It should be pointed out that for the same track width the tractive efficiency for Vehicle A (8 W) is considerably higher than those for Vehicle A and Vehicle A (6 W). For instance, at the track width of 32 cm, the values of the tractive efficiency for Vehicle A, Vehicle A (6 W) and Vehicle A (8 W) are 6.59, 11.01 and 28.23%, respectively. This indicates that the value of the tractive efficiency for Vehicle A (8 W) is 328.4% higher than that for Vehicle A and is 156.4% higher than that for Vehicle A (6 W). At the track width of 56 cm, the values of the tractive efficiency for Vehicle A, Vehicle A (6 W) and Vehicle A (8 W) are 17.90, 20.42 and 36.41%, respectively. This indicates that the value of the tractive efficiency for Vehicle A (8 W) is 103.4% higher than that for Vehicle A and is 78.3% higher than that for Vehicle A (6 W).

B. On clayey soil

Operating on the clayey soil, the track sinkages for the three vehicle configurations with various track widths are less than the vehicle ground clearance and the vehicle belly is not in contact with the terrain surface. Consequently, the belly trim angle no longer has any direct effect on vehicle performance. For a given track contact length, the track width determines the track contact area, which affects the normal pressure exerted on the track–terrain interface, as well as the vehicle thrust on terrain with a cohesive component in its shear strength. Figure 9.41 shows the variations of the mean maximum pressure (MMP) with the track width for the three configurations at 20% slip on the clayey soil. The values of the MMP shown are derived from the pressure distributions on the track–terrain interface predicted by NTVPM. As noted previously, the value of MMP is an indication of the uniformity of the pressure distribution on the track–terrain interface. It is seen that at the same track width the value of the MMP for Vehicle A (8 W) is lowest among the three configurations, whereas that for Vehicle A is the

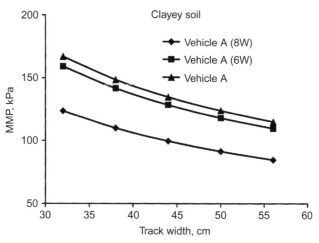

Figure 9.41: Variations of the mean maximum pressure (MMP) with the track width for the three vehicle configurations at 20% slip on the clayey soil

highest. It also shows that the values of the MMP for Vehicle A and Vehicle A (6 W) are close and that for Vehicle A (8 W) is considerably lower. For instance, at the track width of 38 cm, the values of the MMP for Vehicle A, Vehicle A (6 W) and Vehicle A (8 W) are 148.8, 141.9 and 110.0 kPa, respectively.

The track sinkage is related to the normal pressure exerted on the track, which is related to the MMP. Table 9.15 gives the values of the front roadwheel sinkage at various track widths for the three vehicle configurations. It shows that for Vehicle A, increasing the track width from 32 to 56 cm, the front roadwheel sinkage decreases from 11.6 to 8.3 cm, representing a decrease of 28.4%. For Vehicle A (6 W), the variation of front roadwheel sinkage with the track width shows a similar trend to that for Vehicle A. For Vehicle A (8 W), increasing the track width over the same range, the front roadwheel sinkage decreases from 7.4 to 5.1 cm, representing a decrease of 31.1%. Comparing the front roadwheel sinkage of 8.3 cm for Vehicle A with that of 5.1 cm for Vehicle A (8 W) at the same track width of 56 cm, it is seen that the reduction is 38.6%. Table 9.15 also gives the values of the rear roadwheel sinkage at various track widths for the three vehicle configurations. It shows that for Vehicle A, increasing the track width from 32 to 56 cm, the rear roadwheel sinkage decreases from 20.3 to 14.6 cm, representing a decrease of 28.1%. For Vehicle A (6 W), the variation of rear roadwheel sinkage with the track width shows similar trend to that for Vehicle A. For Vehicle A (8 W), increasing the track width over the same range, the rear roadwheel sinkage decreases from 15.1 to 10.8 cm, representing a decrease of 28.5%. Comparing the rear roadwheel sinkage of 14.6 cm for Vehicle A with that of 10.8 cm for Vehicle A (8 W) at the same track width of 56 cm, it is seen that the reduction is 26%.

The track motion resistance is related to roadwheel sinkage. Table 9.15 shows the variations of the track motion resistance coefficient with the track width for the three vehicle configurations.

Table 9.15: Effects of track width on performance parameters of Vehicle A, Vehicle A (6W) and Vehicle A (8W) with initial track tension coefficient of 10% at 20% slip on clayey soil

Vehicle configuration	Track width (cm)	Belly load coeff. %	Belly drag coeff. %	Track motion resistance coeff. %	Total motion resistance coeff. %	Thrust coeff. %	Drawbar pull coeff. %	Tractive efficiency %	Front roadwheel sinkage (cm)	Rear roadwheel sinkage (cm)	MMP (kPa)
Vehicle A (baseline vehicle)	32	0	0	8.83	8.83	17.02	8.19	38.49	11.6	20.3	166.8
	38	0	0	8.25	8.25	17.65	9.40	42.58	10.5	18.4	148.8
	44	0	0	7.77	7.77	18.31	10.54	46.04	9.6	16.8	135.0
	50	0	0	7.36	7.36	19.01	11.65	49.01	8.9	15.6	124.1
	56	0	0	7.01	7.01	19.73	12.72	51.57	8.3	14.6	115.1
Vehicle A (6W)	32	0	0	7.63	7.63	16.57	8.94	43.17	10.9	19.0	159.1
	38	0	0	7.10	7.10	17.18	10.08	46.93	9.8	17.1	141.9
	44	0	0	6.66	6.66	17.83	11.17	50.12	9.0	15.7	128.7
	50	0	0	6.29	6.29	18.53	12.24	52.85	8.3	14.5	118.2
	56	0	0	5.97	5.97	19.24	13.27	55.20	7.7	13.6	109.7
Vehicle A (8W)	32	0	0	4.50	4.50	17.67	13.17	59.64	7.4	15.1	123.8
	38	0	0	4.12	4.12	18.30	14.18	61.99	6.6	13.6	110.0
	44	0	0	3.82	3.82	18.97	15.15	63.89	6.0	12.4	99.6
	50	0	0	3.57	3.57	19.69	16.12	65.48	5.5	11.5	91.5
	56	0	0	3.37	3.37	20.42	17.05	66.81	5.1	10.8	84.8

It shows that for a given vehicle configuration, the track motion resistance coefficient decreases with the increase of the track width. This is because the front and rear roadwheel sinkages decrease with the increase of track width. For instance, for Vehicle A increasing the track width from 32 to 56 cm, the track motion resistance coefficient decreases from 8.83 to 7.01%, representing a decrease of 20.6%. For Vehicle A (8 W), increasing the track width over the same range, the track motion resistance coefficient decreases from 4.50 to 3.37%, representing a decrease of 25.1%. It should be pointed out that in general, the reduction in the track motion resistance coefficient is not in proportion to the increase of the track width. For the same track width, the track motion resistance coefficient for Vehicle A (8 W) is lower than those for Vehicle A and Vehicle A (6 W). For instance, at the track width of 56 cm, the track motion resistance coefficients for Vehicle A, Vehicle A (6 W) and Vehicle A (8 W) are 7.01, 5.97 and 3.37%, respectively. This indicates that the track motion resistance coefficient for Vehicle A (8 W) is 51.9% lower than that for Vehicle A and is 43.6% lower than that for Vehicle A (6 W).

Table 9.15 shows the variations of the thrust coefficient with the track width for the three vehicle configurations. As shown in the table, for a given vehicle configuration, the thrust coefficient increases noticeably with the increase of the track width. It should be noted that the thrust developed by the shearing action of the track on the terrain is derived from the frictional and cohesive components of the shear strength of the terrain, as discussed previously. For the clayey soil, as shown in Table 9.4, it has a significant cohesive component in its shear strength. The part of the thrust derived from the cohesive component of the shear strength of the terrain is the product of the cohesion of the terrain and the track contact area which is proportional to the track width, while the part of the thrust derived from the frictional component of the shear strength of the terrain is independent of the track contact area. As a result, the thrust coefficient increases with the increase of the track width, although the increase is not proportional to the increase of the track width. For instance, for Vehicle A increasing the track width from 32 to 56 cm, representing an increase of 75%, the thrust coefficient increases from 17.02 to 19.73%, representing an increase of 15.9%. For Vehicle A (8 W), increasing the track width over the same range, the thrust coefficient increases from 17.67 to 20.42%, representing an increase of 15.6%. It should be pointed out that for the same track width the thrust coefficient for Vehicle A (8 W) is higher than those for Vehicle A and Vehicle A (6 W), and that the thrust coefficient of Vehicle A (6 W) is slightly lower than that for Vehicle A. For instance, at the track width of 56 cm, the thrust coefficients for Vehicle A, Vehicle A (6 W) and Vehicle A (8 W) are 19.73%, 19.24 and 20.42%, respectively. This indicates that the thrust coefficient for Vehicle A (8 W) is 3.5% higher than that for Vehicle A and is 6.1% higher than that for Vehicle A (6 W).

Figure 9.42 shows the variations of the drawbar pull coefficient with the track width for the three vehicle configurations. Figure 9.42 and Table 9.15 show that for a given vehicle configuration, the drawbar pull coefficient increases noticeably with the increase of the track width. This is because when the track width increases, the thrust coefficient increases, while the track motion resistance coefficient decreases. For instance, for Vehicle A increasing the

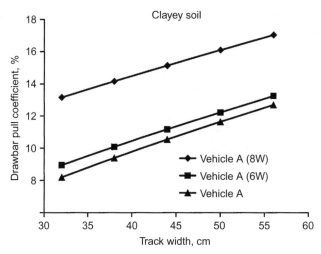

Figure 9.42: Variations of the drawbar pull coefficient with the track width for the three vehicle configurations at 20% slip on the clayey soil

track width from 32 to 56 cm, the drawbar pull coefficient increases from 8.19 to 12.72%, representing an increase of 55.3%. For Vehicle A (8 W), increasing the track width over the same range, the drawbar pull coefficient increases from 13.17 to 17.05%, representing an increase of 29.5%. It should be pointed out that for the same track width the drawbar pull coefficient for Vehicle A (8 W) is considerably higher than those for Vehicle A and Vehicle A (6 W). For instance, at the track width of 32 cm, the values of the drawbar pull coefficients for Vehicle A, Vehicle A (6 W) and Vehicle A (8 W) are 8.19, 8.94 and 13.17%, respectively. This indicates that the value of the drawbar pull coefficient for Vehicle A (8 W) is 60.8% higher than that for Vehicle A and is 47.3% higher than that for Vehicle A (6 W). At the track width of 56 cm, the drawbar pull coefficients for Vehicle A, Vehicle A (6 W) and Vehicle A (8 W) are 12.72, 13.27 and 17.05%, respectively. This indicates that the drawbar pull coefficient for Vehicle A (8 W) is 34% higher than that for Vehicle A, and is 28.5% higher than that for Vehicle A (6 W).

Figure 9.43 shows the variations of the tractive efficiency with the track width for the three vehicle configurations. Figure 9.43 and Table 9.15 show that for a given vehicle configuration, the tractive efficiency increases with the increase of the track width. This is similar to the variations of the drawbar pull coefficient with the track width discussed above. For instance, for Vehicle A increasing the track width from 32 to 56 cm, the tractive efficiency increases from 38.49 to 51.57%, representing an increase of 34%. For Vehicle A (8 W), increasing the track width over the same range, the tractive efficiency increases from 59.64 to 66.81%, representing an increase of 12%. This indicates that the increase in the track width has more significant effects on the improvement of the tractive efficiency for Vehicle A than for Vehicle A (8 W). It should be pointed out, however, that for the same track width the tractive efficiency for Vehicle A (8 W) is considerably higher than those for Vehicle A and Vehicle A

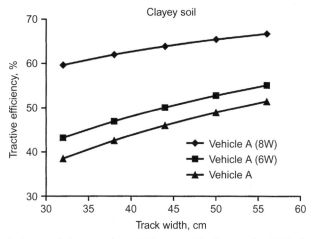

Figure 9.43: Variations of the tractive efficiency with the track width for the three vehicle configurations at 20% slip on the clayey soil

(6 W). For instance, at the track width of 32 cm, the values of the tractive efficiency for Vehicle A, Vehicle A (6 W) and Vehicle A (8 W) are 38.49, 43.17 and 59.64%, respectively. This indicates that the value of the tractive efficiency for Vehicle A (8 W) is 54.9% higher than that for Vehicle A and is 38.2% higher than that for Vehicle A (6 W). At the track width of 56 cm, the values of the tractive efficiency for Vehicle A, Vehicle A (6 W) and Vehicle A (8 W) are 51.57, 55.20 and 66.81%, respectively. This indicates that the value of the tractive efficiency for Vehicle A (8 W) is 29.6% higher than that of Vehicle A and is 21% higher than that for Vehicle A (6 W).

C. Summary

For a given track contact length, the track width determines the track contact area. When the track sinkage is less the vehicle ground clearance, the vehicle belly is not in contact with the terrain surface and the vehicle weight is entirely supported by the track. The track width affects the pressure exerted on the track–terrain interface, hence the track sinkage and track motion resistance. With the increase of the track width, track sinkage and track motion resistance coefficient decrease. As discussed previously, vehicle thrust is developed by the shearing action of the track and is derived from the frictional and cohesive components of the shear strength of the terrain. For the clayey soil, it has a significant cohesive component in its shear strength, as shown in Table 9.4. The part of the thrust derived from the cohesive component of its shear strength increases proportionally with the increase of track contact area. This indicates that the thrust coefficient generally increases with the increase of track width. Because of the decrease of track motion resistance coefficient and the increase of the thrust coefficient, the drawbar pull coefficient and the tractive efficiency generally increase with the increase of the track width. It should be pointed out that the effects of track width on performance are related to vehicle configuration. In general, increasing the track width has more significant effects on the improvement in performance of Vehicle A with five roadwheels than on that of Vehicle A (8 W) with eight overlapping roadwheels. For instance, for Vehicle A increasing the track width from 32 to 56 cm, the drawbar pull coefficient increases from 8.19 to 12.72%, representing an increase of 55.3%. For Vehicle A (8 W), increasing the track width over the same range, the drawbar pull coefficient increases from 13.17 to 17.05%, representing an increase of 29.5%. It should be noted, however, that for the same track width the performance of Vehicle A (8 W) with eight overlapping roadwheels is the highest among the three vehicle configurations examined. For instance, at the track width of 56 cm, the drawbar pull coefficients for Vehicle A, Vehicle A (6 W) and Vehicle A (8 W) are 12.72, 13.27 and 17.05%, respectively. This indicates that the drawbar pull coefficient for Vehicle A (8 W) is 34% higher than that for Vehicle A, and is 28.5% higher than that for Vehicle A (6 W). Furthermore, at the track width of 56 cm, the values of the tractive efficiency for Vehicle A, Vehicle A (6 W) and Vehicle A (8 W) are 51.57, 55.20 and 66.81%, respectively. This indicates that the value of the tractive efficiency for Vehicle A (8 W) is 29.6% higher than that of Vehicle A and is 21% higher than that for Vehicle A (6 W).

On highly compressible terrain, such as the Hope Valley snow, where the vehicle belly comes into contact with the terrain surface, the track width affects not only the normal pressure exerted on the track–terrain interface, but also vehicle belly–terrain interaction. On the Hope Valley snow, the belly load coefficient and the belly drag coefficient decrease with the increase of track width, while the load applied to the track and the track motion resistance coefficient increase with the increase of the track width. However, the decrease in the belly drag coefficient more or less compensates for the increase in the track motion resistance coefficient. As a result, the total motion resistance coefficient varies within a narrow range with the increase of track width for Vehicle A and Vehicle A (6 W), while that for Vehicle A (8 W) decreases slightly with the increase of track width. On the other hand, as the proportion of the load applied to the track and the track contact area increase with the increase of track width, the thrust coefficient increases accordingly. As a result, the drawbar coefficient and tractive efficiency increase with the increase of track width for the three vehicle configurations. It should be pointed out that the effects of track width on performance are related to vehicle configuration. For instance, for Vehicle A increasing the track width from 32 to 56 cm, the drawbar pull coefficient increases from 1.58 to 5.11%, representing an increase of 223.4%. For Vehicle A (8 W), increasing the track width over the same range, the drawbar pull coefficient increases from 7.73 to 11.01%, representing an increase of 42.4%. This indicates that the increase of track width has more significant impact on the improvement in performance of Vehicle A with five roadwheels than on that of Vehicle A (8 W) with eight overlapping roadwheels. It should be pointed out that for the same track width the drawbar pull coefficient for Vehicle A (8 W) is considerably higher than those for Vehicle A and Vehicle A (6 W). For instance, at the track width of 56 cm, the values of the drawbar pull coefficients for Vehicle A, Vehicle A (6 W) and Vehicle A (8 W) are 5.11, 5.75 and 11.01%, respectively. This indicates that the value of the drawbar pull coefficient for Vehicle A (8 W) is 115.5% higher than that for Vehicle A and is 91.5% higher than that for Vehicle A (6 W). At the track width of 56 cm, the values of the tractive efficiency for Vehicle A, Vehicle A (6 W) and Vehicle A (8 W) are 17.90, 20.42 and 36.41%, respectively. This indicates that the value of the tractive efficiency for Vehicle A (8 W) is 103.4% higher than that for Vehicle A and is 78.3% higher than that for Vehicle A (6 W).

On highly compressible terrain, such as the Hope Valley snow, the increase in track width has a more significant impact on the improvement in vehicle performance than on firmer terrain, such as the clayey soil, where the track sinkage is less than the vehicle ground clearance and the vehicle belly does not come into contact with the terrain surface. Furthermore, the increase of track width has more significant effects on improving the performance of Vehicle A with five roadwheels than on that of Vehicle A (8 W) with eight overlapping roadwheels.

9.6 Effects of Sprocket Location

The sprocket location of a tracked vehicle, in the front or at the rear, depends to a great extent on the function (or mission) of the vehicle. For instance, for an armoured personnel carrier, to provide better protection and more convenient access for the personnel to enter (or exit) the vehicle, a large door is usually installed at the rear, while the engine is mounted in the front of the vehicle. With mechanical transmissions, the sprocket is constrained to locate in the front of the vehicle. For fighting vehicles, the main gun is normally mounted in the front and the engine is installed at the rear. As a result, it is more suitable to locate the sprocket at the rear of the vehicle. For conventional industrial and agricultural tracked vehicles, the sprocket is normally located at the rear. The proposed use of electric drive systems to replace mechanical transmissions for future generations of military vehicles would offer more flexibility in vehicle layout and in selecting the sprocket location.

The location of the sprocket affects the track tension distribution. For instance, for a vehicle with a sprocket located in the front, in forward motion the track segments between the sprocket and the rear idler and those between the idler and the rear roadwheel are subject to higher tension, whereas the track segments between the front roadwheel and the sprocket are subject to lower tension. On the other hand, for a vehicle with rear sprocket, only the track segments between the rear sprocket and the rear roadwheel are subject to higher tension, whereas the track segments between the front roadwheel and the front idler and those between the front idler and the rear sprocket are subject to lower tension. It should be noted that a track with rubber bushings, commonly used in military vehicles, elongates when subject to tension. Figure 9.44 shows the relationship between elongation and tension for the track of a widely used armoured personnel carrier, upon which Vehicle A is based. For a vehicle with a front sprocket, the total elongation of the track is larger than that for a vehicle with a rear sprocket. With larger elongation, more track length is available for the deflection

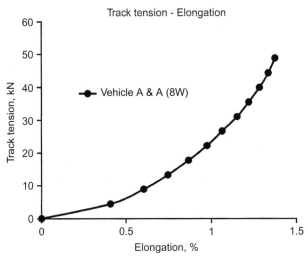

Figure 9.44: The track tension–elongation relationship for Vehicle A and Vehicle A (8W) used in the study of the effects of sprocket location on vehicle performance

of track segments between roadwheels on the track–terrain interface, and these track segments become looser and support less load. As a result, the sprocket location affects the performance of a tracked vehicle, if the effect of track tension on its elongation is significant.

To quantitatively examine the effects of sprocket location, the performances of Vehicle A and Vehicle A (8W) with front and rear sprockets were simulated using NTVPM on the Hope Valley snow and on the clayey soil. As noted previously, NTVPM takes into account the relationship between track elongation and track tension. To evaluate the effects of sprocket location on vehicle performance on a common basis, the relationship between track elongation and track tension is taken to be the same for both Vehicle A and Vehicle A (8W), as shown in Figure 9.44. In addition, the initial track tension coefficient of 10%, the longitudinal location of the CG at a distance of 198 cm from the centre of the sprocket, the vehicle total weight of 110.57 kN, and the track width of 38 cm are kept the same for the two vehicle configurations.

A. *On Hope Valley snow*

Table 9.16 shows that on the Hope Valley snow, the front and rear roadwheel sinkages for the two vehicle configurations with the sprocket located at the rear are lower than those with the sprocket located in the front. This is primarily due to the track elongation being higher with the sprocket located in the front than with the sprocket located at the rear. As a result, the load supported by the track segments between roadwheels with the front sprocket is less than that with the rear sprocket. The vehicle bellies for the two configurations with the sprocket either in the front or at the rear at 20% slip are either partially or fully in contact with the snow surface. For instance, for Vehicle A with the sprocket either in the front or at the rear, the track sinkages under both the front and rear roadwheels are greater than the ground clearance, hence its belly is in full contact with the snow surface. On the other hand, for Vehicle A (8 W) with the sprocket either in the front or at the rear, the track sinkage under the front roadwheel is less than the ground clearance, while that under the rear roadwheel is greater than the ground clearance, hence the belly is in partial contact with the snow surface. With the vehicle belly in contact with the terrain, it supports part of the vehicle weight, while the remainder is applied to the track. Table 9.16 shows the belly trim angles for the two configurations with the sprocket either in the front or at the rear on the Hope Valley snow. It shows that the belly trim angles for the two vehicle configurations slightly increase with the change of sprocket location from the front to the rear. This is primarily due to a greater load transfer from the front to the rear, as a result of the higher drawbar pull with the rear sprocket than with the front sprocket.

Table 9.16 shows that for a given vehicle configuration, the belly load coefficient decreases with the change of sprocket location from the front to the rear. For instance, for Vehicle A, changing the sprocket location from the front to the rear, the belly load coefficient decreases from 24.02 to 21.13%, representing a decrease of 12%. For Vehicle A (8 W), changing the sprocket location from the front to the rear, the belly load coefficient decreases from 14.97 to 12.58%, representing a decrease of 15.8%. This indicates that changing the sprocket location from the front to the rear, the belly load coefficient decreases proportionally more for Vehicle A (8 W) than that for Vehicle A. It should be pointed out that for the same sprocket location, in the front or at the rear, the belly load coefficient for Vehicle A (8 W) is lower than that for Vehicle A. For instance, with the sprocket located at the rear, the belly load coefficients for Vehicle A and Vehicle A (8 W) are 21.13% and 12.58%, respectively. This indicates that the belly load coefficient for Vehicle A (8 W) is 40.5% lower than that for Vehicle A.

Table 9.16 shows that for a given vehicle configuration, the belly drag coefficient decreases with the change of the sprocket location from the front to the rear. For instance, for Vehicle A, changing the sprocket location from the front to the rear, the belly drag coefficient decreases from 5.26 to 4.73%, representing a decrease of 10.1%. For Vehicle A (8 W), changing the sprocket location from the front to the rear, the belly drag coefficient decreases from 3.49 to 3.05%, representing a decrease of 12.6%. This indicates that changing the sprocket location from the front to the rear, the belly drag coefficient decreases proportionally more for Vehicle A (8 W) than that for Vehicle A. It should be pointed out that for the same sprocket location, in the front or at the rear, the belly drag coefficient for Vehicle A (8 W) is lower than that for Vehicle A. For instance, with the sprocket located at the rear, the belly drag coefficients for Vehicle A and Vehicle A (8 W) are 4.73 and 3.05%, respectively. This indicates that the belly drag coefficient for Vehicle A (8 W) is 35.5% lower than that for Vehicle A.

Table 9.16 shows that for Vehicle A and Vehicle A (8 W), the total motion resistance coefficient decreases with the change of the sprocket location from the front to the rear. For instance, for Vehicle A, changing the sprocket location from the front to the rear, the total motion resistance coefficient decreases from 17.68 to 16.38%, representing a decrease of

Table 9.16: Effects of sprocket location on performance parameters of Vehicle A and Vehicle A (8W) with initial track tension coefficient of 10% at 20% slip on Hope Valley snow

Vehicle configuration	Sprocket Location	Belly load coeff. %	Belly drag coeff. %	Track motion resistance coeff. %	Total motion resistance coeff. %	Thrust coeff. %	Drawbar pull coeff. %	Tractive efficiency %	Front roadwheel sinkage (cm)	Rear roadwheel sinkage (cm)	Belly trim angle (degrees)
Vehicle A (baseline vehicle)	Front	24.02	5.26	12.42	17.68	20.37	2.69	10.58	45.6	59.7	2.35
	Rear	21.13	4.73	11.65	16.38	21.82	5.44	19.94	43.0	58.2	2.62
Vehicle A (8W)	Front	14.97	3.49	10.44	13.93	22.65	8.72	30.78	39.1	55.8	3.14
	Rear	12.58	3.05	9.43	12.48	25.36	12.88	40.64	35.4	53.7	3.63

7.4%. For Vehicle A (8 W), changing the sprocket location from the front to the rear, the total motion resistance coefficient decreases from 13.93 to 12.48%, representing a decrease of 10.4%. This indicates that changing the sprocket location from the front to the rear, the total motion resistance coefficient decreases proportionally more for Vehicle A (8 W) than that for Vehicle A. It should be pointed out that for the same sprocket location, in the front or at the rear, the total motion resistance coefficient for Vehicle A (8 W) is lower than that for Vehicle A. For instance, with the sprocket located at the rear, the total motion resistance coefficients for Vehicle A and Vehicle A (8 W) are 16.38 and 12.48%, respectively. This indicates that the total motion resistance coefficient for Vehicle A (8 W) is 23.8% lower than that for Vehicle A.

Table 9.16 shows that for a given vehicle configuration, the thrust coefficient increases with the change of the sprocket location from the front to the rear. This is because changing the sprocket location from the front to the rear, the belly load coefficient decreases. As a result, the proportion of vehicle weight applied to the track increases. Since the Hope Valley snow has a significant frictional component in its shear strength (see Table 9.4), the increase in the proportion of the load applied to the track causes an increase in the thrust coefficient. For instance, for Vehicle A, changing the sprocket location from the front to the rear, the thrust coefficient increases from 20.37 to 21.82%, representing an increase of 7.1%. For Vehicle A (8 W), changing the sprocket location from the front to the rear, the thrust coefficient increases from 22.65 to 25.36%, representing an increase of 12%. This indicates that changing the sprocket location from the front to the rear, the thrust coefficient increases proportionally more for Vehicle A (8 W) than for Vehicle A. It should be pointed out that for the same sprocket location the thrust coefficient for Vehicle A (8 W) is higher than that for Vehicle A. For instance, with the sprocket located at the rear, the thrust coefficients for Vehicle A and Vehicle A (8 W) are 21.82 and 25.36%, respectively. This indicates that the thrust coefficient for Vehicle A (8 W) is 16.2% higher than that for Vehicle A.

Figure 9.45 shows that variations of drawbar pull coefficient with slip for Vehicle A and Vehicle A (8 W) with the sprocket located in the front (FSD) and with the sprocket located at the rear (RSD) on the Hope Valley snow. It shows that for Vehicle A (8 W) with the sprocket located at the rear it becomes self-propelled at a slip of approximately 1.5%, and with the sprocket in the front it becomes self-propelled at a slip of approximately 2%. For Vehicle A with the sprocket located at the rear it becomes self-propelled at a slip of approximately 7%, and with the sprocket located in the front it becomes self-propelled at a slip of approximately 10%. Table 9.16 shows that for a given vehicle configuration, the drawbar pull coefficient at 20% slip increases with the change of the sprocket location from the front to the rear. This is because with the change of the sprocket location from the front to the rear, the thrust coefficient increases, while the total motion resistance coefficient decreases. For instance, for Vehicle A changing the sprocket location from the front to the rear, the drawbar pull coefficient at 20% slip increases from 2.69 to 5.44%, representing an increase of 102.2%. For

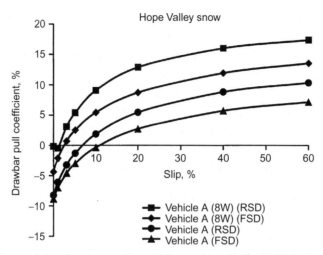

Figure 9.45: Variations of the drawbar pull coefficient with slip for Vehicle A and Vehicle A (8 W) with front and rear sprocket drives on the Hope Valley snow

Vehicle A (8 W) changing the sprocket location from the front to the rear, the drawbar pull coefficient at 20% slip increases from 8.72 to 12.88%, representing an increase of 47.7%. It should be pointed out that for the same sprocket location the drawbar pull coefficient for Vehicle A (8 W) is considerably higher than that for Vehicle A. For instance, with the sprocket located at the rear, the drawbar pull coefficients for Vehicle A and Vehicle A (8 W) are 5.44 and 12.88%, respectively. This indicates that the drawbar pull coefficient for Vehicle A (8 W) is 136.8% higher than that for Vehicle A. The results indicate that in general the change of the sprocket location from the front to the rear improves vehicle performance on highly compressible terrain.

Figure 9.46 shows the variations of the tractive efficiency with slip for Vehicle A and Vehicle A (8 W) with the sprocket located in the front (FSD) and with the sprocket located at the rear (RSD) on the Hope Valley snow. The figure shows that the tractive efficiency of the two vehicle configurations with the sprocket located at the rear is higher than that with the sprocket located in the front at the same slip. Table 9.16 shows that for a given vehicle configuration, the tractive efficiency at 20% slip increases with the change of the sprocket location from the front to the rear. For instance, for Vehicle A, changing the sprocket location from the front to the rear, the tractive efficiency at 20% slip increases from 10.58 to 19.94%, representing an increase of 88.5%. For Vehicle A (8 W), changing the sprocket location from the front to the rear, the tractive efficiency increases from 30.78 to 40.64%, representing an increase of 32%. It should be pointed out that for the same sprocket location the tractive efficiency for Vehicle A (8 W) is considerably higher than that for Vehicle A. For instance, with the sprocket located at the rear, the values of the tractive efficiency for Vehicle A and Vehicle A (8 W) are 19.94 and 40.64%, respectively. This indicates that the value of the tractive efficiency for Vehicle A (8 W) is 103.8% higher than that for Vehicle A.

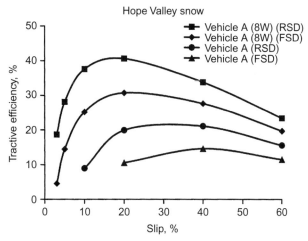

Figure 9.46: Variations of the tractive efficiency with slip for Vehicle A and Vehicle A (8W) with front and rear sprocket drives on the Hope Valley snow

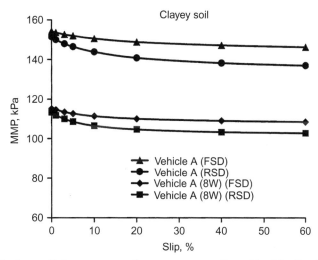

Figure 9.47: Variations of the mean maximum pressure (MMP) with slip for Vehicle A and Vehicle A (8W) with front and rear sprocket drives on the clayey soil

B. *On clayey soil*

Operating on the clayey soil, the track sinkages for the two vehicle configurations, with the sprocket located in the front or at the rear, are less than the vehicle ground clearance and the vehicle belly is not in contact with the terrain surface. Consequently, the belly trim angle no longer has any direct effect on vehicle performance. Figure 9.47 shows the variations of the mean maximum pressure (MMP) with slip for Vehicle A and Vehicle A (8W) with the sprocket located in the front (FSD) and with the sprocket located at the rear (RSD) on the clayey soil. The values of the MMP shown are derived from the pressure distributions on the track–terrain interface predicted by NTVPM. As noted previously, the value of MMP is an indication of the uniformity of the pressure distribution on the track–terrain interface. It is seen that for a given vehicle configuration, the value of the MMP for the sprocket located at the rear is lower than that for the sprocket located in the front. For instance, for Vehicle A, changing the sprocket location from the front to the rear, the value of MMP at 20% slip decreases from 148.8 to 140.8 kPa. For Vehicle A (8 W), changing the sprocket location from the front to the rear, the value of MMP at 20% slip decreases from 110.0 to 104.6 kPa. It also shows that for the same sprocket location the value of the MMP for Vehicle A (8 W) is considerably lower than that for Vehicle A at the same slip.

The track sinkage is related to the normal pressure exerted on the track, which is related to the MMP. Table 9.17 shows the front roadwheel sinkages for Vehicle A and Vehicle A (8 W) with sprocket location in the front and at the rear. It shows that for Vehicle A, changing the sprocket location from the front to the rear, the front roadwheel sinkage decreases from 10.5 to 9.5 cm. For Vehicle A (8 W), changing the sprocket location from the front to the rear, the front roadwheel sinkage decreases from 6.6 to 5.6 cm. Comparing the front roadwheel sinkage of 9.5 cm for Vehicle A with the sprocket located at the rear with that of 5.6 cm for Vehicle A (8 W) with the same sprocket location, it is seen that the reduction is 41.1%. Table 9.17 also gives the values of the rear roadwheel sinkage for the two vehicle configurations with sprocket location in the front and at the rear. It shows that for Vehicle A, changing the sprocket location from the front to the rear, the rear roadwheel sinkage decreases from 18.4 to 17.5 cm. For Vehicle A (8 W), changing the sprocket location from the front to the rear, the rear roadwheel sinkage decreases from 13.6 to 13.0 cm. Comparing the rear roadwheel sinkage of 17.5 cm for Vehicle A with the sprocket located at the rear with that of 13.0 cm for Vehicle A (8 W) with the same sprocket location, it is seen that the reduction is 25.7%.

Table 9.17 shows that for Vehicle A and Vehicle A (6 W), the track motion resistance coefficient decreases with the change of the sprocket location from the front to the rear. For instance, for Vehicle A, changing the sprocket location from the front to the rear, the track motion resistance coefficient decreases from 8.25 to 7.39%, representing a decrease of 10.4%. For Vehicle A (8 W), changing the sprocket location from the front to the rear, the track motion resistance coefficient decreases from 4.12 to 3.69%, representing a decrease of 10.4%. This indicates that changing the sprocket location from the front to the rear, the track motion resistance coefficient for Vehicle A decreases with the same percentage as that for Vehicle A (8 W). It should be pointed out that for the same sprocket location, in the front or at the rear, the track motion resistance coefficient for Vehicle A (8 W) is lower than that for Vehicle A. For instance, with the sprocket located at the rear, the track motion resistance coefficients for Vehicle A and Vehicle A (8 W) are 7.39 and 3.69%, respectively. This indicates that the track motion resistance coefficient for Vehicle A (8 W) is 50.1% lower than that for Vehicle A.

Table 9.17 shows that for a given vehicle configuration, the thrust coefficient increases with the change of the sprocket location from the front to the rear. For instance, for Vehicle A,

Table 9.17: Effects of sprocket location on performance parameters of Vehicle A and Vehicle A (8W) with initial track tension coefficient of 10% at 20% slip on clayey soil

Vehicle configuration	Sprocket location	Belly load coeff. %	Belly drag coeff. %	Track motion resistance coeff. %	Total motion resistance coeff. %	Thrust coeff. %	Drawbar pull coeff. %	Tractive efficiency %	Front roadwheel sinkage (cm)	Rear roadwheel sinkage (cm)	MMP (kPa)
Vehicle A (baseline vehicle)	Front	0	0	8.25	8.25	17.65	9.40	42.58	10.5	18.4	148.8
	Rear	0	0	7.39	7.39	18.07	10.68	47.26	9.5	17.5	140.8
Vehicle A (8W)	Front	0	0	4.12	4.12	18.30	14.18	61.99	6.6	13.6	110.0
	Rear	0	0	3.69	3.69	19.05	15.36	64.50	5.6	13.0	104.6

changing the sprocket location from the front to the rear, the thrust coefficient increases slightly from 17.65 to 18.07%, representing an increase of 2.4%. For Vehicle A (8 W), changing the sprocket location from the front to the rear, the thrust coefficient increases from 18.3 to 19.05%, representing an increase of 4.1%. It should be pointed out that for the same sprocket location the thrust coefficient for Vehicle A (8 W) is higher than that for Vehicle A. For instance, with the sprocket located at the rear, the thrust coefficients for Vehicle A and Vehicle A (8 W) are 18.07 and 19.05%, respectively. This indicates that the thrust coefficient for Vehicle A (8 W) is 5.4% higher than that for Vehicle A.

Figure 9.48 shows the variations of drawbar pull coefficient with slip for Vehicle A and Vehicle A (8 W) with the sprocket located in the front (FSD) and with the sprocket located at the rear (RSD) on the clayey soil. It shows that for Vehicle A, changing the sprocket location from the front to the rear, the drawbar pull coefficient increases. The same trend is observed for Vehicle A (8 W). As shown in Table 9.17, for Vehicle A, changing the sprocket location from the front to the rear, the drawbar pull coefficient at 20% slip increases from 9.40 to 10.68%, representing an increase of 13.6%. For Vehicle A (8 W), changing the sprocket location from the front to the rear, the drawbar pull coefficient at 20% slip increases from 14.18 to 15.36%, representing an increase of 8.3%. It should be pointed out that for the same sprocket location the drawbar pull coefficient for Vehicle A (8 W) is considerably higher than that for Vehicle A. For instance, with the sprocket located at the rear, the drawbar pull coefficients for Vehicle A and Vehicle A (8 W) are 10.68 and 15.36%, respectively. This indicates that the drawbar pull coefficient for Vehicle A (8 W) is 43.8% higher than that for Vehicle A. This shows that the change of the sprocket location from the front to the rear generally improves vehicle performance on the clayey soil.

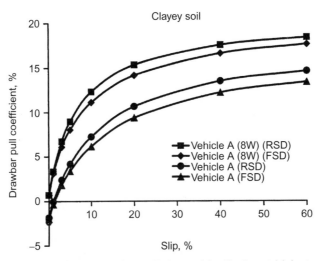

Figure 9.48: Variations of the drawbar pull coefficient with slip for Vehicle A and Vehicle A (8 W) with front and rear sprocket drives on the clayey soil

Figure 9.49 shows the variations of the tractive efficiency with slip for Vehicle A and Vehicle A (8 W) with the sprocket located in the front (FSD) and with the sprocket located at the rear (RSD) on the clayey soil. The figure shows that the tractive efficiency of the two vehicle configurations with the sprocket located at the rear is higher than that with the sprocket located in the front at the same slip. Table 9.17 shows that for a given vehicle configuration, the tractive efficiency at 20% slip increases with the change of the sprocket location from the front to the rear. For Vehicle A, changing the sprocket location from the front to the rear, the tractive efficiency at 20% slip increases from 42.58 to 47.26%, representing an increase of 11%. For Vehicle A (8 W), changing the sprocket location from the front to the rear, the tractive efficiency increases from 61.99 to 64.50%, representing an increase of 4.1%. It should be pointed out that for the same sprocket location the tractive efficiency for Vehicle A (8 W) is considerably higher than that for Vehicle A. For instance, with the sprocket located at the rear, the values of the tractive efficiency for Vehicle A and Vehicle A (8 W) are 47.26 and 64.50%, respectively. This indicates that the value of the tractive efficiency for Vehicle A (8 W) is 36.5% higher than that for Vehicle A.

C. Summary

For a vehicle with tracks that have a certain level of longitudinal elasticity (stiffness), such as those commonly used in military vehicles with rubber bushings, the location of the sprocket has noticeable effects on vehicle performance on soft terrain.

With the sprocket located in the front of the vehicle, in forward motion, the segments in the top run of the track between the front sprocket and the rear idler and those between the idler

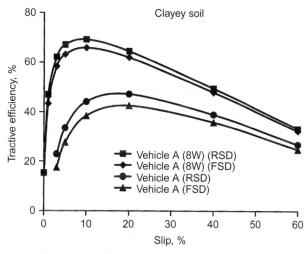

Figure 9.49: Variations of the tractive efficiency with slip for Vehicle A and Vehicle A (8 W) with front and rear sprocket drives on the clayey soil

and the rear roadwheel are subject to higher tension, whereas the track segments between the front roadwheel and the sprocket are subject to lower tension. On the other hand, with the sprocket located at the rear, only the track segments between the rear sprocket and the rear roadwheel are subject to higher tension, whereas the segments in the top run of the track between the front idler and the rear sprocket and those between the front roadwheel and the front idler are subject to lower tension. Consequently, with the sprocket located in the front the total elongation of the track is larger than that with the sprocket located at the rear. With larger elongation, more track length is available for the deflection of track segments between roadwheels on the track–terrain interface and these track segments become looser and support less load. As a result, the sprocket location affects the performance of a tracked vehicle, if the effect of track tension on its elongation is significant.

On the clayey soil, changing the sprocket location from the front to the rear causes a decrease in the value of MMP, hence lower track sinkage and track motion resistance coefficient, and higher performance. For instance, for Vehicle A, changing the sprocket location from the front to the rear, the drawbar pull coefficient at 20% slip increases by 13.6%. For Vehicle A (8 W), changing the sprocket location from the front to the rear, the drawbar pull coefficient at 20% slip increases by 8.3%. This shows that the change of the sprocket location from the front to the rear generally improves vehicle performance on the clayey soil. It is also shown that for the same sprocket location the drawbar pull coefficient for Vehicle A (8 W) is considerably higher than that for Vehicle A. For instance, with the sprocket located at the rear, the drawbar pull coefficient at 20% slip for Vehicle A (8 W) is 43.8% higher than that for Vehicle A.

On highly compressible terrain, such as the Hope Valley snow, changing the sprocket location from the front to the rear has more significant effects on the improvement in performance than on the clayey soil. For instance, for Vehicle A, changing the sprocket location from the front to the rear, the drawbar pull coefficient at 20% slip increases by 102.2%. For Vehicle A (8 W), changing the sprocket location from the front to the rear, the drawbar pull coefficient at 20% slip increases by 47.7%. It is also shown that for the same sprocket location the drawbar pull coefficient for Vehicle A (8 W) is considerably higher than that for Vehicle A on the Hope Valley snow. For instance, with the sprocket located at the rear, the drawbar pull coefficient at 20% slip for Vehicle A (8 W) is 136.6% higher than that for Vehicle A. This shows that with the sprocket located at the rear, together with the increase of the number of roadwheels from five to eight, vehicle performance can be noticeably improved on the Hope Valley snow.

As noted previously, the sprocket location is often constrained by the layout of the vehicle. For instance, for an armoured personnel carrier, to provide better protection and more convenient access for the personnel to enter (or exit) the vehicle, a large door is usually installed at the rear, while the engine is mounted in the front of the vehicle. With mechanical transmissions, the sprocket is constrained to be located in the front of the vehicle. With the proposed electrical drive system for future military vehicles, it will provide more flexibility in vehicle layout, as well as the potential of locating the sprocket at the rear for an armoured personnel carrier.

9.7 Concept of a High-Mobility Tracked Vehicle for Operation on Soft Ground

Based on the results on the investigation on the effects of design features on performance presented above, a concept for a high-mobility tracked vehicle, from the traction perspective on soft ground, emerges. Using Vehicle A as a reference, a high-mobility version will have eight overlapping roadwheels (with the same roadwheel–suspension system parameters as those of Vehicle A (8 W) shown in Table 9.2), rear sprocket drive (with electric drivetrain), track (rubber belt or band type) with width of 44 cm, and an initial track tension regulating system, with which the driver may set the initial track tension in accordance with terrain conditions. They are in contrast with the design features of Vehicle A, which has five roadwheels, front sprocket drive, track (segmented metal type with rubber pads) with width of 38 cm, and an initial track tension coefficient of 10%. The other basic parameters of the high-mobility version are the same as those of Vehicle A, in order to provide a common basis for comparing their performances.

Table 9.18 shows the performance parameters of Vehicle A and those of its high-mobility version on the Hope Valley snow, predicted by NTVPM. It shows that the values of the drawbar pull coefficients at 20% slip for Vehicle A and its high-mobility version are 2.69 and 20.28%, respectively. This indicates that the drawbar pull coefficient at 20% slip of the high-mobility version is 653.9% higher than that of Vehicle A. The values of the tractive efficiency at 20% slip for Vehicle A and its high-mobility version at 20% slip are 10.58 and 53.09%, respectively. This indicates that the tractive efficiency at 20% slip of the high-mobility version is 401.8% higher than that of Vehicle A.

Table 9.19 shows the performance parameters of Vehicle A and those of its high-mobility version on the clayey soil, predicted by NTVPM. It shows that the values of the drawbar pull coefficient at 20% slip for Vehicle A and its high-mobility version are 9.40 and 19.09%, respectively. This indicates that the drawbar pull coefficient at 20% slip of the high-mobility version is 103.1% higher than that of Vehicle A. The values of the tractive efficiency at 20% slip for Vehicle A and its high-mobility version at 20% slip are 42.58 and 69.83%, respectively. This indicates that the tractive efficiency at 20% slip of the high-mobility version is 64% higher than that of Vehicle A.

The results show that the performance of the high-mobility version is significantly higher than that of Vehicle A, particularly on the Hope Valley snow. It also demonstrates that the computer-aided method NTVPM is a useful and valuable tool for the vehicle designer in evaluating design concepts and in optimizing vehicle design parameters.

Table 9.18: Performance parameters of Vehicle A and its high-mobility version at 20% slip on Hope Valley snow

Vehicle configuration	Belly load coeff. %	Belly drag coeff. %	Track motion resistance coeff. %	Total motion resistance coeff. %	Thrust coeff. %	Drawbar pull coeff. %	Tractive efficiency %	Front roadwheel sinkage (cm)	Rear roadwheel sinkage (cm)	Belly trim angle (degrees)
Vehicle A (baseline vehicle)	24.02	5.26	12.42	17.68	20.37	2.69	10.58	45.6	59.7	2.35
High-mobility version*	10.59	2.58	7.70	10.28	30.56	20.28	53.09	29.6	49.7	3.68

*Eight overlapping roadwheels, rear sprocket drive, initial track tension coefficient of 40% and track width of 44 cm.

Table 9.19: Performance parameters of Vehicle A and its high-mobility version at 20% slip on clayey soil

Vehicle configuration	Belly load coeff. %	Belly drag coeff. %	Track motion resistance coeff. %	Total motion resistance coeff. %	Thrust coeff. %	Drawbar pull coeff. %	Tractive efficiency %	Front roadwheel sinkage (cm)	Rear roadwheel sinkage (cm)	MMP (kPa)
Vehicle A (baseline vehicle)	0	0	8.25	8.25	17.65	9.40	42.58	10.5	18.4	148.8
High-mobility version*	0	0	2.78	2.78	21.88	19.09	69.83	3.0	11.0	87.3

*Eight overlapping roadwheels, rear sprocket drive, initial track tension coefficient of 40%, and track width of 44 cm.

9.8 Effects of Design Features on the Performance of Two-Unit Articulated Vehicles

Two-unit articulated tracked vehicles, such as that shown in Figure 9.50, have been in commercial production for decades. They have been used as all-terrain transport vehicles, as well as military logistic vehicles in many countries for troop carrying, command and repair, and recovery. A unique feature of this type of vehicle is its method of steering, which is accomplished by rotating one unit against the other in the yaw plane using an articulation joint. The joint for steering usually consists of two hydraulic cylinders, as shown in Figure 9.51, which are servo-controlled by a steering wheel similar to that of an automobile. By appropriately activating the hydraulic rams through the steering wheel, the front and rear units of the vehicle follow a prescribed curved path. This type of steering method is commonly known as articulated steering. It is different in principle from the conventional skid steering used in single-unit

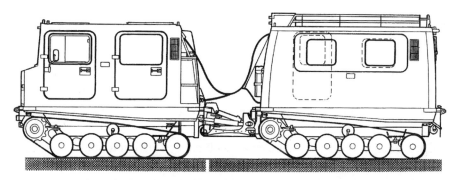

Figure 9.50: Two-unit articulated tracked vehicle – Hagglunds Bv206 (Courtesy BAE Systems Hagglunds AB)

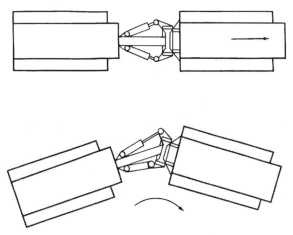

Figure 9.51: Schematic diagram for articulated steering (Reproduced by permission of *ISTVS* from Nuttall, 1964)

tracked vehicles, that requires the adjustment of the thrusts on the outside and inside tracks to generate a turning moment to overcome the moment of turning resistance due to skidding of the tracks on the ground. As the moment of turning resistance usually is considerable, braking of the inside track is often required. The braking force applied on the inside track reduces the resultant forward thrust of the vehicle during a turn and on soft terrain it may lead to immobilization. Detailed discussions on the mechanics of skid steering are presented in the references (Wong, 2008; Wong and Chiang, 2001). Using articulated steering, no adjustment of the thrust on the outside or inside track is required. Thus, using articulated steering the resultant forward thrust of the vehicle can be maintained during a turn. This provides tracked vehicles with improved mobility during turning manoeuvres, particularly on soft ground. For articulated steering the power required to execute a turn is usually much less than that for skid steering.

With respect to the performance of the two-unit, articulated tracked vehicle in straight line motion, it is similar to that of two single-unit vehicles joined together. The tracks of the rear unit usually run in the ruts formed by those of the front unit. This generally reduces the motion resistance of the tracks of the rear unit, hence improving the overall performance of the two-unit articulated vehicle.

The effects of vehicle design features, such as track system configuration, initial track tension, suspension setting, location of the centre of gravity, track width, etc., on the performance of two-unit articulated tracked vehicles are generally similar to those on the performance of single-unit tracked vehicles discussed previously. However, since the front and rear units of an articulated tracked vehicle are connected with an articulation joint, its configuration may have an effect on the interaction between the front and rear units, hence the overall performance of the articulated vehicle. As noted previously, the computer-aided method NTVPM is capable of simulating the performance of two-unit articulated vehicles (Wong, 1992a). A study was performed using NTVPM to investigate the possible effects of the articulation joint configuration on performance of a two-unit articulated tracked vehicle in straight line motion. The baseline vehicle used in this study is similar to the one that has been widely used in many parts of the world. Its basic design parameters are given in Table 9.20 and the parameters of its roadwheel–suspension system are presented in Table 9.21.

In addition to its steering function, the articulation joint also allows the two units to have varying degrees of freedom to pitch relative to each other, as described below:

(a) The rotation of the connecting link pivoted between the two units is unrestricted in the pitch plane;
(b) The connecting link is constrained to rotate within a certain range, with respect to the front and/or rear units;
(c) The connecting link is locked in a particular position and the two units are rigidly connected and cannot pitch relative to each other in straight line motion.

Table 9.20: Basic design parameters of the two-unit articulated tracked vehicle used in the study

Parameters	Front unit	Rear unit
Sprung weight, kN	28.4	28.4
Unsprung weight, kN	4.5	4.5
Sprung weight x-coordinate*, cm	134	105
Sprung weight y-coordinate*, cm	−46	−46
Articulation joint x-coordinate*, cm	273	−29
Articulation joint y-coordinate*, cm	3	3
Sprocket radius, cm	19	19
Track type	Reinforced rubber belt	Reinforced rubber belt
Track width, cm	62	62
Track longitudinal stiffness, kN/per cent elongation	10910	10910
Drawbar hitch x-coordinate*, cm		265
Drawbar hitch y-coordinate*, cm		−2
Ground clearance on level, hard ground, cm	39.6	39.6

*Coordinate origin is at the centre of the sprocket of respective unit. The sprocket is located at the front of the unit. Positive x- and y-coordinates are to the rear and down, respectively.

Another pitch control function, which regulates the angle between the two units in the pitch plane using an additional actuator, may also be added to the articulation joint. This function may enhance the obstacle negotiation capability of the vehicle. It may also decrease the moment of turning resistance by setting the pitch angle between the two units to an appropriate value, so as to shorten the effective track contact length on the ground during a turn.

A. On Hope Valley snow

There has been a perception that if in straight line motion the articulation joint is locked and the two units are rigidly connected and form a longer vehicle, it will alleviate the characteristic 'tail-down' attitude and will lead to improvement in the overall performance of the articulated vehicle. The locking of the articulation joint is usually set with the vehicle at rest on a level, hard ground. To examine this issue, the performances of the baseline articulated vehicle were predicted using NTVPM for two articulation joint configurations: unrestricted and locked. Figure 9.52 shows the variations of the drawbar pull coefficient at 20% slip with the initial track tension coefficient for the two articulation joint configurations on the Hope Valley snow. Figure 9.52 and Table 9.22 show that for a given articulated joint configuration, the drawbar pull coefficient at 20% slip increases with the increase of the initial track tension coefficient, similar to that for a single-unit tracked vehicle discussed previously. For

Table 9.21: Basic design parameters of the roadwheel–suspension system for the two-unit articulated tracked vehicle used in the study

						Front unit		
Roadwheel		Torsion arm pivots		Torsional stiffness (kN-m/deg.)	Torsion arm angles (+ is cw from horizontal) (degrees)			Torsion arm length (cm)
No.	Radius (cm)	x*-coordinate (cm)	y*-coordinate (cm)		Rebound limit	Free position	Jounce limit	
1	19	28	7	0.053	–	76	15	25
2	19	68	20	0.053	–	43	–15	25
3	19	111	20	0.053	–	43	–15	25
4	19	155	20	0.053	–	43	–15	25
5	19	214	20	0.053	17	17	–20	20
					Rear unit			
Roadwheel		Torsion arm pivots		Torsional stiffness (kN-m/deg.)	Torsion arm angles, (+ is cw from horizontal) (degrees)			Torsion arm length (cm)
No.	Radius (cm)	x*-coordinate (cm)	y*-coordinate (cm)		Rebound limit	Free position	Jounce limit	
1	19	28	7	0.053	–	76	15	25
2	19	68	20	0.053	–	43	–15	25
3	19	111	20	0.053	–	34	–15	25
4	19	155	20	0.053	–	24	–15	25
5	19	214	20	0.053	17	8	–20	20

*Coordinate origin is at the centre of the sprocket of each unit. The sprocket is located at the front of each unit. Positive x- and y-coordinates are to the rear and down, respectively.

instance, for the vehicle with the unrestricted articulation joint, increasing the initial track tension coefficient from 5 to 30% the drawbar pull coefficient increases from 8.71 to 17.24%, representing an increase of 97.9%. For the vehicle with the locked articulation joint, increasing the initial track tension coefficient over the same range, the drawbar pull coefficient increases from 3.72 to 12.90%, representing an increase of 246.8%. It should be pointed out, however, that for the same initial track tension coefficient the drawbar pull coefficient for the vehicle with the unrestricted joint is considerably higher than that for the vehicle with the locked joint, particularly when the initial track tension coefficient is 25% or higher, as shown in Figure 9.52. For instance, at the initial track tension coefficient of 25%, the values of the drawbar pull coefficient for the vehicle with the unrestricted joint and for the vehicle with the locked joint are 15.76 and 11.23%, respectively. This indicates that the drawbar pull coefficient for the vehicle with the unrestricted joint is 40.3% higher than that for the vehicle with the locked joint. At the initial track tension coefficient of 30%, the values of the drawbar pull coefficient for the vehicle with the unrestricted joint and for the vehicle with the locked joint

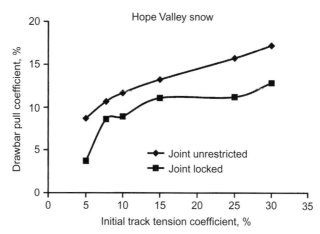

Figure 9.52: Variations of the drawbar pull coefficient with the initial track tension coefficient for a two-unit articulated tracked vehicle with unrestricted and locked articulation joints on the Hope Valley snow

are 17.24 and 12.90%, respectively. This indicates that the value of the drawbar pull coefficient for the vehicle with the unrestricted joint is 33.6% higher than that for the vehicle with the locked joint. This shows that the performance of the vehicle with the unrestricted joint is noticeably higher than that of the vehicle with the locked joint on the Hope Valley snow. This is contrary to the perception noted above that the locking of the articulation joint would lead to improvement in the overall performance of the articulated vehicle. Table 9.22 shows that when the initial track tension coefficient is equal to or higher than 25%, the bellies of the front and rear units are not in contact with the terrain surface for both the unrestricted and locked configurations of the articulation joint, as indicated by the belly load coefficient and belly drag coefficient being zero.

An examination of the predicted normal pressure distributions under the tracks of the front and rear units with the locked joint, shown in Figure 9.53, reveals that at the initial track tension coefficient of 25% or higher, the rear roadwheel of the front unit is actually lifted off the ground. With one roadwheel off the ground, the effective total track contact area is reduced. This leads to the deterioration of vehicle performance, as shown in Figure 9.52 and Table 9.22. The lift-off of the rear roadwheel of the front unit is a result of complex interactions between the front and rear units and between the tracks and the terrain. It is also a function of the track system configuration, suspension characteristics, and terrain conditions.

Figure 9.54 shows the variations of the tractive efficiency with the initial track tension coefficient at 20% slip for the two articulation joint configurations, unrestricted and locked. Figure 9.54 and Table 9.22 show that for a given joint configuration, the tractive efficiency increases with the increase of the initial track tension coefficient. This is similar to the variations of the drawbar pull coefficient with the initial track tension coefficient discussed above. For

Table 9.22: Effects of articulation joint configuration and initial track tension coefficient on performance parameters of the two-unit articulated tracked vehicle at 20% slip on Hope Valley snow

Articulation joint configuration	Initial track tension coeff. %	Belly load coeff. %	Belly drag coeff. %	Track motion resistance coeff. %	Total motion resistance coeff. %	Thrust coeff. %	Drawbar pull coeff. %	Tractive efficiency %	Front roadwheel sinkage* (cm)		Rear roadwheel sinkage* (cm)		Belly trim angle (degrees)	
									Front unit	Rear unit	Front unit	Rear unit	Front unit	Rear unit
Unrestricted	5	0.80	0.19	14.22	14.41	23.12	8.71	30.14	13.4	37.6	31.6	42.4	7.31	3.58
	7.75	0.42	0.10	13.29	13.39	24.09	10.70	35.52	11.8	36.7	30.5	41.4	7.41	3.29
	10	0.22	0.05	12.75	12.80	24.48	11.68	38.16	11.4	36.2	30.1	40.7	7.37	3.03
	15	0.01	0.01	11.81	11.82	25.10	13.28	42.33	10.8	35.4	29.4	39.6	7.25	2.33
	25	0	0	10.59	10.59	26.35	15.76	47.84	9.9	34.0	28.3	38.0	6.96	1.53
	30	0	0	9.84	9.84	27.08	17.24	50.94	9.5	32.9	27.8	37.0	6.78	1.35
Locked	5	2.06	1.32	23.52	24.84	28.56	3.72	10.41	−14.4†	51.1	31.4	49.3	19.78	−4.89‡
	7.75	0.83	0.18	15.95	16.13	24.74	8.61	27.84	13.7	40.2	30.3	44.1	6.35	2.07
	10	1.02	0.20	17.06	17.26	26.20	8.94	27.31	4.1	42.5	28.3	45.7	9.73	1.02
	15	0.27	0.06	14.55	14.61	25.72	11.11	34.56	11.6	38.8	29.0	42.7	6.56	1.83
	25	0	0	11.63	11.63	22.86	11.23	39.31	25.4	34.4	33.8	38.0	1.82	1.75
	30	0	0	10.83	10.83	23.73	12.90	43.48	24.7	33.2	33.2	36.9	1.70	1.67

*Roadwheel sinkages are with respect to the original terrain surface.
†Negative wheel sinkage indicates that the wheel is above ground.
‡Negative belly trim angle indicates the vehicle belly takes a nose-down attitude.

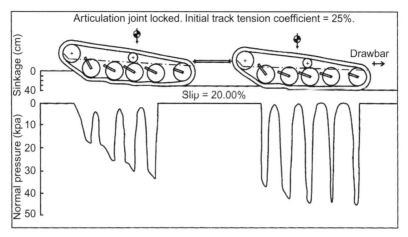

Figure 9.53: Normal pressure distributions under the tracks of a two-unit articulated vehicle with articulation joint locked, at the initial track tension coefficient of 25% and at 20% slip on the Hope Valley snow

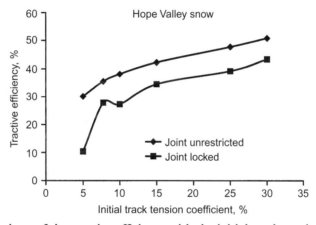

Figure 9.54: Variations of the tractive efficiency with the initial track tension coefficient for a two-unit articulated tracked vehicle with unrestricted and locked articulation joints on the Hope Valley snow

instance, for the vehicle with the unrestricted joint, increasing the initial track tension coefficient from 5 to 30%, the tractive efficiency increases from 30.14 to 50.94%, representing an increase of 69%. For the vehicle with the locked joint, increasing the initial track tension coefficient over the same range, the tractive efficiency increases from 10.41 to 43.48%, representing an increase of 317.1%. It should be pointed out that for the same initial track tension coefficient the tractive efficiency for the vehicle with the unrestricted joint is noticeably higher than that for the vehicle with the locked joint. For instance, at the initial track tension coefficient of 25%, the values of the tractive efficiency for the vehicle with the unrestricted joint and for the vehicle with the locked joint are 47.84 and 39.31%, respectively. This indicates that the value of the tractive efficiency for the vehicle with the unrestricted joint is 27.1% higher than that for the vehicle with the locked joint. At the initial track tension coefficient of 30%, the values of the tractive efficiency for the vehicle with the unrestricted joint and for the vehicle with the locked joint are 50.94 and 43.48%, respectively. This indicates that the value of the tractive efficiency for the vehicle with the unrestricted joint is 17.2% higher than that for the vehicle with the locked joint. This is similar to the difference in drawbar pull coefficient for the two articulation joint configurations discussed previously.

B. On clayey soil

Table 9.23 shows the performance parameters at 20% slip of the baseline articulated vehicle at various initial track tension coefficients with the two different articulation joint configurations, unrestricted and locked, on the clayey soil. It shows that for a given joint configuration, the drawbar pull coefficient increases slightly with the increase of the initial track tension coefficient. For instance, for the vehicle with the unrestricted articulation joint, increasing the initial track tension coefficient from 5 to 30% the drawbar pull coefficient increases from 30.70 to 32.40%, representing an increase of 5.5%. For the vehicle with the locked articulation joint, increasing the initial track tension coefficient over the same range, the drawbar pull coefficient increases from 25.35 to 30.64%, representing an increase of 20.9%. It should be pointed out that for the same initial track tension coefficient the drawbar pull coefficient for the vehicle with the unrestricted joint is slightly higher than that for the vehicle with the locked joint. For instance, at the initial track tension coefficient of 25%, the values of the drawbar pull coefficient for the vehicle with the unrestricted joint and for the vehicle with the locked joint are 32.14 and 30.51%, respectively. This indicates that the drawbar pull coefficient for the vehicle with the unrestricted joint is slightly higher than that for the vehicle with the locked joint by 5.3%. At the initial track tension coefficient of 30%, the values of the drawbar pull coefficient for the vehicle with unrestricted joint and for the vehicle with the locked joint are 32.40 and 30.64%, respectively. This indicates that the value of the drawbar pull coefficient for the vehicle with the unrestricted joint is slightly higher than that for the vehicle with the locked joint by 5.7%. Table 9.23 shows that on the clayey soil, the bellies of the front and rear units are not in contact with the terrain surface for both the unrestricted and locked configurations of the articulation joint, as indicated by the belly load coefficient and belly drag coefficient being zero.

As shown in Table 9.23, for a given joint configuration, the tractive efficiency increases slightly with the increase of the initial track tension coefficient. This is similar to the variations of the drawbar pull coefficient with the initial track tension coefficient discussed above. For instance, for the vehicle with the unrestricted joint, increasing the initial track tension coefficient from 5 to 30%, the tractive efficiency increases from 72.63 to 75.01%, representing an increase of 3.3%. For the vehicle with the locked joint, increasing the initial track tension coefficient over the same range, the tractive efficiency increases from 66.66 to 74.08%,

Table 9.23: Effects of articulation joint configuration and initial track tension coefficient on performance parameters of the two-unit articulated tracked vehicle at 20% slip on clayey soil

Articulation joint configuration	Initial track tension coeff. %	Belly load coeff. %	Belly drag coeff. %	Track motion resistance coeff. %	Total motion resistance coeff. %	Thrust coeff. %	Drawbar pull coeff. %	Tractive efficiency %	Front roadwheel sinkage* (cm)		Rear roadwheel sinkage* (cm)		MMP (kPa)	
									Front unit	Rear unit	Front unit	Rear unit	Front unit	Rear unit
Unrestricted	5	0	0	3.11	3.11	33.81	30.70	72.63	0.9	8.1	8.2	8.3	57.2	68.1
	7.75	0	0	2.86	2.86	34.05	31.19	73.29	0.6	7.9	8.1	8.1	56.0	66.1
	10	0	0	2.73	2.73	34.16	31.43	73.61	0.4	7.9	8.0	8.0	55.4	65.1
	15	0	0	2.51	2.51	34.32	31.81	74.14	0.1	7.7	7.9	7.7	54.5	62.0
	25	0	0	2.26	2.26	34.40	32.14	74.75	−0.5†	7.3	7.8	7.5	66.7	55.6
	30	0	0	2.15	2.15	34.55	32.40	75.01	−0.7†	7.1	7.7	7.4	67.6	53.2
Locked	5	0	0	5.07	5.07	30.42	25.35	66.66	0.0	10.3	8.1	9.7	49.9	82.3
	7.75	0	0	4.94	4.94	30.89	29.95	67.20	0.1	10.1	7.9	9.7	46.6	82.4
	10	0	0	4.32	4.32	31.77	27.45	69.11	1.9	9.2	7.2	9.0	54.8	69.2
	15	0	0	2.53	2.53	32.09	29.56	73.69	3.4	7.0	6.9	7.2	58.5	63.6
	25	0	0	2.25	2.25	32.76	30.51	74.49	3.5	6.1	6.9	6.9	58.6	55.6
	30	0	0	2.45	2.45	33.09	30.64	74.08	1.1	7.3	7.1	7.4	52.3	59.5

*Roadwheel sinkages are with respect to the original terrain surface.
†Negative wheel sinkage indicates that the wheel is above ground.

representing an increase of 11.1%. It should be pointed out that for the same initial track tension coefficient the tractive efficiency for the vehicle with the unrestricted joint is slightly higher than that for the vehicle with the locked joint. For instance, at the initial track tension coefficient of 25%, the values of the tractive efficiency for the vehicle with the unrestricted joint and for the vehicle with the locked joint are 74.75 and 74.49%, respectively. This indicates that the value of the tractive efficiency for the vehicle with the unrestricted joint is marginally higher than that for the vehicle with the locked joint by 0.3%. At the initial track tension coefficient of 30%, the values of the tractive efficiency for the vehicle with the unrestricted joint and for the vehicle with the locked joint are 75.01 and 74.08%, respectively. This indicates that the value of the tractive efficiency for the vehicle with the unrestricted joint is slightly higher than that for the vehicle with the locked joint by 1.3%.

C. Summary

On highly compressible terrain, such as the Hope Valley snow, the performance of the two-unit articulated vehicle with the unrestricted articulation joint is considerably higher than that of the vehicle with the locked articulation joint, particularly at relatively high values of the initial track tension coefficient. For instance, at the initial track tension coefficient of 30%, the drawbar pull coefficient at 20% slip for the vehicle with the unrestricted joint is 33.6% higher than that for the vehicle with the locked joint. As noted previously, this is primarily due to the rear roadwheel of the front unit lifting off the ground for the vehicle with the locked joint. With one of the five roadwheels on a track of the front unit off the ground, the track contact area is reduced. On highly compressible terrain, such as the Hope Valley snow, this leads to the deterioration of vehicle performance. On highly compressible terrain, the increase of the initial track tension coefficient generally leads to the improvement in vehicle performance for a vehicle with either the unrestricted or the locked articulation joint.

On the clayey soil which is firmer than the Hope Valley snow, as indicated by the lower roadwheel sinkages and the vehicle belly not contacting the terrain surface, the performance of the vehicle with the unrestricted joint is only slightly higher than that of the vehicle with the locked joint. This indicates that the articulation joint configuration, whether unrestricted or locked, has less significant effects on vehicle performance on the clayey soil than on the Hope Valley snow. Also, the initial track tension coefficient has less significant effects on vehicle performance on the clayey soil than on the Hope Valley snow.

It should be pointed out that the effects of initial track tension on vehicle performance are dependent on vehicle design features. For instance, the two-unit articulated vehicle examined in this study has a relatively low ratio of vehicle weight to the nominal track contact area on hard ground of 13.49 kN/m^2, as compared with 54.85 kN/m^2 for Vehicle A examined earlier. As a result, the average value of MMP for the front and rear units of the articulated vehicle at the initial track tension coefficient of 10% is 60.25 kPa on the clayey soil (see Table 9.23), which is considerably lower than 148.8 kPa for Vehicle A under the same conditions (see Table 9.6). Consequently, the articulated vehicle has lower roadwheel sinkages and higher performance than Vehicle A. For instance, the drawbar pull coefficient at 20% slip for the articulated vehicle with the initial track tension coefficient of 10% is 31.43% on the clayey soil (see Table 9.23), as against 9.40% for Vehicle A under the same operating conditions (see Table 9.6). Because of the difference in design features, such as the vehicle weight per unit of nominal track contact area, between the articulated vehicle and Vehicle A, increasing the initial track tension coefficient from 10 to 30%, the drawbar pull coefficient at 20% slip on the clayey soil for the articulated vehicle with the unrestricted joint increases from 31.43 to 32.40%, representing an increase of only 3.1% (see Table 9.23). On the other hand, for Vehicle A increasing the initial track tension coefficient over the same range, the drawbar pull coefficient increases from 9.40 to 14.42%, representing a significant increase of 53.4% (see Table 9.6). This demonstrates that the effects of initial track tension on the improvement of vehicle performance are dependent on the design features of the vehicle, as well as terrain conditions.

9.9 Analysis and Evaluation of Detracking Risks

Detracking refers to a phenomenon where the track separates from the sprocket or idler during operation, hence rendering the track system inoperative. The expense to commercial operations and the loss of tempo to the military caused by the delay due to detracking are very significant.

To initiate detracking, normally two conditions must be satisfied. One is the existence of sufficient slacks or excess track lengths (in comparison with the corresponding track lengths under taut conditions) in certain track segments that enable the track to override the sprocket or idler. This is referred to as the necessary condition for detracking. The other is the presence of a sufficiently large side force acting on the track, such as that during a turning manoeuvre at high speeds or with small turning radii. This is referred to as the sufficient condition for detracking. Significant slacks or excess track lengths may appear in certain track segments, such as during a sharp turn with large driving or braking torque applied to the sprocket; the track hitting a sizable obstacle causing significant changes in suspension deflection and track–suspension system geometry; or operating on rough surfaces causing significant vibrations of certain track segments with large amplitudes.

A literature survey reveals that so far little analytical study on detracking has been performed. As an initial attempt to theoretically examine the detracking process, the first part of this section focuses on the analysis of the necessary conditions that enables the track to override the sprocket or idler under steady-state operating conditions, such as in a steady turning manoeuvre on deformable terrain. Based on the analysis, a set of detracking risk indicators is established. To facilitate the evaluation of detracking risks, a special module of NTVPM, known as DETRACK, has been developed. The detracking risks indicators and the module DETRACK are then applied to the evaluation of detracking risks of track systems (Wong and Huang, 2006b).

9.9.1 The Necessary Condition for Detracking

The necessary condition for detracking under steady-state operating conditions will be illustrated through an example of a tracked vehicle undergoing a steady turning manoeuvre. In the following analysis, detracking caused by the ingress of debris into the sprocket/idler–track system is not included.

As mentioned previously, in skid-steering, the thrust of the outside track is increased and that of the inside track is reduced, so as to create a turning moment to overcome the moment of turning resistance due to the skidding of the tracks on the ground. Since the moment of turning resistance is usually considerable, braking of the inside track is often required.

Figure 9.55 shows a proposed band (rubber belt) track system for future combat vehicles with the sprocket in the front. When a driving torque is applied to the sprocket of the outside track, the track segments between the sprocket and idler and those between the idler and rear roadwheel are subject to higher tension and relatively small slacks (sags) in these track segments are expected, as shown in Figure 9.56(a). On the other hand, the track segments between the sprocket and front roadwheel are subject to lower tension and a considerable slack is expected. When a braking torque is applied to the sprocket of the inside track, the

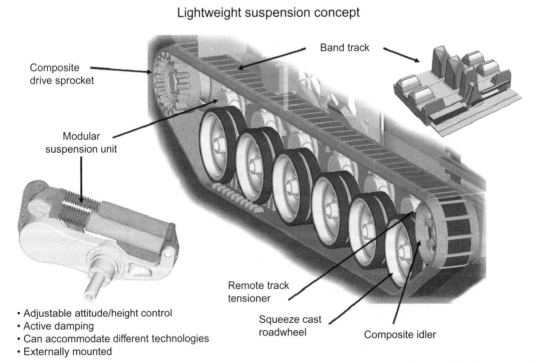

Figure 9.55: A proposed band (rubber belt) track system for a future combat vehicle (Approved for Public Release, Distribution unlimited, TACOM 3 Sept. 2004, FCS Case 04-083)

track segments between the sprocket and the front roadwheel are subject to higher tension and a relatively small slack is expected, as shown in Figure 9.56(b). On the other hand, the track segments between the sprocket and idler and those between the idler and rear roadwheel are subject to lower tension and considerable slacks in these track segments are expected.

Figures 9.57(a) and (b) show schematically the slacks in the track segments of a track system with the sprocket at the rear, when subject to a driving and a braking torque, respectively. In this case, when a driving torque is applied to the sprocket, the track segments between the sprocket and rear roadwheel are subject to higher tension and a relatively small slack is expected, as shown in Figure 9.57(a). On the other hand, the track segments between the sprocket and idler and those between the idler and front roadwheel are subject to lower tension and considerable slacks are expected. When a braking torque is applied to the sprocket, the track segments between the idler and front roadwheel and those between the idler and

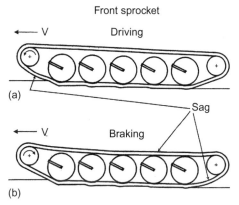

Figure 9.56: A front-sprocket-drive track system (Reprinted by permission of the Council of the Institution of Mechanical Engineers from Wong and Huang, 2006b)

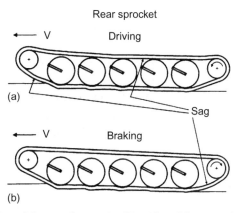

Figure 9.57: A rear-sprocket-drive track system (Reprinted by permission of the Council of the Institution of Mechanical Engineers from Wong and Huang, 2006b)

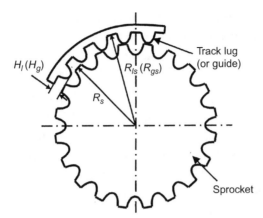

Figure 9.58: Track lugs (or guides) disengaging sprocket teeth (Reprinted by permission of the Council of the Institution of Mechanical Engineers from Wong and Huang, 2006b)

sprocket are subject to higher tension and relatively small slacks in these track segments are expected, as shown in Figure 9.57(b). On the other hand, the track segments between the sprocket and rear roadwheel are subject to lower tension and a considerable slack is expected.

To illustrate the mechanism of detracking, the band track system shown in Figure 9.55 is used as an example. As shown in the top right corner of the figure, there are two rows of drive lugs on the inside surface of the track to engage sprocket teeth and one row of track guides in the middle to provide directional guidance for the roadwheels. In general, there are four possible types of mechanism for detracking, as outlined below:

(a) Track drive lugs overriding the outside diameter of the sprocket, causing track drive lugs to disengage sprocket teeth, as shown in Figure 9.58;

(b) Track guides overriding the outside diameter of the sprocket;

(c) Track drive lugs overriding the outside diameter of the idler, as shown in Figure 9.59;

(d) Track guides overriding the outside diameter of the idler.

For segmented metal tracks (metal link tracks) commonly used in the current generation of fighting vehicles, there are no track drive lugs as there are in band tracks. Transmission of power from the sprocket to the segmented metal track is through engagement of sprocket teeth with drive slots on the track. In this case, track drive slots replace track drive lugs in detracking mechanisms (a) and (c) noted above and in the following analysis of detracking risks.

9.9.2 Detracking Risk Indicators for Track Drive Lugs/Track Guides Overriding the Sprocket/Idler

To assess the risk of detracking, a parameter called Detracking Risk Indicator is proposed. In view of the four possible types of detracking mechanism noted above for either track drive

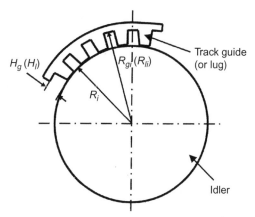

Figure 9.59: Track guides (or lugs) disengaging idler (Reprinted by permission of the Council of the Institution of Mechanical Engineers from Wong and Huang, 2006b)

lugs or track guides overriding either the sprocket or idler, four different Detracking Risk Indicators are introduced and defined as follows:

(a) Detracking Risk Indicator D_{ls} for track drive lugs overriding the sprocket shown in Figure 9.58 is expressed by

$$D_{ls} = (R_{ls} - R_s)/H_l \qquad (9.1)$$

where R_{ls} is the distance of the root of the track drive lug to the centre of the sprocket due to excess track lengths, R_s is the outside radius of the sprocket, and H_l is the height of the track drive lug, as shown in Figure 9.58.

D_{ls} is also an indication of possible interruption of power transmission from the sprocket to the track.

(b) Detracking Risk Indicator D_{gs} for track guides overriding the sprocket shown in Figure 9.58 is expressed by

$$D_{gs} = (R_{gs} - R_s)/H_g \qquad (9.2)$$

where R_{gs} is the distance of the root of the track guide to the centre of the sprocket due to excess track lengths, and H_g is the height of the track guide, as shown in Figure 9.58.

(c) Detracking Risk Indicator D_{li} for track drive lugs overriding the idler shown in Figure 9.59 is expressed by

$$D_{li} = (R_{li} - R_i)/H_l \qquad (9.3)$$

where R_{li} is the distance of the root of the track drive lug to the centre of the idler due to excess track lengths, and R_i is the outside radius of the idler, as shown in Figure 9.59.

(d) Detracking Risk Indicator D_{gi} for track guides overriding the idler shown in Figure 9.59 is expressed by

$$D_{gi} = (R_{gi} - R_i)/H_g \qquad (9.4)$$

where R_{gi} is the distance of the root of the track guide to the centre of the idler due to excess track lengths, as shown in Figure 9.59.

It should be pointed out that the parameters R_{ls}, R_{gs}, R_{li} and R_{gi} are functions of track system configuration, torque applied to the sprocket, initial track tension, track longitudinal stiffness, suspension stiffness, etc. The methods for predicting the values of these parameters, as well as those of various Detracking Risk Indicators, are discussed later.

It should be noted that if the value of a Detracking Risk Indicator is 100%, then either track drive lugs or track guides will override or completely disengage either the sprocket or idler. On the other hand, if the value of a Detracking Risk Indicator is zero (0), there will be no risk for either track drive lugs or track guides to override either the sprocket or idler. If the value of a Detracking Risk Indicator is between zero (0) and 100%, there will be partial disengagement of either track drive lugs or track guides with either the sprocket or idler.

It should also be mentioned that for most band tracks, the height of the track guide is usually much greater than that of the track drive lug, as shown in Figure 9.55. Dependent upon the outside diameter of the sprocket or that of the idler, either the Detracking Risk Indicator D_{gs} or D_{gi}, representing the risk of track guides overriding either the sprocket or idler, respectively, determines the propensity to detrack in practice. The Detracking Risk Indicator D_{ls} can be used to indicate the possible interruption in transmitting power from the sprocket to the track due to partial or complete disengagement of track drive lugs with sprocket teeth.

9.9.3 Risk Indicators for Disengagement of the Leading Track Drive Lug/Track Guide with the Sprocket/Idler

To indicate the onset of disengagement of either the leading track drive lug or track guide with either the sprocket or idler, four Risk Indicators are introduced and defined as follows:

(a) Risk Indicator D'_{ls} for the leading track drive lug disengaging the sprocket shown in Figure 9.60 is expressed by

$$D'_{ls} = (R'_{ls} - R_s)/H_l \qquad (9.5)$$

where R'_{ls} is the distance of the root of the leading track drive lug to the centre of the sprocket due to excess track lengths, as shown in Figure 9.60.

(b) Risk Indicator D'_{gs} for the leading track guide disengaging the sprocket shown in Figure 9.60 is expressed by

$$D'_{gs} = (R'_{gs} - R_s)/H_g \qquad (9.6)$$

where R'_{gs} is the distance of the root of the leading track guide to the centre of the sprocket due to excess track lengths, as shown in Figure 9.60.

(c) Risk Indicator D'_{li} for the leading track drive lug disengaging the idler shown in Figure 9.61 is expressed by

$$D'_{li} = (R'_{li} - R_i)/H_l \tag{9.7}$$

where R'_{li} is the distance of the root of the leading track drive lug to the centre of the idler due to excess track lengths, as shown in Figure 9.61.

(d) Risk Indicator D'_{gi} for the leading track guide disengaging the idler shown in Figure 9.61 is expressed as

$$D'_{gi} = (R'_{gi} - R_i)/H_g \tag{9.8}$$

where R'_{gi} is the distance of the root of the leading track guide to the centre of the idler due to excess track lengths, as shown in Figure 9.61.

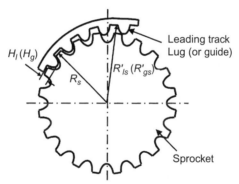

Figure 9.60: Leading track lug (or guide) disengaging sprocket tooth (Reprinted by permission of the Council of the Institution of Mechanical Engineers from Wong and Huang, 2006b)

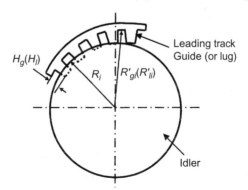

Figure 9.61: Leading track guide (or lug) disengaging idler (Reprinted by permission of the Council of the Institution of Mechanical Engineers from Wong and Huang, 2006b)

It should be pointed out that the parameters R'_{ls}, R'_{gs}, R'_{li} and R'_{gi} are functions of track system configuration, torque applied to the sprocket, initial track tension, track longitudinal stiffness, suspension stiffness, etc. The methods for predicting the values of these parameters, as well as those of various Risk Indicators, are discussed later.

It should be emphasized that the four Risk Indicators noted above are for indicating the onset of disengagement of either the track drive lug or track guide with either the sprocket or idler. As mentioned previously, for most band tracks, the height of the track guide is usually greater than that of the track drive lug, as shown in Figure 9.55. Dependent upon the outside diameter of the sprocket or that of the idler, either the Risk Indicator D'_{gs} or D'_{gi}, representing disengagement of the leading track guide with either the sprocket or idler, respectively, would be more important in indicating the onset of detracking in practice. The Risk Indicator D'_{ls} can be used to indicate the onset of possible interruption in transmitting power from the sprocket to the track due to disengagement of the leading track drive lug with sprocket teeth.

9.9.4 Basic Features of the Module DETRACK for Evaluating Detracking Risks

As noted previously, to compute the Detracking Risk Indicators described in Section 9.9.2 and the Risk Indicators described in Section 9.9.3, a special module of NTVPM, known as DETRACK, has been developed. It evaluates the lengths and corresponding excess track lengths of various track segments. This enables the parameters R_{ls}, R_{gs}, R_{li} and R_{gi} and the Detracking Risk Indicators described in Section 9.9.2, as well as the parameters R'_{ls}, R'_{gs}, R'_{li} and R'_{gi} and the Risk Indicators described in Section 9.9.3, to be predicted.

The major features and functions of the module DETRACK are summarized below:

(a) DETRACK evaluates the lengths of various track segments, including the top run of the track between the sprocket and idler, and segments between the sprocket or idler and the front or rear roadwheel, under driving or braking conditions at various rates of slip or skid (corresponding to various driving or braking torques applied to the sprocket of the outside or inside track, respectively, in a turning manoeuvre). In calculating the lengths of various track segments, the effects of sprocket torque on the elongation of the track (due to track tension and track longitudinal stiffness) and on suspension spring deflection and track–suspension system geometry (due to changes in tension distribution along the track) have been taken into account.

(b) The lengths of various track segments are then compared with those under taut conditions to determine the slacks or excess track lengths. From these, the parameters R_{ls}, R_{gs}, R_{li} and R_{gi} described in Section 9.9.2, as well as the parameters R'_{ls}, R'_{gs}, R'_{li} and R'_{gi} described in Section 9.9.3, are determined. In the analysis, the heights of the track drive lug, track guide, and sprocket tooth and the outside diameters of the sprocket and idler are taken into consideration.

(c) Using Eqns (9.1) to (9.4), Detracking Risk Indicators described in Section 9.9.2 are determined. Based on the design parameters of the track system (such as the heights of the track drive lug, track guide and sprocket tooth, the outside diameters of the sprocket and idler, etc.), DETRACK will compare various Detracking Risk Indicators described in Section 9.9.2. The appropriate Detracking Risk Indicator that determines the propensity to detrack under given operating conditions will be identified and included as part of the output in both tabular and graphic format. For instance, for the track system with a segmented metal track to be evaluated and described later, the Detracking Risk Indicator D_{gi} is found to be the one that determines the propensity to detrack in practice.

(d) Detracking Risk Indicator D_{ls} for track drive lugs disengaging sprocket teeth, given by Eqn (9.1), indicates the possible interruption in transmitting power from the sprocket to the track. It would be of interest to the track or vehicle designer. Its value is therefore included as part of the output of DETRACK in both tabular and graphic format.

(e) Using Eqns (9.5) to (9.8), Risk Indicators described in Section 9.9.3 are determined. The Risk Indicators for indicating the onset of disengagement of either the track drive lug or track guide with either the sprocket or idler, given by Eqns (9.5) to (9.8) and illustrated in Figures 9.60 and 9.61, would be of interest to the track or vehicle designer. The Risk Indicator for indicating the onset of disengagement corresponding to the Detracking Risk Indicator identified in (c) above will be included as part of the output in both tabular and graphic format. For instance, for the track system with the segmented metal track to be evaluated and described later, the Risk Indicator D'_{gi} is included in the output.

9.9.5 Applications of DETRACK to Evaluating Detracking Risks of Track Systems

The module DETRACK of NTVPM has been applied to the evaluation of detracking risks of a segmented metal track (metal link track) system and a band (rubber belt) track system at various values of initial track tension, track longitudinal stiffness and suspension spring stiffness (Wong and Huang, 2006b). In the following, only the results on the effects of initial track tension on the detracking risks of the segmented metal track system of Vehicle A operating on the clayey soil described previously are presented. For details of the study on the effects of track longitudinal stiffness and suspension spring stiffness on detracking risks of the segmented metal track system, as well as the band track system, please refer to Wong and Huang (2006b).

A. When a braking torque is applied

As noted previously, the sprocket of Vehicle A is located in the front. It has five roadwheels with independent torsion bar suspensions and no roller for supporting the top run of the track. The width and pitch of the segmented metal track are 38 and 15 cm, respectively. The height of the track guide and the thickness of the drive slot (for engaging sprocket teeth) are 8.5 and 4 cm, respectively. The outside diameters of the sprocket and idler are 22.7 and 21.9 cm, respectively. Other parameters of Vehicle A have been given previously in Tables 9.1 and 9.2.

The module DETRACK is used to evaluate the detracking risk when a braking torque is applied to the sprocket. This is analogous to examining the detracking risk of the inside track during a steady turning manoeuvre. It is found that the track guide disengaging the idler determines the propensity of this track system to detrack. Figure 9.62 shows the variations of Detracking Risk Indicator D_{gi} for the track guide disengaging the idler, given by Eqn (9.4) and shown in Figure 9.59, with skid and at various values of initial track tension.

For the study of detracking, vehicle skid (or slip) is defined as

$$\text{Skid (or slip)} = 1 - (\text{vehicle actual forward speed } V/\text{vehicle theoretical speed } V_t) \quad (9.9)$$

where V_t is equal to the product of the angular speed and pitch radius of the sprocket. It should be noted that from Eqn (9.9), slip has a positive value, while skid has a negative value. Applying a driving torque to the sprocket results in slip of the track, whereas applying a braking torque to the sprocket results in skid.

It can be seen from Figure 9.62 that the initial track tension has significant effects on the Detracking Risk Indicator D_{gi}. For instance, reducing the value of initial track tension from 10.05 to 5.5 kN, the value of D_{gi} at 60% skid increases from approximately 6.3 to 34.4%, representing an increase of the propensity to detrack by 446%. On the other hand, increasing the value of the initial track tension from 10.05 to 20.10 kN, the value of D_{gi} at 60% skid decreases from approximately 6.3 to 1.2%, representing a decrease of the propensity to detrack by 81%. It is also shown that increasing skid (or the braking torque applied to the sprocket) increases the propensity to detrack. This is particularly significant when the value of initial track tension is low. For instance, at the initial track tension of 5.5 kN, increasing the skid from 5 to 60%, the value of D_{gi} increases from 13.5 to 34.4%, representing an increase of 155%.

Figure 9.63 shows the variations of Risk Indicator D'_{gi} for the leading track guide disengaging the idler, given by Eqn (9.8) and shown in Figure 9.61, with skid and at various values of initial track tension. It can be seen from the figure that the initial track tension has significant effects on the Risk Indicator D'_{gi}. For instance, reducing the value of the initial track tension from 10.05 to 5.5 kN, the value of D'_{gi} at 60% skid increases from approximately 14.7 to 80%, representing an increase of 444%. On the other hand, increasing the value of the initial track tension from 10.05 to 20.10 kN, the value of D'_{gi} at 60% skid decreases from approximately

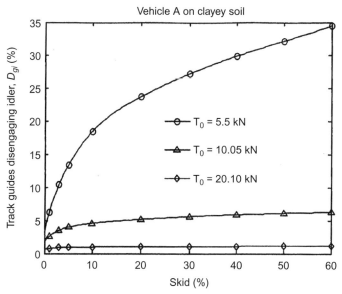

Figure 9.62: Variations of Detracking Risk Indicator D_{gi} for track guides disengaging the idler with skid at various values of initial track tension for Vehicle A on the clayey soil (Reprinted by permission of the Council of the Institution of Mechanical Engineers from Wong and Huang, 2006b)

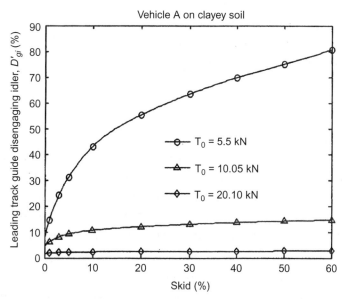

Figure 9.63: Variations of Detracking Risk Indicator D'_{gi} for the leading track guide disengaging the idler with skid at various values of initial track tension for Vehicle A on the clayey soil (Reprinted by permission of the Council of the Institution of Mechanical Engineers from Wong and Huang, 2006b)

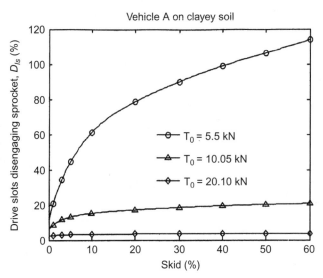

Figure 9.64: Variations of Detracking Risk Indicator D_{ls} for track drive slots disengaging the sprocket with skid at various values of initial track tension for Vehicle A on the clayey soil (Reprinted by permission of the Council of the Institution of Mechanical Engineers from Wong and Huang, 2006b)

14.7 to 3.2%, representing a decrease of 78%. It is also shown that increasing skid increases the risk for the leading track guide disengaging the idler. This is particularly significant when the value of the initial track tension is low. For instance, at the initial track tension of 5.5 kN, increasing the skid from 5 to 60%, the value of D'_{gi} increases from 31 to 80%, representing an increase of 158%.

Figure 9.64 shows the variations of Detracking Risk Indicator D_{ls} for track drive slots disengaging sprocket teeth, given by Eqn (9.1) and shown in Figure 9.58, with skid and at various values of initial track tension. D_{ls} is an indication of possible interruption of power transmission from the sprocket to the track as described in Section 9.9.2. It can be seen from the figure that reducing the value of the initial track tension from 10.05 to 5.5 kN, the value of D_{ls} at 60% skid increases from approximately 20 to 115%, representing an increase of 475%. It should also be pointed out that with $D_{ls} = 115\%$, track drive slots completely disengage sprocket teeth and the track overrides the outside diameter of the sprocket, indicating interruption of transmission of power from the sprocket to the track.

Table 9.24 summarizes, for the segmented metal track system of Vehicle A, the Detracking Risk Indicator D_{gi} for track guides disengaging the idler, Risk Indicator D'_{gi} for the leading track guide disengaging the idler, Detracking Risk Indicator D_{ls} for track drive slots disengaging sprocket teeth, and Risk Indicator D'_{ls} for the leading track drive slot disengaging the sprocket at 60% skid, with the initial track tension of 10.05 kN.

Table 9.24: Detracking risks of the segmented metal track system
of Vehicle A at 60% skid on the clayey soil

Initial track tension	10.05 kN
Detracking risk indicator, D_{gi}	6.3%
Leading track guide disengaging idler, D'_{gi}	14.6%
Track drive slots disengaging sprocket, D_{ls}	20.8%
Leading track drive slot disengaging sprocket, D'_{ls}	48.0%
Source: Wong and Huang (2006b).	

B. When a driving torque applied

Examining the detracking risk when a driving torque is applied to the sprocket is analogous to examining the detracking risk of the outside track during a steady turning manoeuvre. It is found that the track guide disengaging the idler determines the propensity of the track system to detrack. Figure 9.65 shows the variations of Detracking Risk Indicator D_{gi} for track guides disengaging the idler with slip and at various values of initial track tension. Figure 9.66 shows the variations of Risk Indicator D'_{gi} for the leading track guide disengaging the idler with slip and at various values of initial track tension. Figure 9.67 shows the variations of Detracking Risk Indicator D_{ls} for track drive slots disengaging sprocket teeth with slip and at various values of initial track tension.

It can be seen from Figures 9.65, 9.66 and 9.67 that the propensity to detrack decreases with the increase of slip (or the driving torque applied to the sprocket). This is particularly noticeable when the value of the initial track tension is low. For instance, at the initial track tension of 5.5 kN, increasing the slip from 5 to 60%, the value of D_{gi} decreases from 3.6 to 2.0%, representing a decrease of 44%. It is also shown that the initial track tension has significant effects on the Detracking Risk Indicators D_{gi} and D_{ls} and Risk Indicator D'_{gi}. For instance, reducing the value of the initial track tension from 10.05 to 5.5 kN, the value of D_{gi} at 60% slip increases from approximately 1.3 to 2.0%, representing an increase of the propensity to detrack by 54%. On the other hand, increasing the value of the initial track tension from 10.05 to 20.10 kN, the value of D_{gi} decreases from approximately 1.3 to 0.4%, representing a decrease of the propensity to detrack by 69%.

It should be noted that while for a track system with front sprocket the slip and the initial track tension have an influence on detracking risks, the probability of detracking caused by the application of driving torque is very low, as indicated by the very low values of Detracking Risk Indicators D_{gi} and D_{ls} and Risk Indicator D'_{gi}.

C. Summary

Based on the results presented above, it can be concluded that for a track system with the sprocket located in the front, the propensity to detrack when a driving torque is applied to the

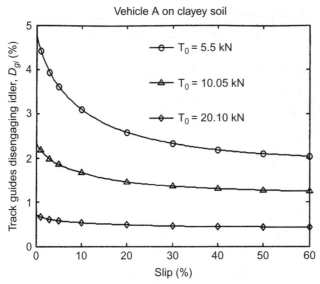

Figure 9.65: Variations of Detracking Risk Indicator D_{gi} for track guides disengaging the idler with slip at various values of initial track tension for Vehicle A on the clayey soil (Reprinted by permission of the Council of the Institution of Mechanical Engineers from Wong and Huang, 2006b)

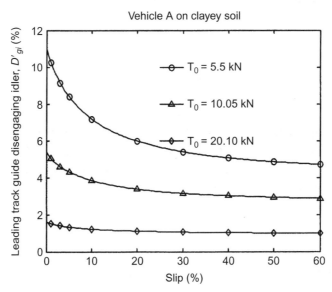

Figure 9.66: Variations of Detracking Risk Indicator D'_{gi} for the leading track guide disengaging the idler with slip at various values of initial track tension for Vehicle A on the clayey soil (Reprinted by permission of the Council of the Institution of Mechanical Engineers from Wong and Huang, 2006b)

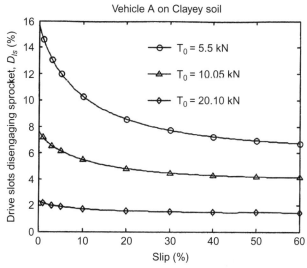

Figure 9.67: Variations of Detracking Risk Indicator D_{ls} for track drive slots disengaging the sprocket with slip at various values of initial track tension for Vehicle A on the clayey soil (Reprinted by permission of the Council of the Institution of Mechanical Engineers from Wong and Huang, 2006b)

sprocket is much lower than that when a braking torque is applied, particularly at high torque. This indicates that for a vehicle with the sprocket located in the front, the detracking risk for the inside track is generally much higher than that for the outside track, during a steady turning manoeuvre. On the other hand, for a vehicle with the sprocket located at the rear, the detracking risk for the outside track with a driving torque applied to its sprocket is generally higher than that for the inside track with a braking torque applied to its sprocket.

It has been found that the track system configuration, including the location of the sprocket and idler, initial track tension, and the magnitude of the braking or driving torque applied to the sprocket have significant effects on the propensity to detrack. The track longitudinal stiffness and suspension stiffness have less significant effects on detracking (Wong and Huang, 2006b).

9.10 Applications of NTVPM to Product Development in the Off-Road Vehicle Industry

NTVPM has been successfully employed to assist off-road vehicle manufacturers in the development of new products, as well as governmental agencies in the evaluation of vehicle candidates. For instance, NTVPM has been used to assist an off-road vehicle manufacturer in a Nordic country in the development of a new combat vehicle and a new version of a two-unit articulated all-terrain vehicle (Wong, 1992a), and in the evaluation of design configurations for a proposed main battle tank. It has been used to assist a military vehicle manufacturer in continental Europe in the development of a high-mobility version of an infantry fighting vehicle (Wong, 1995), as well as a vehicle manufacturer in Asia in the development of an infantry carrier vehicle. NTVPM has been employed to assist a rubber track manufacturer in Canada on the improvement in the design of rubber tracks for recreational and industrial vehicles and in the evaluation of detracking risks of segmented metal track (metal link track) systems and band (rubber belt) track systems. It has also been employed in the assessment of the effects of design modifications on the mobility of a main battle tank for a Canadian government agency and in the evaluation of the mobility of a series of military logistic tracked vehicles for a US government agency.

In this section, the development of a high-mobility version of an infantry fighting vehicle known as ASCOD (Austrian-Spanish Co-Operative Development) is briefly described, to illustrate the applications of NTVPM to new product development in the off-road vehicle industry (Wong, 1995). The original version of the vehicle known as ASCOD PT2 is schematically shown in Figure 9.68(a) (Engeler, 1994). From field tests, it was revealed that its performance on soft ground, particularly on snow-covered terrain, was not as competitive as that of its rivals, and that improvement in its mobility was required. Using NTVPM, the performance of ASCOD PT2 over a variety of terrains was simulated and examined. It was uncovered that over soft terrain, such as deep snow, its rear roadwheel would lift off the ground under certain operating conditions, which led to deterioration of its mobility. A number of new design concepts with different track–roadwheel–suspension configurations were examined, and their performances over various types of terrain were simulated using NTVPM. As a result of a detailed simulation study, a new configuration for the high-mobility version of the ASCOD was recommended, within the design constraints specified by the manufacturer. The new version known as ASCOD PT3 is schematically shown in Figure 9.68(b). Major differences between the design of the original version PT2 and that recommended for the high-mobility version PT3 are highlighted below:

- In PT2, there are six roadwheels on each of the two tracks. On the other hand, in PT3, there are seven roadwheels on each track. The increase in the number of roadwheels would lead to a reduction in peak pressure and a more uniform normal pressure distribution on the track–terrain interface, hence leading to improved performance on soft terrain.

- The roadwheel spacing in PT3 is closer than that in PT2, which improves the load-carrying characteristics of the track segments between adjacent roadwheels. This leads to a reduction in peak pressure and track sinkage, and to improved performance on soft terrain.

- In the suspension system of the PT2, trailing arms (i.e. one end of the torsion arm connected with the centre of the roadwheel being behind the other end of the arm

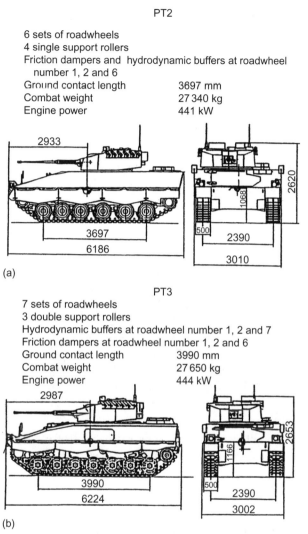

Figure 9.68: Infantry fighting vehicle ASCOD PT2 and its high-mobility version PT3 developed on the basis of simulation results obtained using the computer-aided method NTVPM (Reprinted by permission of *ISTVS* from Engeler, 1994)

connected with the torsion bar) were used for the first five roadwheels and a leading arm (i.e. the end of the torsion arm connected with the centre of the roadwheel being in front of the other end of the arm connected with the torsion bar) was used for the last roadwheel. To minimize the potential for the last roadwheel to lift off the ground, in the suspension system of the PT3, all the torsion arms are of the trailing type.

- New torsion bar suspension settings are selected for the PT3 to provide improvement in load distribution among roadwheels. The measured load distribution for the PT2 is shown in Figure 9.69 (Engeler, 1994), and that for the PT3, together with the predicted load distribution obtained using NTVPM, is shown in Figure 9.70 (Wong, 1995). It can be seen that the load distribution among the intermediate roadwheels for the PT3 is more uniform than that for the PT2. It should be noted that the loads under the front and rear roadwheels are generally lower than those on the intermediate roadwheels, because of the lifting effect of the track tension acting on them. It is seen that the load distribution among the roadwheels for the PT3 predicted using NTVPM is in close agreement with the measured data.

- The longitudinal profile of the vehicle belly (hull) for the PT3 is shaped similar to that of a ski, to minimize the belly drag, in the event that the track sinkage is greater

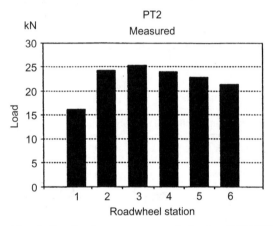

Figure 9.69: Measured load distribution among roadwheels of ASCOD PT2 (Reprinted by permission of *ISTVS* from Engeler, 1994)

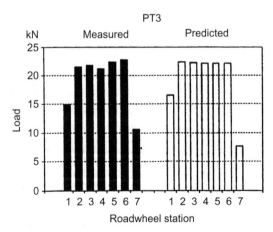

Figure 9.70: Measured load distribution among roadwheels of ASCOD PT3 in comparison with that predicted using NTVPM (Reprinted by permission of *ISTVS* from Engeler, 1994 and Wong, 1995)

than the vehicle ground clearance and the vehicle belly is in contact with the terrain surface.

- To further enhance its mobility on highly compressible terrain, such as deep snow, the suspension settings are tuned in such a way that the vehicle belly takes a nose-down attitude when stationary on level, hard ground. As mentioned previously, this would minimize the load supported by the belly and the associated belly drag and would enhance the traction developed by the tracks when operating on highly compressible terrain.

- For PT3 there is a central initial track tension regulating system remotely controlled by the driver to further enhance its mobility over soft terrain.

The original version PT2 and the new version PT3 were tested in parallel over a variety of terrains by the Military Technology Agency, Austria (Engeler, 1994).

Figure 9.71 shows a comparison of the coefficient of rolling resistance of ASCOD PT3 and that of the PT2 over a variety of surfaces, ranging from asphalt to soft clay. On asphalt, the rolling resistance is primarily due to the internal losses in the track–roadwheel system. The rolling resistance coefficients for both the PT3 and the PT2 on the asphalt are the same, as shown in Figure 9.71. However, on the soft clay or on the field, there is a noticeable difference in the value of the coefficient of rolling resistance between the PT2 and PT3. Assuming that the rolling resistance measured on the asphalt represents the internal resistance of the track–roadwheel system, the difference between the rolling resistance coefficient measured on the soft clay or on the field and that measured on the asphalt represents the rolling resistance

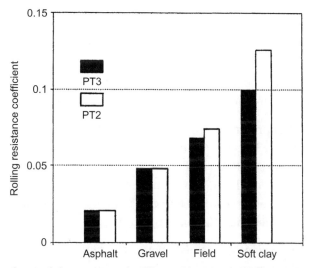

Figure 9.71: Comparison of the measured rolling resistant coefficient of PT2 with that of PT3 over various surfaces (Reprinted by permission of *ISTVS* from Engeler, 1994)

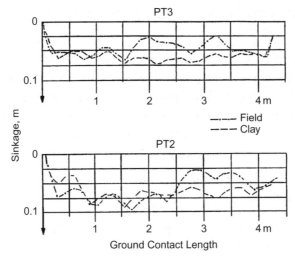

Figure 9.72: Comparison of the measured sinkages under roadwheel stations of PT2 with those of PT3 on clay and on the field (Reprinted by permission of *ISTVS* from Engeler, 1994)

coefficient due to track–terrain interaction. Following this approach, the ratio of the rolling resistance coefficient due to track–terrain interaction of the PT3 to that of the PT2 on the soft clay is 0.76. This is consistent with the simulation results obtained using NTVPM (Wong, 1995). The lower rolling resistance for the PT3 is primarily due to its sinkage being lower than that for the PT2. Figure 9.72 shows the measured sinkages for the PT3 and PT2 on the soft clay and on the field (Engeler, 1994). It shows that on the soft clay the maximum sinkage of the PT3 is approximately 25% lower than that of the PT2.

Figure 9.73 shows a comparison between the measured drawbar pull of the PT3 and that of the PT2 on the soft clay, on the field, and on the asphalt (Engeler, 1994). On the clay, the ratio of measured drawbar pull of the PT3 to that of the PT2 is 1.56. It is consistent with the ratio predicted. On the field, the ratio of the measured drawbar pull of the PT3 to that of the PT2 is 1.13. This is again consistent with the ratio predicted using NTVPM (Wong, 1995).

Based on the field test data presented above, it can be said that the predicted improvement in soft ground mobility of the PT3 over PT2 using NTVPM has been confirmed. The detailed test data on the performances of PT3 and PT2 on snow-covered terrain have not been released. However, it is noted in Engeler (1994) that significant improvement in mobility of the PT3 over PT2 on snow was also observed. Because of its improved performance on soft ground, PT3 has since been in production in Europe (under different trade names).

In summary, through the example presented above, it is demonstrated that NTVPM is a useful tool for the off-road vehicle industry in the evaluation of competing designs and in the development of new products.

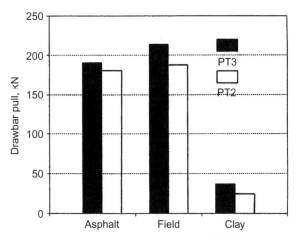

Figure 9.73: Comparison of the measured drawbar pull of PT2 with that of PT3 over various surfaces (Reprinted by permission of *ISTVS* from Engeler, 1994)

9.11 Concluding Remarks

Based on the results of the parametric analyses presented in this chapter and on the example for the applications of NTVPM to product development described above, the capabilities of the computer-aided method NTVPM in performance and design evaluation of tracked vehicles with flexible tracks have been demonstrated. It is shown that NTVPM can be effectively used to evaluate the effects of vehicle design on performance over unprepared terrain, or to assess the impact of terrain conditions on vehicle performance.

From the results presented, the following remarks may be made:

1. The track system configuration is shown to have significant effects on vehicle performance, particularly on highly compressible terrain, such as the Hope Valley snow used in the study.

   For given overall dimensions of a track system, it is desirable, from the traction perspective, to install as many roadwheels as appropriate, giving due consideration to the desirable roadwheel diameter and suspension travel for off-road operations. Among the track system configurations examined in the study, the overlapping roadwheel configuration is shown to exhibit better mobility than the conventional (non-overlapping) arrangement, particularly on soft terrain. It provides a more uniform ground pressure distribution along the contact length of the track, while retaining the desirable characteristics of large diameter roadwheels with appropriate suspension travels for negotiating rough terrain. The improvement in uniformity of normal pressure distribution on the track–terrain interface with the overlapping roadwheel arrangement reduces track sinkage and track motion resistance, resulting in improved vehicle mobility. If its susceptibility to clogging with terrain materials in the spacing between the overlapping roadwheels is alleviated by proper design, this system may have the potential for wide applications to improving tracked vehicle mobility on soft terrain.

It is shown that on firmer terrain, the track system configuration generally has less significant effects on vehicle mobility than on highly compressible terrain.

2. The initial track tension, which is the tension in the track system when the vehicle is stationary on level, hard ground, is shown to have significant effects on vehicle performance, particularly on highly compressible terrain.

   With the increase of initial track tension, the track becomes tighter and the track segments between roadwheels take up more load, hence lowering the peak normal pressure on the track–terrain interface, track sinkage and track motion resistance. With a tighter track, it also enhances the development of vehicle thrust, as the track segments between roadwheels flatten out and the horizontal component of the shear stress on the track–terrain interface becomes larger.

   On highly compressible terrain, such as the Hope Valley snow, the track sinkage exceeds vehicle ground clearance and the vehicle belly comes into contact with the terrain surface. Under these circumstances, increasing the initial track tension has two major effects. Firstly, it reduces the sinkages of both the belly and the track, hence lowering both the belly drag and the track motion resistance. Secondly, it reduces the load supported by the belly, hence increasing the proportion of the load applied to the track. This enables the track to develop higher thrust on most terrains with a noticeable frictional component in their shear strengths. This indicates that when the vehicle belly contacts the terrain surface, increasing the initial track tension has even more significant effects on improving vehicle mobility.

   While increasing the initial track tension generally improves vehicle performance, it may increase the wear and tear of the track system. As a practical measure to accommodate these conflicting factors, it is suggested that a central initial track tension regulating system controlled by the driver be installed. This remotely controlled regulating system enables the driver to conveniently increase the initial track tension, when traversing soft terrain is anticipated. It also enables the driver to reduce the initial track tension when the vehicle operates on firm terrain to minimize the wear and tear of the track system. The central initial track tension regulating system for improving tracked vehicle mobility is analogous to the central tyre inflation system for improving the mobility of off-road wheeled vehicles. It also appears that in many cases retrofitting existing tracked vehicles with the central initial track tension regulating system may be one of the most cost-effective means of enhancing their soft ground mobility, as it is simply an add-on device. Its installation requires only minor modifications to existing vehicles.

For vehicles designed for operation over soft terrain for an extended period of time, a case may be made for the development of an automatic initial track tension regulating system controlled by the slip of the track, in order to further enhance its effectiveness.

It should be mentioned that as the terrain becomes firmer, the effects of the initial track tension on vehicle performance become less significant. Furthermore, the effects of initial track tension are dependent on vehicle design features. For instance, for the two-unit articulated vehicle examined in the study, it has a relatively low weight per unit nominal track contact area, hence relatively low track sinkage and track motion resistance. The increase of initial track tension may not necessarily lead to significant improvement in vehicle performance on relatively firm terrain, such as the clayey soil used in the study. For agricultural or industrial tractors with friction drive between the sprocket and the rubber belt (band) track, to minimize the slippage between the sprocket and the track, the initial track tension coefficient under normal operating conditions is set to a high level. Under these circumstances, further increasing the initial track tension coefficient would not necessarily result in significant improvement in vehicle performance, as its effects may already reach a plateau.

3. For a given track system, the increase in vehicle weight generally has adverse effects on mobility on soft terrain.

   On soft terrain, with the increase of vehicle weight, track sinkage and track motion resistance coefficients generally increase, while the thrust coefficient may decrease, particularly on terrain with a significant cohesive component in its shear strength. As discussed previously, the thrust of a vehicle is developed by the shearing action of the track on the terrain. It is derived from the frictional and cohesive components of the shear strength of the terrain. The part of vehicle thrust derived from the frictional component of the shear strength increases proportionally with the increase of vehicle weight, while the other part derived from the cohesive component is proportional to the track contact area and is independent of vehicle weight. For a given track system, the thrust coefficient, which is the ratio of vehicle thrust to its weight, therefore generally decreases with the increase of vehicle weight, on terrain with a significant cohesive component in its shear strength.

   It is shown that on both the Hope Valley snow and the clayey soil examined in the study, for a given track system configuration, vehicle performance deteriorates with the increase of vehicle weight. It should be pointed out, however, that for the same vehicle weight, the mobility of the track system with overlapping roadwheel arrangement is significantly higher than that of the conventional (non-overlapping) configuration.

4. On soft terrain, it is shown that for a given track system configuration, increasing the width of the track generally has beneficial effects on vehicle performance.

As noted previously, on terrain with both frictional and cohesive components in its shear strength, the part of vehicle thrust derived from the cohesive component increases proportionally with the increase of track contact area which is directly related to the track width.

It is shown that on both the Hope Valley snow and the clayey soil examined in this study, for a given track system configuration, vehicle performance increases with the increase of track width. It should be pointed out, however, that with the same track width the mobility of the track system with overlapping roadwheel arrangement is significantly higher than that of the conventional (non-overlapping) configuration.

5. On highly compressible terrain where the track sinkage exceeds vehicle ground clearance and the vehicle belly comes into contact with the terrain surface, suspension setting has noticeable effects on vehicle mobility.

    Suspension setting affects the load distribution among the roadwheels, as well as vehicle belly attitude, such as 'nose-up' or 'nose-down'. If the belly takes a nose-down attitude in operation, then a large part of the belly (except the front part) will not be in contact with the terrain surface. If the belly takes a nose-up attitude, then the belly comes into contact with the terrain surface, in part or in full. This induces a drag, commonly known as the belly drag, which is caused by the sliding of the belly on the terrain and by its bulldozing effect. Also with the belly in contact with the terrain, it supports part of the vehicle weight and reduces the portion of the load applied to the track. On terrain with a significant frictional component in its shear strength, this reduces the vehicle thrust. This indicates that an unfavourable belly attitude not only induces belly drag but also reduces vehicle thrust.

    Operating on highly compressible terrain, the vehicle belly usually takes a nose-up attitude, caused by load transfer from the front to the rear due to drawbar pull among other factors. It is shown that the suspension setting that can minimize the positive trim angle provides the vehicle with improved mobility.

    On less compressible terrain, where the vehicle belly is not in contact with the terrain surface, the suspension setting only affects the load distribution among the roadwheels. Thus the optimal suspension setting that provides the vehicle with peak performance varies with terrain conditions. In view of this, consideration may be given to the development of a suspension setting regulating system that enables the driver to change suspension settings according to terrain conditions. For vehicles with hydro-pneumatic suspensions, it appears that such a suspension setting regulating system may be readily implemented.

6. On soft terrain, the longitudinal location of the centre of gravity (CG) of the sprung weight has noticeable effects on vehicle mobility.

The longitudinal location of the CG affects the load distribution among the roadwheels, as well as the attitude of the vehicle belly.

On highly compressible terrain where the vehicle belly comes into contact with the terrain surface, the effects of the longitudinal location of the CG are analogous to that of the suspension setting. For a given track system configuration on a particular terrain, an optimal longitudinal location of the CG may be identified, which can provide the vehicle with the highest mobility. It should also be pointed out that if the longitudinal location of the CG is shifted to a position that gives an unfavourable vehicle belly attitude and load distribution among the roadwheels, the vehicle may become immobilized. In general, the optimal longitudinal location of the CG, at which the performance reaches its peak, varies with terrain conditions.

7. For a vehicle with tracks that exhibit a certain level of longitudinal elasticity (stiffness), such as those commonly used in military vehicles with rubber bushings, the location of the sprocket has noticeable effects on vehicle performance on soft terrain.

   With the sprocket located in the front of the vehicle, in forward motion the segments in the top run of the track between the front sprocket and the rear idler and those between the idler and the rear roadwheel are subject to higher tension, whereas the track segments between the front roadwheel and the sprocket are subject to lower tension. On the other hand, with the sprocket located at the rear, only the track segments between the rear sprocket and the rear roadwheel are subject to higher tension, whereas the segments in the top run of the track between the front idler and the rear sprocket and those between the front roadwheel and the idler are subject to lower tension. Consequently, in forward motion a vehicle with the sprocket located in the front, the total elongation of the track is larger than that with the sprocket located at the rear. With larger elongation, more track length is available for the deflection of the track segments between roadwheels and as a result these track segments become looser and support less load. Consequently, on soft terrain the sprocket location affects the performance of a tracked vehicle, if the effect of track tension on its elongation is noticeable.

   On highly compressible terrain, such as the Hope Valley snow, with the sprocket located at the rear, the performance of the baseline vehicle with five roadwheels is noticeably higher than that with the sprocket located in the front. The effects of sprocket location on vehicle mobility, however, become less significant when the terrain becomes firmer or the longitudinal stiffness of the track is higher.

   It should be noted that the sprocket location is often constrained by the function and layout of the vehicle. For instance, for an armoured personnel carrier, to provide better protection and more convenient access for the personnel to enter (or exit) the vehicle, a large door is usually installed at the rear, while the engine is mounted in the front of the vehicle. With mechanical transmissions (or drivelines), the sprocket is constrained to locate in the front. With the proposed electrical drive system for future generation of military vehicles, it may provide more flexibility in vehicle layout, as well as the selection of the location of the sprocket.

8. On highly compressible terrain, such as the Hope Valley snow, the articulation joint configuration in the pitch plane, such as unrestricted or locked, is shown to have noticeable effects on performance of a two-unit articulated vehicle.

    There has been a perception that if the articulation joint is locked in the pitch plane, the two units are rigidly connected and form a longer vehicle, it will alleviate the characteristic 'tail-down' attitude, hence leading to improvement in the overall performance of the articulated vehicle. Results of the study show that in fact the vehicle with the unrestricted articulation joint that allows the two units to pitch freely relative to each other offers better performance than the vehicle with the locked articulation joint. It is found that within a certain range of the initial track tension coefficient, with the locked joint, some of the roadwheels may be lifted off the ground. With a roadwheel on the track off the ground, the effective total track contact area is reduced. This leads to the deterioration of vehicle performance. The lift-off from the ground of a roadwheel in the track system is a result of complex interactions between the front and rear units and between the track and the terrain. It is also a function of the track system configuration, suspension characteristics, and terrain conditions.

9. A special module DETRACK of NTVPM has been developed for evaluating the detracking risks of track systems. It is found that for a front-sprocket-drive track system in a turning manoeuvre, the inside track under braking has a much higher propensity to detrack than the outside track. On the other hand, for a rear-sprocket-drive track system in a turning manoeuvre, the outside track with a driving torque applied to the sprocket has a higher propensity to detrack than the inside track. Both the magnitude of the braking or driving torque applied to the sprocket and the initial track tension have significant effects on the propensity to detrack.

10. It should be realized that the interaction between a tracked vehicle and deformable terrain is a complex phenomenon, and that vehicle mobility is a complicated function of vehicle design and operating parameters and terrain characteristics. To take into account all major design parameters of the vehicle and all pertinent terrain characteristics in predicting tracked vehicle performance on unprepared terrain, a comprehensive and realistic computer simulation model is needed. As demonstrated through the parametric analyses of vehicle design and performance presented in this chapter, the computer-aided method NTVPM can fulfil this need. Consequently, NTVPM is a useful tool for the vehicle engineer to quantitatively evaluate vehicle design concepts and to optimize design parameters, prior to prototyping and hardware testing. This will significantly contribute to

the shortening of the development and design cycle of new products and to the reduction of the associated costs. NTVPM is also useful for the procurement manager to evaluate vehicle candidates prior to fielding testing.

11. It is demonstrated that the applications of the computer-aided method NTVPM to vehicle design study may lead to innovations. For instance, the research findings on the effects of the initial track tension on vehicle performance obtained using NTVPM lead to the innovative concept of the initial track tension regulating system for improving tracked vehicle mobility on soft terrain. The results of the study using NTVPM on the effects of suspension settings on vehicle performance, particularly on highly compressible terrain where track sinkage exceeds vehicle ground clearance, lead to the innovative concept of a suspension setting regulating system to minimize the additional drag caused by the vehicle belly contacting the terrain surface.

CHAPTER 10

Computer-Aided Method RTVPM for Evaluating the Performance of Vehicles with Long-Pitch Link Tracks

The computer-aided method NTVPM described in the previous two chapters is intended for design and performance evaluation of tracked vehicles with rubber belt (band) tracks or link tracks with relatively short track pitch, referred to as flexible tracks. The flexible track allows the vehicle to operate at relatively high speeds without excessive speed fluctuations and associated vibrations. For slow-moving tracked vehicles commonly used in farming, the construction industry, the logging industry, etc., rigid link tracks with relatively long track pitch are employed. The intention is to provide a more uniform ground pressure distribution. For this type of track system, the ratio of roadwheel diameter to track pitch may be as low as 1.2 and the ratio of roadwheel spacing to track pitch is typically 1.5. As a result, the computer-aided method NTVPM would not be appropriate for evaluating track systems with relatively long-pitch tracks. To provide a comprehensive method for design and performance evaluation of tracked vehicles with relatively long-pitch link tracks, a computer-aided method, known as RTVPM, has been developed (Gao and Wong, 1994; Wong and Gao, 1994; Wong, 1998).

10.1 Basic Approach to the Development of the Computer-Aided Method RTVPM

RTVPM treats the track as a system of rigid links connected with frictionless pins, as schematically shown in Figure 10.1. The roadwheels, supporting rollers and sprocket are assumed to be rigidly attached to the track frame. This assumption is realistic, as for most slow-moving tracked vehicles there are no suspensions for the roadwheels. The centre of the idler is considered to be mounted in a pre-compressed spring, as shown in Figure 10.1.

RTVPM takes into account all major design parameters of the vehicle, including the vehicle weight, location of the centre of gravity, number of roadwheels, roadwheel dimensions and spacing, locations of the sprocket and idler, supporting roller arrangements, track width, track pitch, initial track tension, and drawbar hitch location. Terrain characteristics that are taken into account in RTVPM are the same as those in NTVPM.

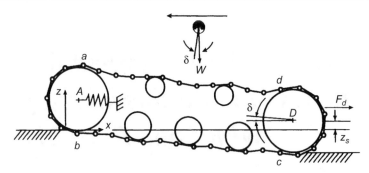

Figure 10.1: Schematic diagram of a track system with long-pitch rigid links (Reprinted by permission of the Council of the Institution of Mechanical Engineers from Gao and Wong, 1994)

RTVPM can be used to predict the normal and shear stress distributions on the track–terrain interface, and the track motion resistance due to track–terrain interaction, thrust, drawbar pull and tractive efficiency of the vehicle as functions of track slip on deformable terrain. As the method takes into account all major design features and pertinent terrain characteristics, it is suited for the evaluation of design concepts and for the assessment of the effects of terrain conditions on vehicle performance. Thus, it is useful to the vehicle engineer in the selection of appropriate vehicle configuration and in the optimization of vehicle design parameters, as well as to the procurement manager in the evaluation of vehicle candidates for a given mission and environment.

The prime objective for the development of RTVPM is to establish an analytical procedure with which the interaction between the long-pitch link track and the terrain under steady-state operating conditions may be predicted in a realistic manner. In the analysis, the track system is divided into four sections: the upper run of the track supported by rollers, sprocket and idler, such as *ad* shown in Figure 10.1; the lower run of the track in contact with the terrain, such as *bc*; the section in contact with the idler, such as *ab*; and the section in contact with the sprocket, such as *cd*. The origin for the coordinate system x, z is located at the front contact point of the track system, as shown in Figure 10.1. The attitude of the track frame is identified by the angle δ, and the vertical position of the centre of the sprocket is defined by z_s, which also indicates the sinkage of the track system on deformable terrain. Since the roadwheels, supporting rollers and sprocket are assumed to be rigidly connected with the track frame, their positions with respect to the coordinate system can be defined as functions of z_s, δ and their locations with respect to the centre of the sprocket. As noted previously, the idler is considered to be mounted on a pre-compressed spring, which allows the centre of the idler to move relative to the track frame within prescribed limits.

10.1.1 Analysis of the Upper Run of the Track

To illustrate the basic approach, a simplified upper run of the track with only one supporting roller, as shown in Figure 10.2, is used in the analysis presented below. It should be pointed

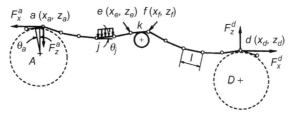

Figure 10.2: The upper run of a track with a supporting roller (Reprinted by permission of the Council of the Institution of Mechanical Engineers from Gao and Wong, 1994)

out, however, that the method of approach can be extended to the analysis of a track system with any number of supporting rollers.

The upper run of the track *ad* shown in Figure 10.2 may be divided into two types of segment: one hanging between two supports, such as *ae*, and the other supported by a roller, such as *ef*. For a track segment hanging between two supports, such as *ae*, the following two equations describing the relative positions of pin *a* and pin *e* may be established:

$$\sum l\cos\theta_j + x_a - x_e = 0 \tag{10.1}$$

$$\sum l\sin\theta_j + z_a - z_e = 0 \tag{10.2}$$

where l is the pitch of the track link, θ_j is the inclination angle of link j, and (x_a, z_a) and (x_e, z_e) are the coordinates of pin *a* and pin *e*, respectively.

For link j in track segment *ae*, shown in Figure 10.3, the following three equations can be established from the equilibrium of the link:

$$\sum F_x = 0, \quad -F_{xl}^j + F_{xr}^j = 0 \tag{10.3}$$

$$\sum F_z = 0, \quad -F_{zl}^j + F_{zr}^j - lw = 0 \tag{10.4}$$

$$\sum M_r = 0, \quad -F_{xl}^j l\sin\theta_j + \left(F_{zl}^j l + \frac{l^2 w}{2}\right)\cos\theta_j = 0 \tag{10.5}$$

where F_{xl}^j and F_{zl}^j are the horizontal and vertical forces at the left-hand side pin of link j, respectively, F_{xr}^j and F_{zr}^j are the horizontal and vertical forces at the right-hand side pin of link j, respectively, and w is the weight of the link per unit length.

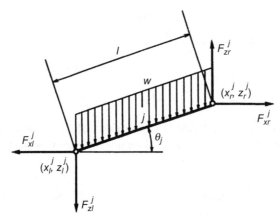

Figure 10.3: The equilibrium of a link in the upper run of the track (Reprinted by permission of the Council of the Institution of Mechanical Engineers from Gao and Wong, 1994)

Based on Eqns (10.3) and (10.4), the following general expressions for the forces at the left- and right-hand side pins of the links in segment ae can be established:

$$F_{xr}^j = F_{xl}^j = F_{xr}^{j-1} = F_{xl}^{j-1} = \cdots = F_x^a \tag{10.6}$$

$$F_{zr}^j = F_{zl}^j + lw = F_{zl}^{j-1} + 2lw = \cdots = F_z^a + jlw \tag{10.7}$$

where F_x^a and F_z^a are the horizontal and vertical forces at pin a, respectively, as shown in Figure 10.2. For pin e in Figure 10.2, $j = 5$, the forces are expressed by

$$F_x^e = F_x^a$$

$$F_z^e = F_z^a + 5lw$$

where F_x^e and F_z^e are the horizontal and vertical forces at pin e, respectively.

From Eqn (10.5), the inclination angle θ_j of link j can be expressed by

$$\theta_j = \tan^{-1}\left(\frac{2F_{zl}^j + lw}{2F_{xl}^j}\right) \tag{10.8}$$

As shown in Eqns (10.6) and (10.7), F_{xl}^j and F_{zl}^j can be expressed as a function of F_x^a and F_z^a, respectively. As a result, for given values of l and w, θ_j can be expressed as a function of F_x^a and F_z^a:

$$\theta_j = \theta_j(F_x^a, F_z^a) \tag{10.9}$$

For track segment *ae* shown in Figure 10.2, Eqns (10.1) and (10.2) can be rewritten as

$$f_1(F_x^a, F_z^a, x_a) - x_e = 0 \tag{10.10}$$

$$f_2(F_x^a, F_z^a, z_a) - z_e = 0 \tag{10.11}$$

where

$$f_1(F_x^a, F_z^a, x_a) = \sum l \cos \theta_j (F_x^a, F_z^a) + x_a$$

$$f_2(F_x^a, F_z^a, z_a) = \sum l \sin \theta_j (F_x^a, F_z^a) + z_a$$

F_x^a and F_z^a are two of the basic unknowns to be determined from the overall equilibrium conditions of the track system, which is discussed later. (x_a, z_a) define the positions of the link in contact with the idler relative to the track frame. (x_e, z_e) are defined by the equilibrium conditions of link k in contact with the supporting roller, as shown in Figure 10.4.

By neglecting the friction between the track link and supporting roller, the following three equations can be established from the equilibrium conditions of link k:

$$\sum F_x = 0, \quad -F_x^e + F_x^f - G_s^k \sin \theta_s^k = 0 \tag{10.12}$$

$$\sum F_z = 0, \quad -F_z^e + F_z^f + G_s^k \cos \theta_s^k - lw = 0 \tag{10.13}$$

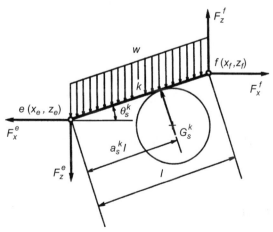

Figure 10.4: The equilibrium of a link in contact with a supporting roller (Reprinted by permission of the Council of the Institution of Mechanical Engineers from Gao and Wong, 1994)

$$\sum M_k = 0, \quad -\left[F_x^e a_s^k + F_x^f (1 - a_s^k)\right] l \sin\theta_s^k + \left[F_z^e a_s^k + F_z^f (1 - a_s^k) + lw(a_s^k - 1/2)\right] \quad (10.14)$$
$$\times l \cos\theta_s^k = 0$$

where F_x^f and F_z^f are the horizontal and vertical forces at pin f of link k, respectively, a_s^k is a parameter defining the position of the contact point between link k and the supporting roller, G_s^k is the normal force exerted on link k by the supporting roller, and θ_s^k is the inclination angle of link k. For each of the links in contact with a supporting roller, three equations similar to Eqns (10.12), (10.13) and (10.14) can be derived.

It should be pointed out that for link k, in the above equations $F_x^e = F_x^a$ and $F_z^e = F_z^a + 5lw$, as noted previously. It should also be mentioned that a_s^k, G_s^k and θ_s^k are unknowns and that the coordinates of pin e, (x_e, z_e), and those of pin f, (x_f, z_f), can be defined in terms of a_s^k and θ_s^k and the coordinates of the centre of the supporting roller, which are functions of z_s, δ and the distance from the centre of the sprocket.

Taking into account that (x_e, z_e) are functions of a_s^k, θ_s^k, z_s and δ, Eqns (10.10) and (10.11) can be rewritten as follows:

$$f_1(z_s, \delta, F_x^a, F_z^a, a_s^k, \theta_s^k, x_a) = 0 \quad (10.15)$$

$$f_2(z_s, \delta, F_x^a, F_z^a, a_s^k, \theta_s^k, z_a) = 0 \quad (10.16)$$

A similar approach to the analysis of track segment ae can be followed to analyse track segment fd shown in Figure 10.2. Two equations similar to Eqns (10.10) and (10.11) can be obtained for segment fd. However, as noted from Eqns (10.12) and (10.13), the forces at pin f, F_x^f and F_z^f, are functions of G_s^k, θ_s^k, F_x^e and F_z^e. Furthermore, the coordinates of pin f, (x_f, z_f), are functions of a_s^k, θ_s^k, z_s and δ. Consequently, for segment fd, the following two functional relationships can be established:

$$f_3(z_s, \delta, F_x^a, F_z^a, a_s^k, G_s^k, \theta_s^k, x_a) - x_d = 0 \quad (10.17)$$

$$f_4(z_s, \delta, F_x^a, F_z^a, a_s^k, G_s^k, \theta_s^k, z_a) - z_d = 0 \quad (10.18)$$

where (x_d, z_d) are coordinated of pin d.

Following a similar approach, Eqn (10.14) can be rewritten as follows:

$$f_5(F_x^a, F_z^a, a_s^k, G_s^k, \theta_s^k) = 0 \quad (10.19)$$

The analysis of the segments of the upper run of the track with one supporting roller described above can be extended to a track system with any number of supporting rollers. For a track system with m supporting rollers, $2 + 3m$ independent equations consisting of Eqns (10.15) and (10.16) and m sets of equations similar to Eqns (10.17), (10.18) and (10.19) can be established. In these equations, z_s, δ, F_x^u, F_z^u, x_a, x_d and z_d, and m sets of a_s^k, G_s^k and θ_s^k ($k = 1, 2, \ldots, m$) are unknowns.

10.1.2 Analysis of the Lower Run of the Track in Contact with the Terrain

To illustrate the basic approach, a simplified lower run of the track in contact with the terrain with only one roadwheel, as shown in Figure 10.5, is used in the analysis presented below. It should be pointed out, however, that the method of approach can be extended to the analysis of a track system with any number of roadwheels.

The approach to the analysis of the lower run of the track in contact with the terrain, such as bc shown in Figure 10.5, is similar to that of the upper run of the track, except the normal pressure and shear stress exerted on the track link by the terrain have to be taken into account. The normal pressure p acting on the link is dependent upon the sinkage z of the link and upon the pressure–sinkage relationship and the response to repetitive loading of the terrain, as described in Chapter 4. The shear stress s is dependent upon the position of the link, x and z, slip of the track i, and the shear stress–shear displacement relationship and the response to repetitive shearing of the terrain, as discussed in Chapter 5.

Similar to the upper run of the track, the lower run of the track may be divided into two different types of segment: one in contact with the terrain only, such as bg in Figure 10.5; and the other in contact with both the terrain and the roadwheel, such as gh shown in the figure. For track segment bg, the following two equations describing the relative positions of b and g can be established:

$$\sum l\cos\theta_j + x_b - x_g = 0 \qquad (10.20)$$

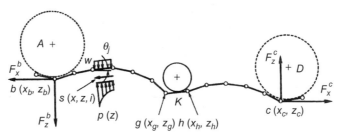

Figure 10.5: The lower run of a track in contact with the terrain (Reprinted by permission of the Council of the Institution of Mechanical Engineers from Gao and Wong, 1994)

$$\sum l\sin\theta_j + z_b - z_g = 0 \tag{10.21}$$

where (x_b, z_b) and (x_g, z_g) are the coordinates of pin b and pin g, respectively, and θ_j is the inclination angle of link j, as shown in Figure 10.6.

By considering the equilibrium conditions for link j in track segment bg shown in Figure 10.6, which is in contact with the terrain only, the following three equations can be established:

$$\sum F_x = 0, \quad -F_{xl}^j + F_{xr}^j - P_j\sin\theta_j - S_j\cos\theta_j = 0 \tag{10.22}$$

$$\sum F_z = 0, \quad -F_{zl}^j + F_{zr}^j + P_j\cos\theta_j - S_j\sin\theta_j - lw = 0 \tag{10.23}$$

$$\sum M = 0, \quad -F_{xl}^j l\sin\theta_j + \left(F_{zl}^j l + \frac{l^2}{2}w\right)\cos\theta_j - P_j l_j = 0 \tag{10.24}$$

where F_{xl}^j and F_{zl}^j are the horizontal and vertical forces at the left-hand side pin of link j, respectively, F_{xr}^j and F_{zr}^j are the horizontal and vertical forces at the right-hand side pin of link j, respectively, P_j and S_j are the resultant normal force and shear force exerted on link j, respectively, and l_j is a parameter defining the location of the point of application of the resultant normal force P_j shown in Figure 10.6.

P_j and S_j can be obtained by integrating the normal pressure p and shear stress s over the contact area of the link, respectively. For a given terrain with known mechanical properties (such as the

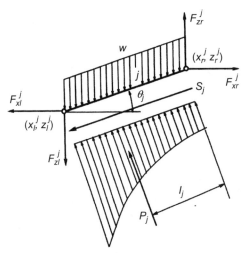

Figure 10.6: The equilibrium of a link in contact with the terrain (Reprinted by permission of the Council of the Institution of Mechanical Engineers from Gao and Wong, 1994)

pressure–sinkage relations, shear strength and shear stress–shear displacement relations), P_j, S_j and l_j can be expressed as functions of the position of the link and track slip i as follows:

$$P_j = P_j(z_l^j, \theta_j) \tag{10.25}$$

$$S_j = S_j(x_l^j, z_l^j, \theta_j, i) \tag{10.26}$$

$$l_j = l_j(z_l^j, \theta_j) \tag{10.27}$$

where (x_l^j, z_l^j) are the coordinates of the left-hand side pin of link j, which can be expressed as functions of x_b, z_b and θ_j. Substituting Eqns (10.25), (10.26) and (10.27) into Eqn (10.24), the following equation is obtained:

$$-F_{xl}^j l \sin\theta_j + \left(F_{zl}^j l + \frac{l^2}{2} w\right)\cos\theta_j - P_j(z_l^j, \theta_j) l_j(z_l^j, \theta_j) = 0 \tag{10.28}$$

From Eqn (10.28), θ_j can be expressed as a function of F_{xl}^j, F_{zl}^j and z_l^j:

$$\theta_j = \theta_j(F_{xl}^j, F_{zl}^j, z_l^j) \tag{10.29}$$

F_{xl}^j and F_{zl}^j are related to the forces (F_x^b, F_z^b) and coordinates (x_b, z_b) at b, and track slip i; z_l^j is a function of (x_b, z_b) and θ_j. Consequently,

$$\theta_j = \theta_j(F_x^b, F_z^b, x_b, z_b, i) \tag{10.30}$$

For track segment bg, substituting Eqn (10.30) into Eqns (10.20) and (10.21), the following equations are obtained:

$$f_6(F_x^b, F_z^b, x_b, z_b, i) - x_g = 0 \tag{10.31}$$

$$f_7(F_x^b, F_z^b, x_b, z_b, i) - z_g = 0 \tag{10.32}$$

where

$$f_6(F_x^b, F_z^b, x_b, z_b, i) = \sum l\cos\theta_j(F_x^b, F_z^b, x_b, z_b, i) + x_b$$

$$f_7(F_x^b, F_z^b, x_b, z_b, i) = \sum l\cos\theta_j(F_x^b, F_z^b, x_b, z_b, i) + z_b$$

Coordinates (x_g, z_g) in the equations above can be defined by considering the equilibrium of link K in contact with both the roadwheel and the terrain as shown in Figure 10.7.

By considering the equilibrium conditions of link K shown in Figure 10.7, the following three equations can be established:

$$\sum F_x = 0, \quad -F_x^g + F_x^h - G_r^K \sin\theta_r^K - P_K \sin\theta_r^K - S_K \cos\theta_r^K = 0 \quad (10.33)$$

$$\sum F_z = 0, \quad -F_z^g + F_z^h - G_r^K \cos\theta_r^K + P_K \cos\theta_r^K - S_K \sin\theta_r^K - lw = 0 \quad (10.34)$$

$$\sum M_K = 0, \quad -\left[F_x^g a_r^K + F_x^h (1 - a_r^K)\right] l \sin\theta_r^K + \left[F_z^g a_r^K + F_z^h (1 - a_r^K) + lw\left(a_r^K - \frac{1}{2}\right)\right] \times l\cos\theta_r^K - P_K l_K = 0 \quad (10.35)$$

where F_x^g and F_z^g are the horizontal and vertical forces at pin g, respectively, F_x^h and F_z^h are the horizontal and vertical forces at pin h, respectively, P_K and S_K are the resultant normal force and shear force exerted on link K by the terrain, respectively, a_r^K is a parameter defining the position of the contact point between the link and the roadwheel, G_r^K is the normal force exerted on the link by the roadwheel, θ_r^K is the inclination angle of link K, and l_K is a parameter defining the location of the point of application of the resultant normal force P_K exerted on the link by the terrain with respect to the centre of the roadwheel, as shown in Figure 10.7. For each link in contact with both the roadwheel and the terrain, three equations similar to Eqns (10.33), (10.34) and (10.35) can be established.

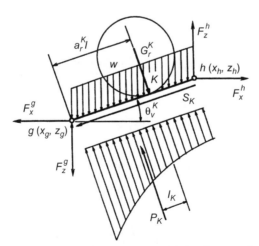

Figure 10.7: The equilibrium of a link in contact with both a roadwheel and the terrain (Reprinted by permission of the Council of the Institution of Mechanical Engineers from Gao and Wong, 1994)

It should be noted that from the analysis described previously, F_x^g and F_z^g can be expressed as a function of F_x^b, F_z^b, x_b, z_b and i. It should be pointed out that a_r^K, G_r^K and θ_r^K are unknowns and that the coordinates of pin g, (x_g, z_g), and those of pin h, (x_h, z_h), can be defined in terms of a_r^K and θ_r^K, and the coordinates of the centre of the roadwheel, which are functions of z_s, δ and the distance from the centre of the sprocket.

Taking into account that (x_g, z_g) are functions of a_r^K, θ_r^K, z_s and δ, Eqns (10.31) and (10.32) can be rewritten as follows:

$$f_6(z_s, \delta, F_x^b, F_z^b, a_r^K, \theta_r^K, x_b, z_b, i) = 0 \tag{10.36}$$

$$f_7(z_s, \delta, F_x^b, F_z^b, a_r^K, \theta_r^K, x_b, z_b, i) = 0 \tag{10.37}$$

A similar approach to the analysis of tack segment bg may be followed to analyse track segment hc shown in Figure 10.5. Two equations similar to Eqns (10.31) and (10.32) can be derived for segment hc. However, as noted from Eqns (10.33) and (10.34), the forces at h are functions of G_r^K, θ_r^K, F_x^g and F_z^g. As pointed out previously, F_x^g and F_z^g are in turn functions of F_x^b, F_z^b, x_b, z_b and i. Furthermore, the coordinates of pin h, (x_h, z_h), are functions of a_r^K, θ_r^K, z_s and δ. Consequently, for segment hc, the following two functional relationships can be established:

$$f_8(z_s, \delta, F_x^b, F_z^b, a_r^K, G_r^K, \theta_r^K, x_b, z_b, i) - x_c = 0 \tag{10.38}$$

$$f_9(z_s, \delta, F_x^b, F_z^b, a_r^K, G_r^K, \theta_r^K, x_b, z_b, i) - z_c = 0 \tag{10.39}$$

Following a similar approach, Eqn (10.35) can be rewritten as follows:

$$f_{10}(z_s, \delta, F_x^b, F_z^b, a_r^K, G_r^K, \theta_r^K, x_b, z_b, i) = 0 \tag{10.40}$$

The analysis of the segments of the lower run of the track with one roadwheel described above can be extended to a track system with any number of roadwheels. For a track system with n roadwheels, $2 + 3n$ independent equations, consisting of Eqns (10.36), (10.37) and n sets of equations similar to Eqns (10.38), (10.39) and (10.40), can be established. In these equations, z_s, δ, F_x^b, F_z^b, x_b, z_b, x_c and z_c, and n sets of a_r^K, G_r^K and θ_r^K ($K = 1, 2, ..., n$) are unknowns. It should be noted that track slip i is considered to be an input parameter to the solution process, at which the performance of the track system is to be evaluated.

10.1.3 Analysis of the Links in Contact with the Idler

The links in contact with the idler are schematically shown in Figure 10.8. In the analysis, it is assumed that each link is in contact with the pitch circle of the idler in the middle of the link. The relative position of the link with respect to the idler is defined by the angle θ_a, which is the angle of the first link in the upper run of the track in contact with the idler. θ_a is taken as an input parameter in the solution process and the interaction between the track system and the terrain varies with θ_a. As noted previously, the coordinates of a, (x_a, z_a), which are used in Section 10.1.1, are functions of θ_a.

As the centre of the front idler is assumed to be mounted on a pre-compressed spring, the location of the centre of the front idler A can be expressed by

$$x_A = x_q - (l_s + \Delta l_s)\cos\delta = x_A(z_s, \delta, \Delta l_s) \tag{10.41}$$

$$z_A = z_q - (l_s + \Delta l_s)\sin\delta = z_A(z_s, \delta, \Delta l_s) \tag{10.42}$$

where (x_A, z_A) and (x_q, z_q) are the coordinates of the centre A of the idler and those of the point of attachment q of the spring on the track frame, respectively, l_s is the original (uncompressed or unstretched) length of the spring, which is a known parameter, and Δl_s is the deformation of the spring. As point q is attached to the track frame, its coordinates are functions of the position of the sprocket, the inclination angle of the track frame, and its distance from the centre of the sprocket, the inclination angle of the track frame, and its distance from the centre of the sprocket:

$$x_q = x_q(z_s, \delta)$$

$$z_q = z_q(z_s, \delta)$$

From Eqns (10.41) and (10.42), the coordinates of pin a, (x_a, z_a), on the upper run of the track shown in Figure 10.8 can be expressed as functions of z_s, δ, θ_a and Δl_s:

$$x_a = x_A - R_i \sin\theta_a + \frac{l}{2}\cos\theta_a = x_a(z_s, \delta, \theta_a, \Delta l_s) \tag{10.43}$$

$$z_a = z_A + R_i \cos\theta_a + \frac{l}{2}\sin\theta_a = z_a(z_s, \delta, \theta_a, \Delta l_s) \tag{10.44}$$

where R_i is the radius of the idler.

As all links in segment ab are assumed to be in contact with the idler in the middle of the links, the coordinates of pin b, (x_b, z_b), in the lower run of the track in contact with the terrain can be expressed by

$$x_b = x_b(z_s, \delta, \theta_a, \Delta l_s) \tag{10.45}$$

$$z_b = z_b(z_s, \delta, \theta_a, \Delta l_s) \tag{10.46}$$

After the positions of the links are defined, the equilibrium conditions of the idler can be examined. From the previous analysis, the forces acting on pin a and pin b are (F_x^a, F_z^a) and (F_x^b, F_z^b), respectively. The equilibrium conditions for the idler are therefore expressed by

$$\sum F_x = 0, \quad F_x^a + F_x^b + F_x^A - \sum(P_j \sin\theta_j + S_j \cos\theta_j) = 0 \tag{10.47}$$

$$\sum F_z = 0, \quad F_z^a + F_z^b + F_z^A + \sum(P_j \cos\theta_j - S_j \sin\theta_j) = 0 \tag{10.48}$$

$$\sum M_A = 0, \quad F_x^a(z_A - z_a) - F_z^a(x_A - x_a) + F_x^b(z_A - z_b) - F_z^b(x_A - x_b)$$
$$+ \sum\left[P_j\left(\frac{l}{2} - l_j\right) - S_j R_i\right] = 0 \tag{10.49}$$

where F_x^A and F_z^A are the horizontal and vertical forces acting at the centre A of the idler, respectively, P_j and S_j are the resultant normal force and shear force acting on link j in contact with the terrain, respectively, l_j is a parameter defining the location of the point of application of P_j, as shown in Figure 10.8; and

$$\theta_j = \theta_a + (j-1)\left[2\tan^{-1}\left(\frac{l}{2R_i}\right)\right]$$

P_j, S_j and l_j can be described in a similar way to that presented in Section 10.1.2, and can be expressed as

$$P_j = P_j(z_s, \delta, \theta_a, \Delta l_s) \tag{10.50}$$

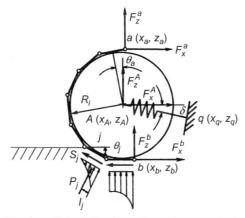

Figure 10.8: The equilibrium of the idler (Reprinted by permission of the Council of the Institution of Mechanical Engineers from Gao and Wong, 1994)

$$S_j = S_j(z_s, \delta, \theta_a, \Delta l_s, i) \tag{10.51}$$

$$l_j = l_j(z_s, \delta, \theta_a, \Delta l_s) \tag{10.52}$$

Substituting Eqns (10.41) to (10.46) into Eqn (10.49), the following equation for F_x^b is obtained:

$$F_x^b = \frac{1}{z_A - z_b}\left[-F_x^a(z_A - z_a) + F_z^a(x_A - x_a) + F_z^b(x_A - x_b)\right.$$
$$\left. - \sum\left[P_j\left(\frac{l}{2} - l_j\right) - S_j R_i\right]\right] = F_x^b(z_s, \delta, F_x^a, F_z^a, F_z^b, \theta_a, i, \Delta l_s) \tag{10.53}$$

Substituting Eqns (10.50), (10.51) and (10.53) into Eqns (10.47) and (10.48), the following two equations are obtained:

$$F_x^A = F_x^A(z_s, \delta, F_x^a, F_z^a, F_z^b, \theta_a, i, \Delta l_s) \tag{10.54}$$

$$F_z^A = F_z^A(z_s, \delta, F_z^a, F_z^b, \theta_a, i, \Delta l_s) \tag{10.55}$$

Knowing the forces acting at the centre of the idler and the spring stiffness, the following expression for the spring deformation Δl_s may be obtaind:

$$\Delta l_s = \Delta l_s(z_s, \delta, F_x^a, F_z^a, F_z^b, \theta_a, i) \tag{10.56}$$

Using Eqn (10.56), Eqns (10.43), (10.44), (10.45), (10.46) and (10.53) may be rewritten as follows:

$$x_a = x_a(z_s, \delta, F_x^a, F_z^a, F_z^b, \theta_a, i) \tag{10.57}$$

$$z_a = z_a(z_s, \delta, F_x^a, F_z^a, F_z^b, \theta_a, i) \tag{10.58}$$

$$x_b = x_b(z_s, \delta, F_x^a, F_z^a, F_z^b, \theta_a, i) \tag{10.59}$$

$$z_b = z_b(z_s, \delta, F_x^a, F_z^a, F_z^b, \theta_a, i) \tag{10.60}$$

$$F_x^b = F_x^b(z_s, \delta, F_x^a, F_z^a, F_z^b, \theta_a, i) \tag{10.61}$$

As noted previously, θ_a and i in the expressions above are considered to be input parameters in the solution process and are not regarded as unknowns.

10.1.4 Analysis of the Links in Contact with the Sprocket

The links in contact with the sprocket are schematically shown in Figure 10.9. In the analysis, the sprocket is considered to be rigidly attached to the track frame. As a result, the coordinates of the centre of the sprocket, (x_D, z_D), are functions of z_s and δ and can be expressed as

$$x_D = x_D(z_s, \delta) \tag{10.62}$$

$$z_D = z_D(z_s, \delta) \tag{10.63}$$

The coordinates of pin c, (x_c, z_c), are given by

$$x_c = x_D + R_s \sin\theta_c - \frac{l}{2}\cos\theta_c = x_c(z_s, \delta, \theta_c) \tag{10.64}$$

$$z_c = z_D - R_s \cos\theta_c - \frac{l}{2}\sin\theta_c = z_c(z_s, \delta, \theta_c) \tag{10.65}$$

where R_s is the pitch radius of the sprocket and θ_c defines the angular position of the link in contact with the sprocket at the bottom, as shown in Figure 10.9.

Similarly, the coordinates of pin d, (x_d, z_d), can be expressed as functions of z_s, δ and θ_c:

$$x_d = x_d(z_s, \delta, \theta_c) \tag{10.66}$$

$$z_d = z_d(z_s, \delta, \theta_c) \tag{10.67}$$

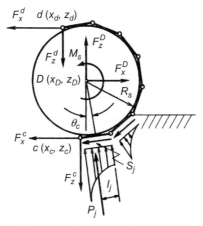

Figure 10.9: The equilibrium of the sprocket (Reprinted by permission of the Council of the Institution of Mechanical Engineers from Gao and Wong, 1994)

Eqns (10.64) to (10.67) indicate that x_c, z_c, x_d and z_d are functions of θ_c. In the solution process, θ_c is regarded as an unknown.

By considering the equilibrium of the sprocket, the following equations can be established:

$$\sum F_x = 0, \quad -F_x^c - F_x^d + F_x^D - \sum(P_j \sin\theta_j + S_j \cos\theta_j) = 0 \quad (10.68)$$

$$\sum F_z = 0, \quad -F_z^c - F_z^d + F_z^D + \sum(P_j \cos\theta_j - S_j \sin\theta_j) = 0 \quad (10.69)$$

$$\sum M_D = 0, \quad -F_x^c(z_D - z_c) + F_z^c(x_D - x_c) - F_x^d(z_D - z_d) + F_z^d(x_D - x_d)$$
$$+ \sum \left[P_j\left(\frac{l}{2} - l_j\right) - S_j R_s\right] + M_s = 0 \quad (10.70)$$

where F_x^c, F_z^c, F_x^d and F_z^d are the forces at pins c and d, respectively, which can be derived from the analysis of the lower and upper runs of the track described in Sections 10.1.1 and 10.1.2, F_x^D and F_z^D are the horizontal and vertical forces at the centre of the sprocket, and M_s is the input torque at the sprocket required to maintain the vehicle operating at a steady-state condition. P_j, S_j and l_j can be derived in the same way as that described in Section 10.1.2

10.1.5 Analysis of the Complete Track System with Rigid Links

Based on the results of the analysis of the four individual segments of the track system described above, the interaction between the entire track system with the terrain can now be examined. From Eqns (10.15) to (10.19), (10.36) to (10.40), (10.57) to (10.61) and (10.64) to (10.67), it can be seen that for a particular track system configuration, operating over a given terrain, at a prescribed relative position of the track links with respect to the track frame defined by the angle θ_a and at a given track slip i, the performance of the vehicle is completely defined by the following unknown parameters: sinkage parameter z_s, track frame inclination angle δ, the horizontal and vertical forces at pin a, F_x^a and F_z^a, the vertical force at pin b, F_z^b, the angle defining the relative position of the track links with respect to the sprocket θ_c, m sets of three parameters relating to supporting roller–track link interaction, a_s^k, G_s^k and θ_s^k; and n sets of three parameters relating to roadwheel–track link–terrain interaction, a_r^K, G_r^K and θ_r^K. As noted in Sections 10.1.1 and 10.1.2, from the analyses of the upper and lower runs of the track, for a track system with m supporting rollers and n roadwheels, a total of $4 + 3m + 3n$ independent equations can be established. Coupled with the vertical force and moment equilibrium conditions for the entire track system, $6 + 3m + 3n$ independent equations can be derived. The vertical force and moment equilibrium of the entire track system can be expressed as

$$\sum F_z = 0, \quad \sum(P_j \cos\theta_j - S_j \sin\theta_j) - W = 0 \quad (10.71)$$

$$\sum M = 0, \quad \sum \left[P_j \cos\theta_j (x_j^p - x_w) + P_j \sin\theta_j (z_j^p - z_w) - S_j \sin\theta_j (x_j^s - x_w) \right.$$
$$\left. + S_j \cos\theta_j (z_j^s - z_w) \right] + \sum (P_j \sin\theta_j + S_j \cos\theta_j) z_{db} = 0 \quad (10.72)$$

where (x_w, z_w) are the coordinates of the centre of gravity of the track system, (x_j^p, z_j^p) and (x_j^s, z_j^s) are the coordinates of the points of application of the resultant normal force P_j and the shear force S_j exerted on the track link by the terrain, respectively, z_{db} is the vertical distance between the drawbar hitch and the centre of gravity of the track system, and W is the normal load on the track system.

From the $6 + 3m + 3n$ independent equations, the unknown parameters, z_s, δ, F_x^a, F_z^a, F_z^b, θ_c; m sets of parameters a_s^k, G_s^k and θ_s^k; and n sets of parameters, a_r^K, G_r^K, and θ_r^K can be uniquely defined. It should be mentioned that these equations are non-linear. The gradient search method and the iteration procedure (or other suitable numerical methods) may be used to obtain solutions.

After the unknown parameters noted above have been obtained, the performance of the track system can then be determined. The thrust (tractive effort) F developed by the track can be obtained from the horizontal component of the resultant shear force S acting on the track–terrain interface. The motion resistance R_t caused by track–terrain interaction can be determined from the horizontal component of the normal force P acting on the track–terrain interface, while the drawbar pull F_d is obtained by the difference between the thrust F and the motion resistance R_t. As mentioned previously, these performance parameters are functions of track slip i.

The analyses presented above are implemented in the computer-aided method RTVPM, running on Microsoft Windows operating systems. Figure 10.10 shows the control centre, as displayed on the computer screen, for the operation of the latest version of RTVPM.

In addition to the prediction of the performance of vehicles with long-pitch link tracks, RTVPM also predicts the normal and shear stress distributions on the track–terrain interface as functions of track slip. Figure 10.11 shows the predicted normal and shear stress distributions at 20% slip on a clayey loam, under a tracked vehicle with total weight of 329 kN, eight roadwheels of diameter of 26 cm, average roadwheel spacing of 34.2 cm, track width of 50.8 cm and track pitch of 21.6 cm.

10.2 Experimental Substantiation

The basic features of RTVPM have been substantiated by field test data. As an example, Figures 10.12 and 10.13 show a comparison of the measured and predicted drawbar pull coefficient (drawbar pull to vehicle weight ratio) and tractive efficiency (the ratio of the product of drawbar pull and vehicle speed to the power input to the drive sprockets) as functions of track slip for a heavy track-type tractor used in construction industry. The tractor has a total weight

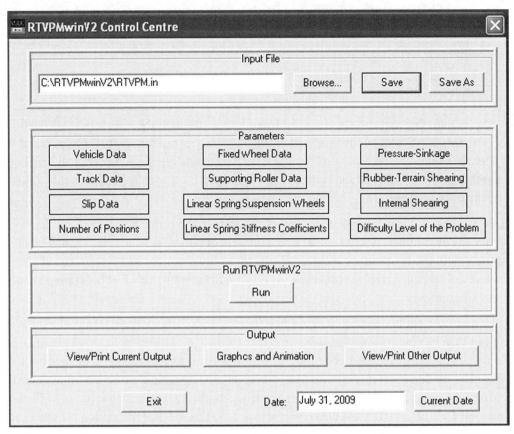

Figure 10.10: Control centre for the operation of the computer-aided method RTVPM, as displayed on the monitor screen

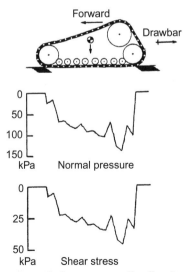

Figure 10.11: The normal pressure and shear stress distributions under a vehicle with long-pitch link tracks predicted by the computer-aided method RTVPM (Reprinted by permission from J.Y. Wong, *Theory of Ground Vehicles*, 4th Ed., Wiley, 2008, copyright © 2008 by John Wiley)

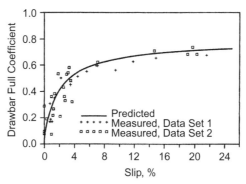

Figure 10.12: Comparison of the measured and predicted drawbar pull coefficient of a vehicle with long-pitch link tracks on a dry, disked sandy loam using the computer-aided method RTVPM (Measured data provided by courtesy of Caterpillar Inc., USA. Reprinted by permission of the Council of the Institution of Mechanical Engineers from Gao and Wong, 1994)

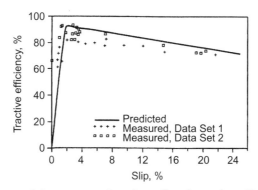

Figure 10.13: Comparison of the measured and predicted tractive efficiency of a vehicle with long-pitch link tracks on a dry, disked sandy loam using the computer-aided method RTVPM (Measured data provided courtesy of Caterpillar Inc., USA. Reprinted by permission of the Council of the Institution of Mechanical Engineers from Gao and Wong, 1994)

of 329 kN, eight roadwheels of diameter of 26 cm on each of the two tracks, roadwheel spacing (the distance between the centres of two adjacent roadwheels) varying from 31.7 to 38.2 cm with an average of 34.2 cm, track width of 50.8 cm and track pitch of 21.6 cm. The terrain on which the performance tests were conducted was a dry, disked sandy loam, with an angle of shearing resistance of 40.1°, and cohesion of 0.55 kPa. The vehicle data, terrain data and measured performance data were provided courtesy of Caterpillar Inc., Peoria, Illinois, USA. The predicted performance was obtained using RTVPM. It can be seen from Figures 10.12 and 10.13 that there is a reasonably close agreement between the measured drawbar performance and the predicted one obtained using RTVPM.

The predictive capability of RTVPM may be further demonstrated through comparisons of the measured and predicted normal and shear stress distributions on the track–terrain

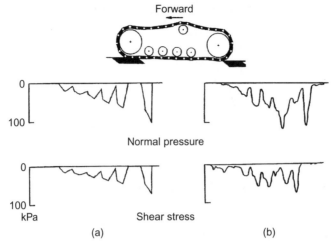

Figure 10.14: Comparison of (a) the predicted and (b) the measured normal pressure and shear stress distributions under a track system with long-pitch rigid links and four roadwheels on a loosely cultivated sand (Measured data from Wills, 1963. Reprinted by permission of the Council of the Institution of Mechanical Engineers from Wong, 1998)

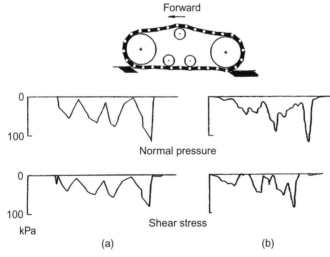

Figure 10.15: Comparison of (a) the predicted and (b) the measured normal pressure and shear stress distributions under a track system with long-pitch rigid links and two roadwheels on a loosely cultivated sand (Measured data from Wills, 1963. Reprinted by permission of the Council of the Institution of Mechanical Engineers from Wong, 1998)

interface. Figures 10.14 and 10.15 show the measured and predicted stress distributions under two experimental track systems, one with four roadwheels and the other with two roadwheels, respectively, on loose cultivated sand in a sand quarry. The measured data were obtained and reported by Wills (1963). The predicted performance data were obtained using RTVPM

(Wong, 1998). The nominal contact lengths for the track systems with four and two roadwheels were 1270 mm and 952.5 mm, respectively. Both systems had the same normal load of 13.12 kN, the same track width of 254 mm, and the same track pitch of 149 mm. Some of the vehicle and terrain parameters required as input to RTVPM, such as the initial track tension and the parameters characterizing the responses of terrain to repetitive normal and shear loadings, were not given in Wills (1963). For these parameters, estimated values based on those of similar track systems and terrain conditions in the author's data bank were used in the simulations using RTVPM. It can be seen that both measured and predicted normal pressure and shear stress distributions exhibit similar characteristics.

In summary, based on the comparisons of the measured and predicted drawbar performance and the normal pressure and shear stress distributions on the track–terrain interface, it is shown that RTVPM can provide realistic predictions of the performances of vehicles with long-pitch link tracks in the field.

10.3 Applications of the Computer-Aided Method RTVPM to Parametric Analysis

With the basic features of the computer-aided method RTVPM substantiated by test data, use can then be made of this method in parametric analyses of vehicle design and performance. In the following, the effects of the number of roadwheels, roadwheel spacing, track pitch and initial track tension on the performances of a baseline vehicle and its two variants on two types of terrain are examined (Wong, 1998).

The basic design parameters of the baseline vehicle used in this study are given in Table 10.1. The track system is similar to that commonly found in agricultural and industrial tractors. The roadwheels and supporting rollers are rigidly mounted on the track frame, while the front idlers are mounted on pre-compressed springs, as shown in Figure 10.1.

Three track system configurations shown in Figure 10.16 were examined. Configuration A, which is the track system of the baseline vehicle, has five roadwheels of radius 16 cm on each of the two tracks, and roadwheel spacing (the distance between the centres of two adjacent roadwheels) of 39.5 cm. Configuration B has seven roadwheels of radius 13 cm and roadwheel spacing of 30 cm; and Configuration C has eight roadwheels of radius 11 cm and roadwheel spacing of 26.5 cm. To study the effects of track pitch and the ratio of roadwheel spacing to track pitch on the performance of Configuration A, the track pitch is varied from 14 to 30 cm, with the corresponding ratio of roadwheel spacing to track pitch (S_{rw}/P_{tr}) from 2.82 to 1.32. For Configuration B, track pitch ranging from 14 to 30 cm and the corresponding ratio of roadwheel spacing to track pitch from 2.14 to 1 are examined. For Configuration C, track pitch ranging from 14 to 26 cm and the corresponding ratio of roadwheeel spacing to track pitch from 1.89 to 1.02 are evaluated. To provide a common basis for comparison, the track contact lengths of the three configurations are kept the same, while the number of roadwheels

Table 10.1: Basic parameters of the baseline vehicle

Total weight, kN	248.27
Centre of gravity, x-coordinate*	−142
Centre of gravity, y-coordinate*	−75
Sprocket pitch radius, cm	40
Idler radius, cm	40
Idler centre, x-coordinate*	−284
Idler centre, y-coordinate*	0
Idler spring stiffness, kN/m	582
Number of supporting rollers	2
Supporting roller radius, cm	13
Initial track tension, kN	15
Drawbar hitch, x-coordinate*	80
Drawbar hitch, y-coordinate*	8.3
Track pitch, cm	16
Roadwheel spacing/track pitch	2.47
Track width, cm	50.8
Track weight per unit length, kN/m	1.973

*Coordinate origin is at the centre of the sprocket. Positive x- and y-coordinates are to the rear and down, respectively.

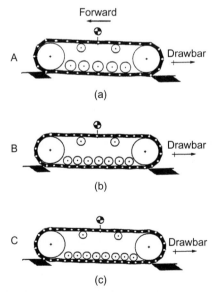

Figure 10.16: Track system configurations with (a) five roadwheels, (b) seven roadwheels, and (c) eight roadwheels used in parametric analyses (Reprinted by permission of the Council of the Institution of Mechanical Engineers from Wong, 1998)

Table 10.2: Parameters for the medium soil

Pressure–sinkage parameters			Repetitive loading parameters		Shear strength parameters		
n	k_c (kN/m^{n+1})	k_ϕ (kN/m^{n+2})	k_o (kN/m^3)	A_u (kN/m^4)	ϕ (degrees)	c (kPa)	K (cm)
0.8	16.54	911.4	0	86 000	29	6.89	2.5

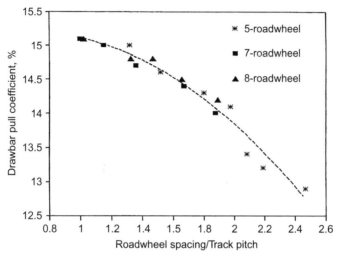

Figure 10.17: Variations of the drawbar pull coefficient at 20% slip with the ratio of roadwheel spacing to track pitch at an initial track tension coefficient of 6.04% on the clayey soil (Reprinted by permission of the Council of the Institution of Mechanical Engineers from Wong, 1998)

and their radii for the three configurations are different. The tractive performances of these three configurations at two values of the ratio of the initial track tension to vehicle weight (the initial track tension coefficient), 6.04 and 12.08%, were predicted using RTVPM on two types of terrain, designated as the clayey soil and medium soil. The initial track tension coefficient of 6.04% is recommended for the baseline vehicle for normal use. The clayey soil used in this study is the same as that described in Chapter 9. Its pressure–sinkage and repetitive loading parameters are given in Table 9.3, while its shear strength parameters are presented in Table 9.4. The medium soil used in this study represents a soil condition found during the wet season in Germany. Its parameters are given in Table 10.2.

10.3.1 Effects of the Ratio of Roadwheel Spacing to Track Pitch

The tractive performances of the three track system configurations with tracks of various pitches were predicted using RTVPM. It was found that for given overall dimensions of a track system, the ratio of roadwheel spacing to track pitch is one of the design parameters that have significant effects on its tractive performance. Figures 10.17 and 10.18 show the

356 Chapter 10

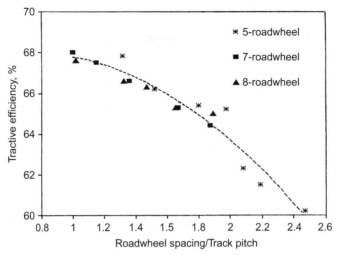

Figure 10.18: Variations of the tractive efficiency at 20% slip with the ratio of roadwheel spacing to track pitch at an initial track tension coefficient of 6.04% on the clayey soil (Reprinted by permission of the Council of the Institution of Mechanical Engineers from Wong, 1998)

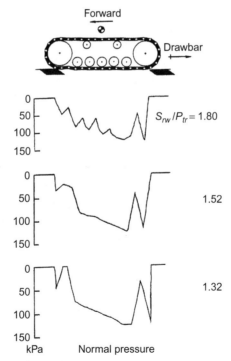

Figure 10.19: Normal pressure distributions under a track system with five roadwheels at various ratios of roadwheel spacing to track pitch on the clayey soil (Reprinted by permission of the Council of the Institution of Mechanical Engineers from Wong, 1998)

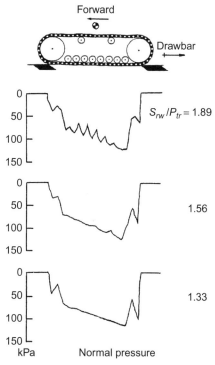

Figure 10.20: Normal pressure distributions under a track system with eight roadwheels at various ratios of roadwheel spacing to track pitch on the clayey soil (Reprinted by permission of the Council of the Institution of Mechanical Engineers from Wong, 1998)

variations of the drawbar pull coefficient and tractive efficiency at 20% slip with the ratio of roadwheel spacing to track pitch, at the initial track tension coefficient of 6.04% on the clayey soil, respectively. It is shown that as long as the ratio of roadwheel spacing to track pitch remains the same, the tractive performances of the three track system configurations with different number of roadwheels ranging from five to eight will be similar. This indicates that the number of roadwheels, roadwheel spacing, or track pitch alone is not necessarily the determining factor for the tractive performance and that the ratio of roadwheel spacing to track pitch is one of the most significant design parameters affecting its performance. This conclusion is further supported by the observations that the normal pressure distributions under the track systems with different number of roadwheels are similar for similar ratios of roadwheel spacing to track pitch. Figures 10.19 and 10.20 show the normal pressure distributions under the track systems with five and eight roadwheels, respectively, at different ratios of roadwheel spacing to track pitch (S_{rw}/P_{tr}) on the clayey soil. It can be seen that for similar ratios of roadwheel spacing to track pitch, the normal pressure distributions under the track system with five roadwheels have similar characteristics to those of the track system with eight roadwheels. It should also be pointed out that as the ratio of roadwheel spacing to track

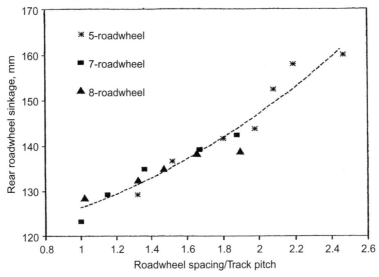

Figure 10.21: Variations of rear roadwheel sinkage at 20% slip with the ratio of roadwheel spacing to track pitch at an initial track tension coefficient of 6.04% on the clayey soil (Reprinted by permission of the Council of the Institution of Mechanical Engineers from Wong, 1998).

pitch decreases, the fluctuation of normal pressure under the track decreases. This indicates that by lowering the ratio of roadwheel spacing to track pitch, the normal pressure under the track system is more uniformly distributed, which leads to improvements in the tractive performance of the vehicle. It should be noted that the normal pressure distributions under long-pitch link tracks shown in Figures 10.19 and 10.20 are generally more uniform than those under flexible tracks shown in Figures 9.13, 9.14 and 9.15.

As can be seen from Figures 10.17 and 10.18, on the clayey soil by reducing the ratio of roadwheel spacing to track pitch from 2.47 to 1, the drawbar pull coefficient at 20% slip increases from 12.9 to 15.1%, and the tractive efficiency increases from 60.2 to 67.6%, which represent improvements of 17.1% in drawbar pull and 12.3% in tractive efficiency, respectively. The improvements in performance are due to the reduction in the track motion resistance coefficient (the ratio of the track motion resistance caused by track–terrain interaction to vehicle weight), which is closely related to the rear roadwheel sinkage, as it determines the rut depth (or deformation of the terrain) after the passage of the vehicle. Figures 10.21 and 10.22 show the variations of the rear roadwheel sinkage and track motion resistance coefficient with the ratio of roadwheel spacing to track pitch, respectively.

Figures 10.23 and 10.24 show the variations of the drawbar pull coefficient and tractive efficiency at 20% slip with the ratio of roadwheel spacing to track pitch, at the initial track tension coefficient of 6.04% on the medium soil, respectively. Similar to the situation on the clayey soil, it can be seen that as long as the ratio of roadwheel spacing to track pitch

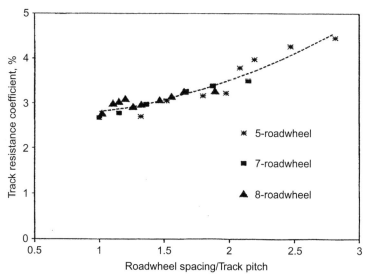

Figure 10.22: Variations of the track motion resistance coefficient at 20% slip with the ratio of roadwheel spacing to track pitch at an initial track tension coefficient of 6.04% on the clayey soil (Reprinted by permission of the Council of the Institution of Mechanical Engineers from Wong, 1998)

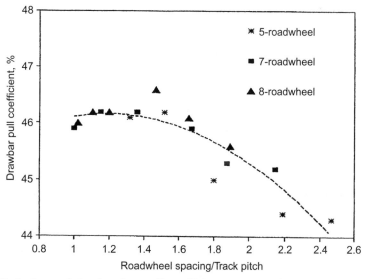

Figure 10.23: Variations of the drawbar pull coefficient at 20% slip with the ratio of roadwheel spacing to track pitch at an initial track tension coefficient of 6.04% on the medium soil (Reprinted by permission of the Council of the Institution of Mechanical Engineers from Wong, 1998)

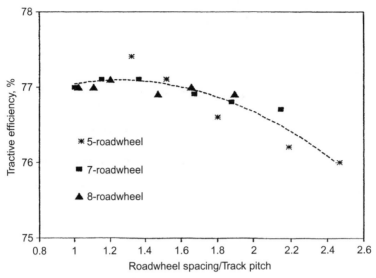

Figure 10.24: Variations of the tractive efficiency at 20% slip with the ratio of roadwheel spacing to track pitch at an initial track tension coefficient of 6.04% on the medium soil (Reprinted by permission of the Council of the Institution of Mechanical Engineers from Wong, 1998)

remains the same, the tractive performances of the three track systems with different numbers of roadwheels will be similar. This is again supported by the observations that the normal pressure distributions under the track systems with different numbers of roadwheels are similar for similar ratios of roadwheel spacing to track pitch (S_{rw}/P_{tr}), as shown in Figures 10.25 and 10.26.

As can be seen from Figures 10.23 and 10.24, on the medium soil reducing the ratio of roadwheel spacing to track pitch from 2.47 to 1, the drawbar pull coefficient at 20% slip increases from 44.3 to 46% and the tractive efficiency increases from 76 to 77%, which represent improvements of 3.8% in drawbar pull and 1.3% in tractive efficiency, respectively. The improvements in performance are again due to the reduction in rear roadwheel sinkage and hence the track motion resistance coefficient, as shown in Figures 10.27 and 10.28, respectively.

It should be pointed out that the medium soil used in this study is firmer than the clayey soil. This is indicated by the roadwheel sinkage of the vehicle on the medium soil being less than that on the clayey soil, as shown in Figures 10.21 and 10.27. As a result, the ratio of roadwheel spacing to track pitch has less significant effects on vehicle drawbar performance on the medium soil than on the clayey soil.

It should also be pointed out that within a certain range of the ratio of roadwheel spacing to track pitch, the drawbar pull coefficient and tractive efficiency vary only slightly. For instance, the drawbar pull coefficient and tractive efficiency at 20% slip change marginally,

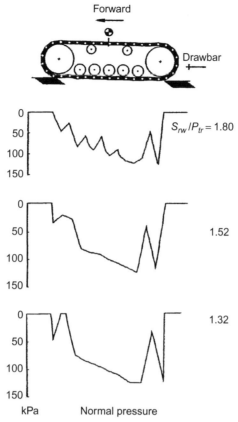

Figure 10.25: Normal pressure distributions under a track system with five roadwheels at various ratios of roadwheel spacing to track pitch on the medium soil (Reprinted by permission of the Council of the Institution of Mechanical Engineers from Wong, 1998)

if the ratio of roadwheel spacing to track pitch varies from 1.3 to 1 on the clayey soil or from 1.6 to 1 on the medium soil, as shown in Figures 10.17, 10.18, 10.23 and 10.24. This implies that the designer would have a certain flexibility in selecting the appropriate track pitch or roadwheel spacing that on the one hand can ensure good tractive performance and on the other hand can minimize the fluctuation of vehicle speed due to the polygon (or chordal) effect of the sprocket–track engagement. It can be shown that the vehicle speed fluctuation ΔV due to the polygon effect is given by

$$\Delta V = 1 - \sqrt{1 - P_{tr}/D_s} \qquad (10.73)$$

where P_{tr} is the track pitch, and D_s is the sprocket pitch diameter.

For current agricultural and industrial tractors with long-pitch link tracks, the ratio of sprocket pitch diameter to track pitch varies from 3.7 to 4.3 approximately. Therefore, the vehicle

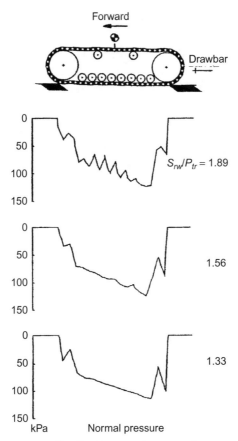

Figure 10.26: Normal pressure distributions under a track system with eight roadwheels at various ratios of roadwheel spacing to track pitch on the medium soil (Reprinted by permission of the Council of the Institution of Mechanical Engineers from Wong, 1998)

speed fluctuation will be in the range from 3.72 to 2.75%. If the speed fluctuation is limited to 2.75% and the sprocket pitch diameter is 92.8 cm, then the track pitch should be 21.6 cm. On the clayey soil, to ensure good drawbar performance, the ratio of roadwheel spacing to track pitch should be in the range from 1 to 1.3. Consequently, the roadwheel spacing should be in the range of 21.6 to 28.1 cm. On the medium soil, to ensure good drawbar performance, the ratio of roadwheel spacing to track pitch should be in the range of 1 to 1.6. Consequently, the roadwheel spacing should be in the range of 21.6 to 34.6 cm. This would allow the vehicle designer to have the flexibility to select the appropriate roadwheel diameter and the number of roadwheels. It should be noted that from the vehicle vibration point of view, it is desirable to have a slight variation in roadwheel spacing from one pair of roadwheels to another on a track system. This could avoid the development of a scalloped profile with a fixed wavelength on the track link rails due to wear.

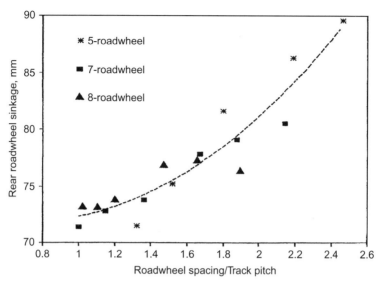

Figure 10.27: Variations of rear roadwheel sinkage at 20% slip with the ratio of roadwheel spacing to track pitch at an initial track tension coefficient of 6.04% on the medium soil (Reprinted by permission of the Council of the Institution of Mechanical Engineers from Wong, 1998)

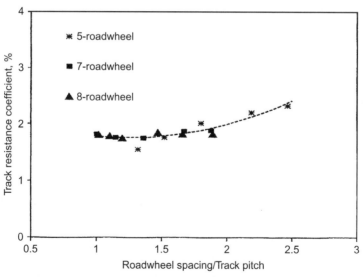

Figure 10.28: Variations of the track motion resistance coefficient at 20% slip with the ratio of roadwheel spacing to track pitch at an initial track tension coefficient of 6.04% on the medium soil (Reprinted by permission of the Council of the Institution of Mechanical Engineers from Wong, 1998)

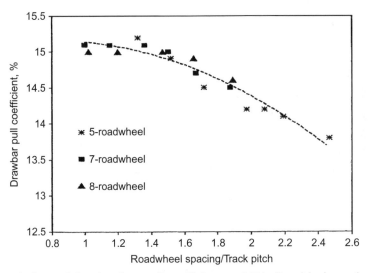

Figure 10.29: Variations of the drawbar pull coefficient at 20% slip with the ratio of roadwheel spacing to track pitch at an initial track tension coefficient of 12.08% on the clayey soil (Reprinted by permission of the Council of the Institution of Mechanical Engineers from Wong, 1998)

A review of current designs of agricultural and industrial tractors with long-pitch link tracks reveals that their ratios of roadwheel spacing to track pitch mostly fall into the range from 1.5 to 1.8. This indicates that they would have adequate drawbar performance on terrain similar to the medium soil, but not necessarily on terrain similar to the clayey soil used in this study.

10.3.2 Effects of Initial Track Tension

Figures 10.29 and 10.30 show the variations of the drawbar pull coefficient and tractive efficiency, respectively, at 20% slip on the clayey soil with the ratio of roadwheel spacing to track pitch, at the initial track tension coefficient of 12.08%, which is double that recommended for normal use noted previously. It can be seen that at high values of the ratio of roadwheel spacing to track pitch, the initial track tension coefficient has considerable effects on tractive performance. Its effects, however, decrease with the decrease of the ratio of roadwheel spacing to track pitch. For instance, on the clayey soil, at a ratio of roadwheel spacing to track pitch of 2.47, the drawbar pull coefficient at 20% slip increases from 12.9 to 13.8%, when the initial track tension coefficient increases from 6.04 to 12.08% (see Figures 10.17 and 10.29). However, at a ratio of roadwheel spacing to track pitch of 1, the increase in the initial track tension coefficient has little effect on the drawbar pull coefficient. On the medium soil, the variations of the drawbar pull coefficient and tractive efficiency at 20% slip with the ratio of roadwheel spacing to track pitch, at the initial track tension coefficient of 12.08%, are shown in Figures 10.31 and 10.32, respectively. It can be seen from Figures 10.23 and 10.31

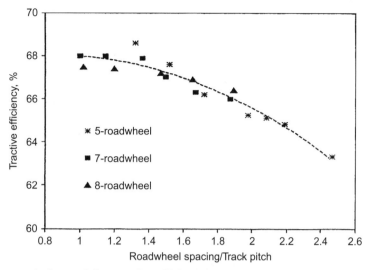

Figure 10.30: Variations of the tractive efficiency at 20% slip with the ratio of roadwheel spacing to track pitch at an initial track tension coefficient of 12.08% on the clayey soil (Reprinted by permission of the Council of the Institution of Mechanical Engineers from Wong, 1998)

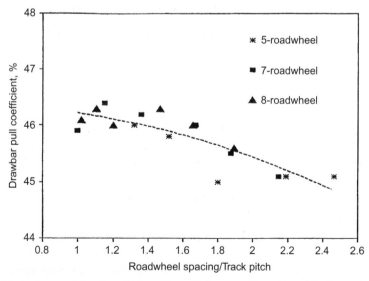

Figure 10.31: Variations of the drawbar pull coefficient at 20% slip with the ratio of roadwheel spacing to track pitch at an initial track tension coefficient of 12.08% on the medium soil (Reprinted by permission of the Council of the Institution of Mechanical Engineers from Wong, 1998)

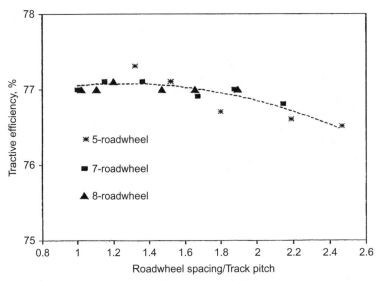

Figure 10.32: Variations of the tractive efficiency at 20% slip with the ratio of roadwheel spacing to track pitch at an initial track tension coefficient of 12.08% on the medium soil (Reprinted by permission of the Council of the Institution of Mechanical Engineers from Wong, 1998)

that at a ratio of roadwheel spacing to track pitch of 2.47, the drawbar pull coefficient at 20% slip increases from 44.4 to 45.1%, when the initial track tension coefficient increases from 6.04 to 12.08%. Again, at low values of the ratio of roadwheel spacing to track pitch, the initial track tension coefficient has little effect on tractive performance.

It should be pointed out that for a given ratio of roadwheel spacing to track pitch, the effects of the initial track tension coefficient are more significant on the clayey soil than on the medium soil. This is primarily due to the fact that the medium soil is firmer than the clayey soil, as noted previously. This indicates that the effects of the initial track tension on tractive performance are more significant on softer terrain than on firmer terrain, similar to the finding for vehicles with flexible tracks described in Section 9.1.

10.3.3 Concept of a Vehicle with Enhanced Performance on Soft Ground

The results presented above indicate that for a given vehicle configuration, its tractive performance on the clayey soil is significantly lower than that on the medium soil. For instance, with the ratio of roadwheel spacing to track pitch ratio of 1 and at the initial track tension coefficient of 6.04%, the drawbar pull coefficient on the clayey soil is 15.1%, which is much lower that that on the medium soil of 46%. It is therefore of interest to explore the concept of a vehicle configuration that could offer improved performance on the clayey soil.

Based on the results of parametric analyses presented above, a concept for an enhanced version of the baseline vehicle emerges. It has seven roadwheels with radius of 12.5 cm, track pitch of

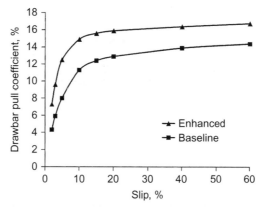

Figure 10.33: Comparison of the predicted drawbar pull coefficient at various slips of the baseline vehicle with that of the enhanced version on the clayey soil using the computer-aided method RTVPM (Reprinted by permission of the Council of the Institution of Mechanical Engineers from Wong, 1998)

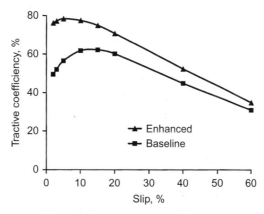

Figure 10.34: Comparison of the predicted tractive efficiency at various slips of the baseline vehicle with that of the enhanced version on the clayey soil using the computer-aided method RTVPM (Reprinted by permission of the Council of the Institution of Mechanical Engineers from Wong, 1998)

21 cm, the ratio of roadwheel spacing to track pitch of 1.45, the initial track tension coefficient of 10%, and the location of the centre of gravity at 40 cm ahead of the midpoint of the track contact length. A comparison of the drawbar pull coefficient at various slips of the enhanced version with that of the baseline vehicle on the clayey soil is shown in Figure 10.33. It can be seen that the drawbar pull coefficient at 20% slip of the enhanced version is 15.9%, whereas that of the baseline vehicle is 12.9%. Figure 10.34 shows that the tractive efficiency at 20% slip of the enhanced version is 70.8%, whereas that of the baseline vehicle is 60.2%.on the clayey soil. This indicates that the drawbar pull coefficient and tractive efficiency at 20% slip of the enhanced version are 23.3 and 17.6% higher than those of the baseline vehicle, respectively.

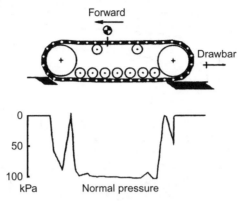

Figure 10.35: Normal pressure distribution under the track of the enhanced version at 20% slip on the clayey soil predicted using the computer-aided method RTVPM (Reprinted by permission of the Council of the Institution of Mechanical Engineers from Wong, 1998)

The higher performance of the enhanced version, in comparison with that of the baseline vehicle, is due to the effects of a combination of factors, including the ratio of roadwheel spacing to track pitch, initial track tension and location of the centre of gravity, which give a more favourable normal pressure distribution on the track–terrain interface. Figure 10.35 shows the normal pressure distribution of the enhanced version at 20% slip on the clayey soil. It shows that except for immediately under the idler and sprocket, the normal pressure is more or less uniformly distributed on the track–terrain interface.

10.4 Concluding Remarks

The results of the parametric analysis presented in this chapter have demonstrated the capabilities of the simulation model RTVPM in performance and design evaluation of vehicles with long-pitch link tracks, commonly used in low-speed agricultural and industrial tractors. It is shown that RTVPM can be used to evaluate the effects of vehicle design on performance over unprepared terrain or the impact of terrain conditions on vehicle performance.

Based on the results of the parametric analyses presented, the following remarks are made:

1. It is shown that the number of roadwheels, track pitch, or roadwheel spacing alone is not necessarily a determining factor for the performance of vehicles with long-pitch link tracks. Rather, the ratio of the roadwheel spacing to track pitch is one of the most significant parameters affecting vehicle performance, particularly on soft terrain. This indicates that for a given roadwheel spacing, track pitch should be carefully selected to ensure adequate tractive performance on the one hand, and to minimize fluctuation of vehicle speed and the associated vehicle vibration on the other hand.

2. At high ratios of roadwheel spacing to track pitch, the initial track tension has noticeable effects on tractive performance over soft terrain. However, at low ratios of roadwheel spacing to track pitch, the effects of the initial track tension on performance become less significant. It is also shown that the effects on performance of the ratio of roadwheel spacing to track pitch are more significant on soft terrain than on firm ground.

3. In comparison with vehicles having flexible tracks discussed in Chapter 9, vehicles with long-pitch link tracks, with appropriate ratio of roadwheel spacing to track pitch, generally have more uniform normal pressure distributions on the track–terrain interface, leading to higher tractive performance on soft terrain..

4. Similar to a vehicle with flexible tracks discussed in Chapters 8 and 9, the interaction between a vehicle with long-pitch link track and deformable terrain is complex. Its performance is a complicated function of vehicle design and operating parameters and terrain behaviour. To take into account all major design features of the vehicle and all pertinent terrain characteristics in predicting vehicle performance, a comprehensive and realistic computer-aided method is needed. As demonstrated through parametric analyses presented in this chapter, RTVPM can fulfil this need. Consequently, RTVPM is a useful tool for the engineer to evaluate vehicle design concepts and to optimize design parameters for vehicles with long-pitch link tracks. RTVPM has been successfully employed in assisting vehicle manufacturers in the development of new products, including underwater tracked vehicles for the mining industry.

CHAPTER 11

Methods for Evaluating Wheeled Vehicle Performance

Close to 5500 years have elapsed since the invention of the wheel. However, human's quest for a better understanding of this seemingly simple invention and his fascination for its refinement appear to be unabated even to this date (Wong, 1984a, c).

The problem of wheel–terrain interaction may seem antiquated in this technologically advanced era of space exploration, nanotechnology, robotics and microelectronics. It is, nevertheless, a complex problem, at least from an analytical viewpoint, for the characteristics of wheel–terrain interaction are influenced by a large number of design and operational factors, as well as terrain characteristics. This is illustrated by the characteristics of soil flow patterns in the longitudinal plane under a driven, towed, 100% slipping, and locked rigid wheel shown in Figures 2.17–2.20, respectively (Wong and Reece, 1966; Wong, 1967). As a result of wheel–terrain interaction, normal and shear stresses develop on the interface, and their distributions on the contact patch vary with the design and operational parameters of the wheel, as well as terrain conditions. Figure 1.13 shows the measured normal and shear stress distributions on the interface of a pneumatic tyre on loose sandy loam (Krick, 1969). All of these indicate that the wheel–terrain interaction is a complex phenomenon.

There are two principal objectives of the study of wheel–terrain interaction. One is to formulate realistic methods for predicting the performance of a wheel in relation to its design parameters and terrain conditions. This is of prime interest to wheeled vehicle designers and users. The other is to establish reliable procedures for predicting changes in terrain conditions caused by the passage of the wheel (or roller). This is of great interest to agricultural engineers in the evaluation of soil compaction caused by farm vehicles and machinery and to road construction equipment engineers in the assessment of the effectiveness of roller compactors.

Since the performance of a wheel, as well as the stress and strain fields in the terrain induced by the passage of the wheel, is directly related to the normal and shear stress distributions on the wheel–terrain interface, one of the central issues in the study of the mechanics of wheel–terrain interaction is the formulation of a mathematical model for predicting the geometry of the contact patch and the stresses on the interface, in terms of the design and operational parameters of the wheel and the mechanical properties of the terrain (Wong, 1984a, c). This provides the technological basis for the prediction of the performance of off-road wheeled vehicles and for the evaluation of terrain compaction due to vehicular traffic.

A variety of methods of approach to the study of wheel–terrain interaction have been developed over the years. They range from entirely empirical to highly analytical (Wong, 1984c). The development of a particular method of approach is greatly influenced by its intended purposes and the state of technological development. A brief review of some of the methods is given below.

11.1 Empirical Methods

One of the better known empirical methods for predicting and evaluating off-road wheeled vehicle performance is that developed by the US Army Corps of Engineers Waterways Experiment Station (WES) (Rula and Nuttall, 1971). Similar to the development of the empirical method for tracked vehicle performance by WES described in Section 7.1, a number of representative off-road wheeled vehicles were tested in a range of terrains, primarily fine- and coarse-grained soils. Terrain conditions were identified using a cone penetrometer. These two sets of measurements were then empirically correlated, and a model known as the WES VCI model was developed for predicting wheeled vehicle performance (Rula and Nuttall, 1971). This model, which was developed in the 1960s, forms part of the basis for the subsequent developments of the AMC 71 and AMM 75 mobility models and the NATO Reference Mobility Model (NRMM).

In the WES VCI method, the mobility index (MI) of a wheeled vehicle is first calculated using an empirical equation. The mobility index for a self-propelled, all-wheel-drive vehicle is expressed by

$$\text{Mobility Index} = \left(\frac{\text{contact pressure factor} \times \text{weight factor}}{\text{tyre factor} \times \text{grouser factor}} + \text{wheel load factor} - \text{clearance factor} \right) \times \text{engine factor} \times \text{transmission factor} \quad (11.1)$$

where

$$\text{Contact pressure factor} = \frac{\text{gross weight, lb}}{\text{tyre section width, in.} \times \text{outside radius of tyre, in.} \times \text{no. of tyres}}$$

Weight factor:

weight range	weight factor equations
< 8.89 kN (2000 lb)	$\overline{Y} = 0.553\,\overline{X}$
8.89 to 60.05 kN (2,000 to 13,500 lb)	$\overline{Y} = 0.033\,\overline{X} + 1.050$
60.05 to 88.96 kN (13,501 to 20,000 lb)	$\overline{Y} = 0.142\,\overline{X} - 0.420$
> 88.96 kN (20,000 lb)	$\overline{Y} = 0.0278\,\overline{X} - 3.115$

where $\overline{Y}$ = weight factor; $\overline{X} = \dfrac{\text{gross weight, lb}}{1000 \times \text{No. of axles}}$

Tyre factor $= \dfrac{10 + \text{tyre width, in.}}{100}$

Grouser factor: with chains = 1.05

without chains = 1.00

Wheel load factor $= \dfrac{\text{gross weight, lb}}{1000 \times \text{No. of axles}/2}$

Clearance factor $= \dfrac{\text{clearance, in.}}{10}$

Engine factor: ≥8.2 kW/tonne (10 hp/ton) of vehicle weight = 1.0

< 8.2 kW/tonne (10 hp/ton) of vehicle weight = 1.05

Transmission factor: automatic = 1.00; manual = 1.05

Similar to the empirical method developed by WES for predicting tracked vehicle performance, the mobility index of a wheeled vehicle is used to determine a vehicle cone index (VCI). The VCI represents the minimum soil strength of the critical layer that is required for a vehicle to successfully make a specific number of passes.

For a self-propelled, wheeled vehicle, the vehicle cone index is related to the mobility index by the following empirical equations:

For one pass

$$VCI_1 = 11.48 + 0.2MI - \left(\dfrac{39.2}{MI + 3.74}\right) \quad (11.2)$$

and for 50 passes

$$VCI_{50} = 28.23 + 0.43MI - \left(\dfrac{92.67}{MI + 3.67}\right) \quad (11.3)$$

For fine-grained soils, after the VCI of a vehicle and the strength of the soil (i.e. rating cone index RCI) have been determined, the performance parameters of a wheeled vehicle, such as the net maximum drawbar pull coefficient and the towed motion resistance coefficient, are predicted as a function of the excess of RCI over VCI (or excess soil strength), as illustrated in Figures 7.1 and 7.2.

Lately, a correction factor taking into account the effect of tyre deflection has been introduced into the calculation of VCI. It is expressed by $\sqrt[4]{0.15(\delta/h)}$, where δ is the deflection and h is the unloaded section height of the tyre. The corrected values of VCI for one pass and for 50 passes are obtained by multiplying Eqn (11.2) and Eqn (11.3) by the correction factor, respectively.

The methodology of the WES VCI model for predicting wheeled vehicle performance on fine-grained soils has been adopted by the NATO Reference Mobility Model Edition II. Similar to the empirical method for predicting tracked vehicle performance developed by WES, it is uncertain that the empirical model presented above, based on test data collected years ago, is still valid for predicting the current or new generations of wheeled vehicles.

For the performance of a single tyre, an empirical model based on soil–tyre numeric was developed at WES (Freitag, 1965; Turnage, 1972, 1978). The clay–tyre numeric N_c is for tyres operating in purely cohesive soil (near-saturated clay), while the sand–tyre numeric N_s is for tyres operating in purely frictional soil (air-dry sand). These two numerics are defined as

$$N_c = \frac{Cbd}{W} \times \left(\frac{\delta}{h}\right)^{1/2} \times \frac{1}{1 + (b/2d)} \tag{11.4}$$

and

$$N_s = \frac{G(bd)^{3/2}}{W} \times \frac{\delta}{h} \tag{11.5}$$

where b is the tyre section width, C is the cone index, d is the tyre diameter, G is the sand penetration resistance gradient, h is the unloaded tyre section height, W is the tyre load, and δ is the tyre deflection.

For tyres operating in cohesive-frictional soils, a soil–tyre numeric N_{cs} was proposed by Wismer and Luth (1972) and is defined as

$$N_{cs} = \frac{Cbd}{W} \tag{11.6}$$

Based on test results obtained primarily in laboratory soil bins, these soil–tyre numerics have been empirically correlated with two tyre performance parameters, the drawbar coefficient μ and the drawbar efficiency η at 20% slip. The drawbar coefficient is defined as the ratio of drawbar pull to the normal load on the tyre, while the drawbar efficiency is defined as the ratio of the drawbar power (i.e. the product of drawbar pull and wheel forward speed) to the power input to the tyre. Figure 11.1 shows the empirical relations between μ and η at 20% slip and the clay–tyre numeric N_c. These relations were obtained on cohesive clays with tyres ranging from 4.00–7 to 31 × 15.50–13, with loads from 0.23 to 20 kN and with ratios of tyre deflection to section height from 0.08 to 0.35. The cone index values of these clays in the top 15 cm ranged from 55 to 390 kPa. Figure 11.2 shows the relations between the two tyre performance parameters at 20% slip and the sand–tyre numeric N_s. These empirical relations were based on test results obtained on a particular type of sand known as desert Yuma sand, with tyres similar to those for Figure 11.1, with loads from 0.19 to 20 kN and with ratios of

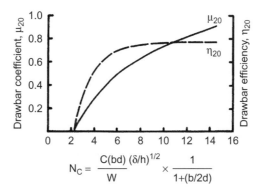

Figure 11.1: Empirical relations between drawbar coefficient and drawbar efficiency at 20% slip and the clay–tyre numeric N_c (Reprinted by permission of *ISTVS* from Turnage, 1978)

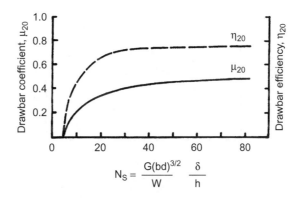

Figure 11.2: Empirical relations between drawbar coefficient and drawbar efficiency at 20% slip and the sand–tyre numeric N_s (Reprinted by permission of *ISTVS* from Turnage, 1978)

tyre deflection to section height from 0.15 to 0.35. The values of the penetration resistance gradient for the desert Yuma sand ranged from 0.9 to 5.4 MPa/m. Figure 11.3 shows the empirical relations between μ and η at 20% slip and the soil–tyre numeric N_{cs}. These relations were obtained on cohesive-frictional soils, with tyres ranging from 36 to 84 cm in width and from 84 to 165 cm in diameter, and loads from 2.2 to 28.9 kN. These soils ranged from a tilled soil with an average before-traffic cone index value of 130 kPa in a layer 15 cm deep to an untilled soil with an average cone index value of 3450 kPa (Turnage, 1978).

As pointed out earlier, the empirical relation between the sand–tyre numeric N_s given by Eqn (11.5) and tyre performance parameters shown in Figure 11.2 was developed from tests entirely conducted in one type of frictional soil, namely, desert Yuma sand. Subsequent tyre tests performed in a quite different frictional soil, a washed sand from an alluvial plain termed mortar sand, indicated that the influence of the sand penetration resistance gradient G on tyre performance depends on the particular type of frictional soil under consideration (Turnage, 1978). As a result, Turnage proposed a revised sand–tyre numeric N_{se} to replace N_s given in Eqn (11.5). N_{se} is defined as

$$N_{sc} = \frac{G_e(bd)^{3/2}}{W} \times \frac{\delta}{h} \tag{11.7}$$

where G_e is an effective sand penetration resistance gradient normalized with respect to Yuma sand. It takes into account a tyre shape factor b/d, soil moisture content, before-traffic relative density D_{rb} and effective relative density D_{re} (Turnage, 1978). The relative density D_r is generally defined as

$$D_r = \frac{e_{max} - e}{e_{max} - e_{min}} \tag{11.8}$$

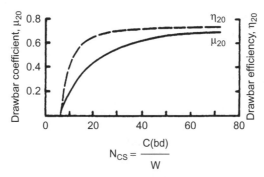

Figure 11.3: Empirical relations between drawbar coefficient and drawbar efficiency at 20% slip and the numeric N_{cs} for cohesive-frictional soil (Reprinted by permission of *ISTVS* from Turnage, 1978)

where e is the void ratio of the sand at the condition of interest, and e_{max} and e_{min} are the void ratios in the loosest and densest states, respectively.

Figure 11.4 shows the empirical relation between the revised sand–tyre numeric N_{se} and the drawbar coefficient μ at 20% slip obtained with tyres similar to those for Figure 11.2.

In 1981, Reece and Peca (1981) reported that the empirical equations relating the drawbar coefficient μ and the drawbar efficiency η at 20% slip to the revised sand–tyre numeric N_{se} did not yield reasonable predictions of tyre performance on an air-dry Cresswell sand. In view of these findings, an extensive reassessment of the original methodology was conducted at WES by Turnage. As a result, he proposed a further revised sand–tyre numeric N_{sey} to replace N_{se} (Turnage, 1984). N_{sey} is defined as

$$N_{sey} = \frac{G_{ey}(bd)^{3/2}}{W} \times \frac{\delta}{h} \tag{11.9}$$

In addition to the factors that have been taken into account in obtaining G_e in Eqn (11.7), G_{ey} includes the effects of sand compatibility (i.e. the ratio of the difference in void ratio for the loosest and densest state to the void ratio for the densest state) and the sand grain median diameter (i.e. the sand grain diameter for which 50% of the sand sample is finer by weight).

It was reported by Turnage (1984) that the use of the further revised sand–tyre numeric N_{sey} improves the prediction of tyre (or wheeled vehicle) performance over a broader range of

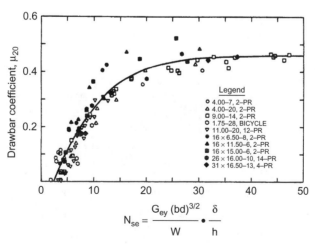

Figure 11.4: Measured data and empirical relation between drawbar coefficient at 20% slip and the revised sand–tyre numeric N_{se} obtained in air-dry Yuma sand (Reprinted by permission of *ISTVS* from Turnage, 1984)

sand types, as shown in Figure 11.5. However, the exact range of sand conditions for which N_{sey} is applicable remains to be determined.

The introduction of the further revised sand–tyre numeric N_{sey} indicates that the original concept of using the simple measurements obtained by a cone penetrometer to describe the conditions of a terrain is inadequate and that a series of field measurements and laboratory evaluations, including the analysis of grain size distribution and compatibility, have to be conducted to adequately define the properties of a given sand.

In summary, it is fair to say that while empirical relations are useful in estimating the performance of tyres (or wheeled vehicles) with design features similar to those that have been tested under similar conditions, it is by no means certain that they can be extrapolated beyond the conditions upon which they were based. This is exemplified by the processes in which various sand–tyre numerics have been developed and modified, as described previously. It, therefore, appears uncertain that an entirely empirical model can play a significant role in the evaluation of new design concepts or in the prediction of tyre (or wheeled vehicle) performance in new operating environments.

11.2 Methods for Parametric Analysis

One of the well-known methods for predicting wheel performance on unprepared terrain is that developed by Bekker (Bekker, 1956 and 1960). His method was later refined (Bekker, 1983, 1985) and extended to predicting flexible tyre performance.

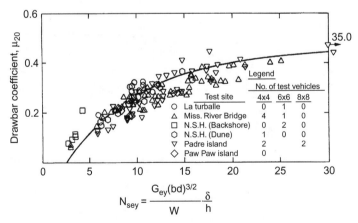

Figure 11.5: Measured data and empirical relation between drawbar coefficient at 20% slip and the further revised sand–tyre numeric N_{sey} obtained from test results with a variety of wheeled vehicles on six sandy fields (Reprinted by permission of *ISTVS* from Turnage, 1984)

11.2.1 Rigid Wheel–Terrain Interaction

While pneumatic tyres have long replaced rigid wheels as the running gear of off-road wheeled vehicles in normal operation, the mechanics of a rigid wheel on unprepared terrain is still of interest, as a pneumatic tyre may behave like a rigid rim, if the inflation pressure is sufficiently high and the terrain is relatively soft, as shown in the diagram in the middle of Figure 11.6. Furthermore, rigid wheels are still in use under certain circumstances, such as in vehicles for operation in the paddy field, or in robotic rovers for the exploration of the Moon, Mars and beyond.

In the method originally developed by Bekker for predicting the performance of a rigid wheel, it is assumed that the terrain reaction on the contact patch is purely radial, as shown in Figure 11.7, and that the radial pressure is equal to the normal pressure p beneath a horizontal sinkage plate (rectangular or circular) at the same depth in the pressure–sinkage test discussed in Chapter 4. It is further assumed that there is no shear stress on the wheel–terrain interface.

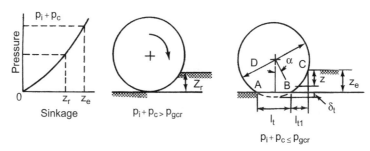

Figure 11.6: Behaviour of a tyre in the rigid and elastic operating modes

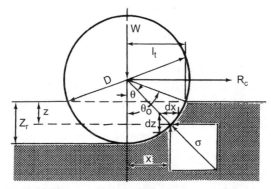

Figure 11.7: Simplified wheel–soil interaction model (Reprinted by permission from M.G. Bekker, *Theory of Land Locomotion*, University of Michigan Press, 1956, copyright © by the University of Michigan)

A. Compaction Resistance of Rigid Wheels

Considering the equilibrium of the horizontal and vertical forces acting on a towed rigid wheel shown in Figure 11.7, the following equations can be established (Bekker, 1956):

$$R_c = b_{ti} \int_0^{\theta_0} \sigma r \sin\theta \, d\theta \tag{11.10}$$

$$W = b_{ti} \int_0^{\theta_0} \sigma r \cos\theta \, d\theta \tag{11.11}$$

where R_c is the motion resistance, W is the vertical load on the wheel, σ is the radial (normal) pressure on the wheel–terrain interface, b_{ti} and r are the width and radius of the wheel, respectively, and θ_0 is the contact angle of the wheel, as shown in Figure 11.7.

As mentioned above, the radial pressure σ on the wheel–terrain interface is assumed to be equal to the normal pressure p beneath a sinkage plate at the same depth z in the pressure–sinkage test, the following relations can be established (Figure 11.7):

$$\sigma r \sin\theta \, d\theta = p \, dz \tag{11.12}$$

$$\sigma r \cos\theta \, d\theta = p \, dx \tag{11.13}$$

If the pressure–sinkage relation is described by Eqn (4.1) proposed by Bekker, then together with Eqn (11.12) given above, Eqn (11.10) may be rewritten as

$$R_c = b_{ti} \int_0^{z_r} \left(\frac{k_c}{b} + k_\phi\right) z^n \, dz = b_{ti} \left[\left(\frac{k_c}{b} + k_\phi\right) \frac{z_r^{n+1}}{n+1}\right] \tag{11.14}$$

where z_r is the sinkage of the wheel, as shown in Figure 11.7.

It should be noted that according to Bekker, in the above equation the parameter b in the term of $(k_c/b + k_\phi)$ is the smaller dimension of the contact patch of the wheel. For a wide wheel, the smaller dimension of the contact patch may be the contact length l_t shown in Figure 11.7, and not the wheel width b_{ti}. This is further discussed later in this section.

The value of R_c calculated by Eqn (11.14) is equivalent to the vertical work done per unit length in pressing a sinkage plate, with the same width as that of the wheel, into the terrain to a depth of z_r. This implies that the motion resistance R_c of a rigid wheel is caused by the vertical work done in making a rut of depth z_r. Accordingly, the motion resistance R_c given by Eqn (11.14) is usually referred to as the compaction resistance.

Using Eqn (11.14) to calculate the compaction resistance R_c, the sinkage (or rut depth) z_r expressed in terms of wheel parameters and of pressure–sinkage parameters of the terrain has to be determined. Equations (11.11) and (11.13) show that the wheel sinkage z_r may be derived from the following equation for vertical force equilibrium:

$$W = -b_{ti}\int_0^{z_r} p\, dx = -b_{ti}\int_0^{z_r}\left(\frac{k_c}{b} + k_\phi\right)z^n\, dx \qquad (11.15)$$

From the geometry shown in Figure 11.7, one obtains

$$x^2 = [D - (z_r - z)](z_r - z) \qquad (11.16)$$

where D is the wheel diameter.

For small sinkage,

$$x^2 = D(z_r - z) \qquad (11.17)$$

and

$$2x\, dx = -D\, dz \qquad (11.18)$$

Substituting Eqn (11.18) into Eqn (11.15), one obtains

$$W = b_{ti}(k_c/b + k_\phi)\int_0^{z_r} \frac{z^n \sqrt{D}}{2\sqrt{z_r - z}}\, dz \qquad (11.19)$$

Let $z_r - z = t^2$, then $dz = -2t\, dt$ and

$$W = b_{ti}\left(\frac{k_c}{b} + k_\phi\right)\sqrt{D}\int_0^{\sqrt{z_r}} (z_r - t^2)^n\, dt \qquad (11.20)$$

Expending $(z_r - t^2)^n$ into a series

$$(z_r - t^2)^n = z_r^n - nz_r^{n-1}t^2 + n(n-1)z_r^{n-2}t^4/2 - n(n-1)(n-2)z_r^{n-3}t^6/6$$
$$+ n(n-1)(n-2)(n-3)z_r^{n-4}t^8/24 + \ldots$$

and taking only the first two terms in the integration in Eqn (11.20), one obtains

$$W = \frac{b_{ti}(k_c/b + k_\phi)\sqrt{z_r D}}{3}z_r^n(3 - n) \qquad (11.21)$$

Rearranging the above equation, one obtains

$$z_r^{(2n+1)/2} = \frac{3W}{b_{ti}(k_c/b + k_\phi)(3-n)\sqrt{D}}$$

or

$$z_r = \left[\frac{3W}{b_{ti}(3-n)(k_c/b + k_\phi)\sqrt{D}}\right]^{2/(2n+1)} \tag{11.22}$$

Substituting Eqn (11.22) into Eqn (11.14), the compaction resistance R_c is expressed by

$$R_c = \frac{1}{(3-n)^{(2n+2)/(2n+1)}(n+1)b_{ti}^{1/(2n+1)}(k_c/b + k_\phi)^{1/(2n+1)}} \left(\frac{3W}{\sqrt{D}}\right)^{(2n+2)/(2n+1)} \tag{11.23}$$

From Eqn (11.23), it appears that to reduce the compaction resistance R_c, it is more effective to increase the wheel diameter D than the wheel width b_{ti}, as D enters the equation in higher power than b_{ti}. Note that Eqn (11.23) is derived from Eqn (11.21) using only the first two terms of the series representing $(z_r - t^2)^n$ in Eqn (11.20). As a result, Eqn (11.23) works well only for values of n up to approximately 1.3. Beyond that, the error in predicting the compaction resistance R_c increases. When the value of n approaches 3, the value of R_c approaches infinity — an obvious anomaly. For values of n greater than 1.3, the first five terms in the series representing $(z_r - t^2)^n$ should be used in the integration in Eqn (11.20). This will greatly improve the accuracy in predicting the compaction resistance R_c.

It should be mentioned that for a wide wheel (or a wide tyre, such as the Terratire or Rolligon, when its inflation pressure is sufficiently high and it behaves like a rigid rim), the smaller dimension of the projected contact patch may be the contact length l_t shown in Figure 11.7 and not the width b_{ti}. Under these circumstances, contact length l_t should substitute for b in the term $(k_c/b + k_\phi)$ in the equations for calculating the sinakge z_r and the compaction resistance R_c presented above. The compaction resistance R_c in this case is expressed by

$$R_c = \frac{1}{(3-n)^{(2n+2)/(2n+1)}(n+1)b_{ti}^{1/(2n+1)}(k_c/l_t + k_\phi)^{1/(2n+1)}} \left(\frac{3W}{\sqrt{D}}\right)^{(2n+2)/(2n+1)} \tag{11.24}$$

Example 11.1

A rigid wheel with diameter of 1 m and width of 0.15 m is to be used to carry a load of 10 kN on a sandy loam. The pressure–sinkage relationship of the terrain may be described by Eqn (4.1). Its pressure–sinkage parameters are: $n = 1.0$, $k_c = 5.7\,\text{kN/m}^2$, and $k_\phi = 2293\,\text{kN/m}^3$. To reduce the compaction resistance R_c on this terrain, one may consider increasing the wheel diameter or wheel width. Evaluate the effect of increasing the wheel diameter or the wheel width by 20% on the compaction resistance of the wheel.

Solution

(a) To determine whether the wheel width or the contact length is the smaller dimension of the contact patch, the sinkage z_r is first calculated using Eqn (11.22) by assuming the wheel width is the smaller dimension of the contact patch:

$$z_r = \left[\frac{3W}{b_{ti}(3-n)(k_c/b_{ti} + k_\phi)\sqrt{D}}\right]^{2/(2n+1)} = \left[\frac{3 \times 10}{0.15 \times 2 \times (5.7/0.15 + 2293)\sqrt{1}}\right]^{2/3}$$
$$= 0.123\,\text{m} = 12.3\,\text{cm}$$

From Figure 11.7, the contact length l_t can be determined as follows:

$$l_t = \sqrt{(D/2)^2 - (D/2 - z_r)^2} = \sqrt{0.25 - (0.5 - 0.123)^2} = 0.328\,\text{m} = 32.8\,\text{cm}$$

Since the contact length is greater than the wheel width, the wheel width is the smaller dimension of the contact patch and should be used in the calculation of the term $(k_c/b + k_\phi)$.

(b) Using Eqn (11.23), the compaction resistance R_c of the wheel with diameter of 1 m, width of 0.15 m and a load of 10 kN, can be predicted as follows:

$$R_c = \frac{1}{(3-n)^{(2n+2)/(2n+1)}(n+1)b_{ti}^{1/(2n+1)}(k_c/b_{ti} + k_\phi)^{1/(2n+1)}}\left(\frac{3W}{\sqrt{D}}\right)^{(2n+2)/(2n+1)}$$

$$= \frac{1}{2^{4/3} \times 2 \times (0.15)^{1/3}(5.7/0.15 + 2293)^{1/3}}\left(\frac{3 \times 10}{\sqrt{1}}\right)^{4/3}$$
$$= 2.627\,\text{kN}$$

(c) Increasing the wheel diameter by 20%, the value of D is 1.2 m. The corresponding compaction resistance R_{c1} is

$$R_{c1} = \frac{1}{2^{4/3} \times 2 \times (0.15)^{1/3}(5.7/0.15 + 2293)^{1/3}} \left(\frac{3 \times 10}{\sqrt{1.2}}\right)^{4/3}$$
$$= 2.325 \text{ kN}$$
$$R_{c1}/R_c = 88.54\%$$

(d) Increasing the wheel width by 20%, the value of b_{ti} is 0.18 m. The sinkage z_r can be calculated by assuming the wheel width b_{ti} is the smaller dimension of the contact patch:

$$z_r = \left[\frac{3W}{b_{ti}(3-n)(k_c/b_{ti} + k_\phi)\sqrt{D}}\right]^{2/(2n+1)} = \left[\frac{3 \times 10}{0.18 \times 2 \times (5.7/0.18 + 2293)\sqrt{1}}\right]^{2/3}$$
$$= 0.109 \text{ m} = 10.9 \text{ cm}$$

The corresponding contact length l_t can be calculated as follows:

$$l_t = \sqrt{(D/2)^2 - (D/2 - z_r)^2} = \sqrt{0.25 - (0.5 - 0.109)^2} = 0.312 \text{ m} = 31.2 \text{ cm}$$

Since the contact length is again greater than the wheel width, the wheel width is the smaller dimension of the contact patch and should be used in the calculation of the term $(k_c/b + k_\phi)$.

The corresponding compaction resistance R_{c2} is

$$R_{c2} = \frac{1}{2^{4/3} \times 2 \times (0.18)^{1/3}(5.7/0.18 + 2293)^{1/3}} \left(\frac{3 \times 10}{\sqrt{1}}\right)^{4/3}$$
$$= 2.473 \text{ kN}$$
$$R_{c2}/R_c = 94.17\%$$

The results indicate that to reduce the compaction resistance, increasing the wheel diameter by 20% is more effective than by increasing the wheel width by 20%. ∎

It was pointed out by Bekker that acceptable predictions may be obtained using Eqn (11.23) for moderate sinkage (i.e. $z_r \leq D/6$), and that the larger the wheel diameter and the smaller the sinkage, the more accurate the predictions will be (Bekker, 1969). It should also be noted that predictions obtained using Eqn (11.23) for wheels with diameters less than 50 cm become less accurate, and that predictions of wheel sinkage in dry sand obtained using Eqn (11.23) will not be accurate, if there is significant slip–sinkage (Bekker, 1969).

The basic assumption used in the above analysis of rigid wheel–terrain interaction implies that the maximum radial (normal) pressure on the contact patch should occur at the lowest point of wheel contact (or the so-called 'bottom-dead-centre'), where the sinkage is a maximum. It also implies that the distribution of radial stress on the contact patch and the compaction resistance R_c is independent of slip. Experimental evidence shows that on clay the normal pressure distribution beneath a rigid wheel seems to be quite uniform and not significantly affected by slip (Wong and Reece, 1967a). A number of experiments on sand, however, show that the radial pressure distribution is different from that assumed in the above analysis.

B. Stress Distributions on the Contact Patch of Driven Rigid Wheels

Figure 11.8 shows the measured radial (normal) pressure distributions on the contact patch of a driven rigid wheel (with diameter of 1.25 m and width of 0.15 m) on compact sand at different slips (Onafeko and Reece, 1967; Wong, 1967). The slip is defined by Eqn (6.10). It shows that the measured maximum radial pressure does not occur at the bottom-dead-centre, as would be expected from the plate sinkage analogy. Rather, it occurs in front (or forward) of the bottom-dead-centre. The location of the measured maximum radial pressure is a function of slip and shifts forward with increasing slip. It is found that the location of the maximum radial pressure is in fact at the junction of the two flow zones in the soil under a driven wheel, as shown in Figure 2.17 (Wong, 1967; Wong and Reece, 1967a, b). A review of several sets of experimental data on sand, such as those shown in Figure 11.9, indicates that the location of the maximum radial stress may be expressed by the following equation (Wong and Reece, 1967a):

$$\theta_m/\theta_0 = c_1 + c_2 i \qquad (11.25)$$

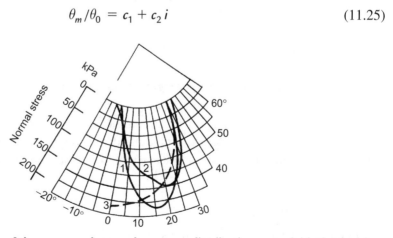

Figure 11.8: Comparison of the measured normal pressure distribution on a rigid wheel with the predicted one using the simplified wheel–soil interaction model on compact sand; curve 1 – measured at 3.1% slip; curve 2 – measured at 35.1%; and curve 3 – predicted (Reprinted by permission of *ISTVS* from Onafeko and Reece, 1967)

Table 11.1: Coefficients c_1 and c_2 for defining the relative position of the maximum radial stress point on a rigid wheel on sand under two types of condition: compact and loose

Soil	Angle of shearing resistance, $\phi°$	Cohesion, c (kPa)	Density (kg/m³)	c_1	c_2
Compact sand	33.3°	0.69	1592	0.43	0.32
Loose sand	31.1°	0.83	1329	0.18	0.32

Source: Wong and Reece (1967a).

where θ_m is the angular position of the maximum radial stress with respect to the vertical, θ_0 is the contact angle of the wheel with the terrain, as shown in Figure 11.7, c_1 and c_2 are coefficients that are related to soil conditions, and i is the slip of the wheel. Table 11.1 gives the values of c_1 and c_2 for a driven rigid wheel with diameter of 1.25 m and width of 0.15 m and those for a driven rigid wheel with diameter of 1.25 m and width of 0.3 m on sand under two different conditions: compact and loose (Onafeko and Reece, 1967; Wong and Reece, 1967a). It is interesting to note that even though the conditions of the sand are different, the values of the coefficient c_2 are the same for both compact and loose sand. This is also shown by the two lines in Figure 11.9 having the same slope in characterizing the relationship between the relative position of the maximum radial stress (θ_m/θ_0) and slip i on both the compact and loose sand. It should be pointed out, however, that for the loose sand the value of coefficient c_1 is lower than that for the compact sand. This indicates that the location of the maximum radial stress on loose sand is closer to the bottom-dead-centre than that on compact sand. It also shows that the position of the maximum radial stress relative to the wheel contact angle, θ_m/θ_0, depends on the compressibility of the sand. It is interesting to point out that the maximum radial stress point moves closer to the bottom-dead-centre as the slip decreases.

Figure 11.8 also indicates that as the wheel slip increases, the magnitude of the maximum radial stress decreases. This is primarily due to the fact that when wheel slip increases, the shear stress on the wheel–terrain interface increases accordingly. The vertical component of the shear stress plays an increasingly significant role in supporting the normal load on the wheel. This leads to the decrease in the magnitude of the radial stress on the contact patch in general and in the magnitude of the maximum radial stress in particular (Onafeko and Reece, 1967; Wong and Reece, 1967a). All of these indicate that in the analysis of the mechanics of wheel–terrain interaction, the shear stress on the wheel–terrain interface must be taken into consideration.

C. Stress Distributions on the Contact Patch of Towed Rigid Wheels

The radial and shear stress distributions under a towed rigid wheel with diameter of 1.25 m and width of 0.15 m on compact sand are shown in Figure 11.10 (Onafeko and Reece, 1967; Wong and Reece, 1967b). It shows that the maximum radial stress occurs at the point where the shear stress changes direction, which may be called the transition point. It is found that

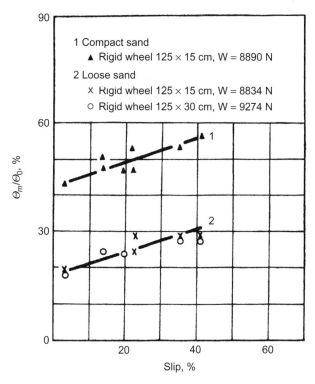

Figure 11.9: Variations of the relative angular position of the maximum radial (normal) stress θ_m/θ_0 with slip (Reprinted by permission of *ISTVS* from Wong and Reece, 1967a)

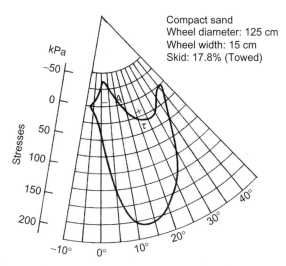

Figure 11.10: Measured radial (normal) and shear stress distributions on the contact patch of a towed rigid wheel on compact sand (Reprinted by permission of *ISTVS* from Onafeko and Reece, 1967)

the transition point is located at the junction of the two flow zones under a towed rigid wheel shown in Figure 2.18 (Wong, 1967; Wong and Reece, 1967b). Under the action of section AD of the rim of a towed rigid wheel, the soil in the region ABD moves upward while the rim rotates around the instantaneous centre I, as shown in Figure 2.18. The soil therefore slides along section AD of the rim in such a way as to produce shear stresses in the direction opposite to that of wheel rotation, which is defined as positive, as shown in Figure 11.10. In the section between A and E in Figure 2.18, the soil moves forward slowly while the wheel rim moves forward relatively fast. In this region, the shear stress therefore acts in the direction of wheel rotation, which is defined as negative, as shown in Figure 11.10. As the shear stress on part of the contact patch is acting in one direction (positive), while that on the other part acting in the other direction (negative), the resultant torque acting on a towed wheel is zero. This indicates that the towed wheel is in free rolling condition.

The location of the transition point may be estimated based on the theory of plastic equilibrium discussed in Section 2.2 of Chapter 2. Since at the transition point, shear stress is zero, the radial (normal) stress at that point must be the major principal stress at the boundary of the soil mass. According to the theory of passive earth pressure, there must be two slip lines on either side of the axis of the major principal stress with an angle of $45° − \phi/2$, as AF and AG shown in Figure 11.11. ϕ is the angle of internal shearing resistance of the soil. It therefore follows that if a point on the wheel–soil interface can be found where the slip lines have an angle of $45° − \phi/2$ with the radial, then this point will be the transition point. This indicates that to locate the transit point, the slip line field on the boundary of the soil mass adjacent to the rim should be examined. It is found that the tangent of the trajectory of soil particles at the junction of the two flow zones beneath a towed rigid wheel coincides with the

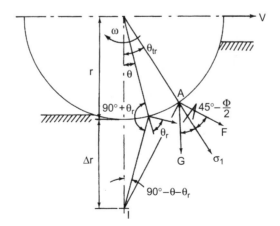

Figure 11.11: Diagram illustrating the angle between the direction of the absolute velocity vector and the radial at a point on the rim of a towed rigid wheel (Reprinted by permission of *ISTVS* from Wong and Reece, 1967b)

direction of the absolute velocity of the corresponding point on the rim. In solving problems in soil mechanics, it is usually assumed that the trajectories of soil particles in the flow zones represent one set of the slip lines. If this assumption is adopted, then it follows that at the transition point the direction of the absolute velocity of the rim coincides with one of the slip lines at the boundary of the soil mass. Based on the above consideration, it becomes evident that the problem of locating the transition point is now reduced to determining where the direction of the absolute velocity on the rim has an angle of $45° - \phi/2$ with the radial.

Referring to Figure 11.11, it can be seen that the angle θ_r between the direction of the absolute velocity and the radial at different points on the wheel rim can be determined from the kinematics of the wheel. For a towed rigid wheel of radius r with skid i_s, the instantaneous centre I is at a distance of Δr below the bottom-dead-centre, as shown in Figure 11.11. In the following analysis, the skid i_s of a wheel is defined as follows:

$$i_s = 1 - \frac{V_t}{V} = 1 - \frac{r\omega}{r_e \omega} = 1 - \frac{r}{r + \Delta r} \tag{11.26}$$

where r is the radius of the rigid wheel, r_e is usually called the effective rolling radius of the wheel, which is equal to $r + \Delta r$ shown in Figure 11.11, and $\Delta r = r i_s/(1 - i_s)$, ω is the angular speed of the wheel, V_t is the theoretical speed of the wheel centre and is equal to $r\omega$, and V is the actual forward speed of the wheel centre and is equal to $r_e \omega$. For a wheel completely locked up during braking (i.e. its angular speed is zero, while the forward speed of the wheel centre is not zero), the skid i_s is 100%, following the definition of i_s given by Eqn (11.26).

From the geometry shown in Figure 11.11, the following relationships can be established (Wong and Reece, 1967b):

$$\frac{r}{r + \Delta r} = \frac{\sin\left[90° - (\theta + \theta_r)\right]}{\sin(90° + \theta_r)} = \frac{\cos(\theta + \theta_r)}{\cos\theta_r}$$

$$1 - i_s = \cos\theta - \sin\theta \tan\theta_r$$

$$\tan\theta_r = \frac{\cos\theta - (1 - i_s)}{\sin\theta} \tag{11.27}$$

Since at the transition point $\theta_r = 45° - \phi/2$, the angular position of the transition point θ_{tr} can be obtained by solving the following equation:

$$\tan(45° - \phi/2) = \frac{\cos\theta_{tr} - (1 - i_s)}{\sin\theta_{tr}} \tag{11.28}$$

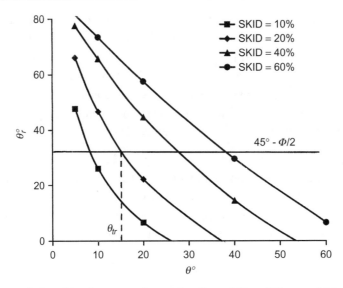

Figure 11.12: Relationships between the angles θ_r and θ at different skids (Reprinted by permission of ISTVS from Wong and Reece, 1967b)

Figure 11.12 shows the variations of θ_r with θ at various values of skid i_s. The angular position θ_{tr} of the transition point on the wheel rim at a given skid i_s is determined by the intersection of the curve for a given skid i_s and the horizontal line for $\theta_r = +(45° - \phi/2)$, as shown in Figure 11.12. Incidentally, another horizontal line representing $\theta_r = -(45° - \phi/2)$ may also be drawn in Figure 11.12 (not shown in the figure) and it may have another intersection point on each curve. This point occurs at an angle much greater than that shown in the figure. Experimental evidence indicates that the actual transition point locates at $\theta_r = +(45° - \phi/2)$.

Table 11.2 shows a comparison between the measured angular positions of the transition point of two towed rigid wheels with different widths and the corresponding predicted ones obtained using the above-noted method on compact and loose sand (Wong and Reece, 1967b). It shows that while there are differences in the absolute value between the predicted and measured angular positions of the transition point, the variation of the predicted value shows a similar trend to that of the measured. The differences are probably due to the fact that the prediction method is based on the consideration of a two-dimensional soil flow beneath a wheel, while the relatively narrow wheels used in the experiments actually cause a three-dimensional soil flow.

In summary, in the original method developed by Bekker, only the radial (normal) stress on the contact patch is taken into account and the shear stress is neglected. Experimental evidence indicates that the shear stress is an essential factor that should be taken into

Table 11.2: Comparison of the measured and predicted locations of the transition points of towed rigid wheels of different dimensions on compact and loose sand

Soil type	Compact sand	Loose sand					
Wheel dimensions	Diameter – 1.25 m Width – 0.15 m	Diameter – 1.25 m Width – 0.15 m		Diameter – 1.25 m Width – 0.30 m			
Skid, %	17.8	30.6	31.1	31.3	33.9	34.9	35.3
Measured location of the transition point, θ_{tr}	11.6°	18.1°	18.1°	20.0°	17.9°	17.9°	17.9°
Predicted location of the transition point, θ_{tr}	15.3°	23.3°	23.6°	23.8°	25.5°	26.0°	26.2°

Source: Wong and Reece (1967b).

consideration in the analysis of wheel–terrain interaction. It also shows that the mechanics of interaction between a wheel and the terrain is much more complex than that assumed in the Bekker method. This suggests that in a more comprehensive analysis of the mechanics of wheel–terrain interaction, both the radial (normal) and shear stresses on the contact patch should be taken into account. This is discussed in detail in the next chapter.

11.2.2 Flexible Tyre–Terrain Interaction

When the inflation pressure of a pneumatic tyre is relatively low and the terrain is relatively firm, the tyre deflects. Under these circumstances, it may be assumed that the lower portion of the tyre in contact with the terrain, section AB shown in the right diagram of Figure 11.6, is flattened, and that the ground contact pressure on the flat portion of the tyre is equal to the sum of the inflation pressure p_i and the pressure due to the stiffness of the tyre carcass p_c. Methods for estimating the conditions under which a tyre deflects and exhibits behaviour shown in the right diagram of Figure 11.6 are presented later in this section.

A. Compaction Resistance of a Flexible Tyre

If the tyre deflects as shown in the right diagram of Figure 11.6 and the pressure–sinkage relation of the terrain is described by Eqn (4.1), the sinkage of a flexible tyre z_e is expressed by

$$z_e = \left(\frac{p_i + p_c}{k_c/b + k_\phi} \right)^{1/n} \tag{11.29}$$

As noted previously, the parameter b in the term $(k_c/b + k_\phi)$ is the smaller dimension of the contact patch of the tyre. For a wide tyre, such as the Terratire or Rolligon, the contact length l_t, and not the tyre width b_{ti}, may well be the smaller dimension of the tyre contact patch. By considering the vertical equilibrium of the tyre, the approximate contact length l_t of a flexible tyre may be estimated as follows:

$$l_t = \frac{W}{b_{ti}\,(p_i + p_c)} \qquad (11.30)$$

If l_t is greater than the tyre width b_{ti}, then b_{ti} is the smaller dimension of the contact patch. On the other hand, if l_t is smaller than the tyre width b_{ti}, then l_t is the smaller dimension of the contact patch.

By substituting z_e given by Eqn (11.29) for z_r in Eqn (11.14), the compaction resistance R_c of a flexible tyre is expressed by

$$R_c = b_{ti}\left[\left(\frac{k_c}{b} + k_\phi\right)\frac{z_e^{n+1}}{n+1}\right] = \frac{b_{ti}\,(p_i + p_c)^{(n+1)/n}}{(n+1)(k_c/b + k_\phi)^{1/n}} \qquad (11.31)$$

where b_{ti} is the tyre width and b is the smaller dimension of the contact patch, as noted previously.

It should be noted that the deflected shape of a pneumatic tyre assumed in the above analysis is only an approximation to that observed in practice. In reality, the lower portion of the tyre is not flat, but deflects gradually. However, it has been shown that in many cases, the difference between the assumed and measured deflected shape is within an acceptable range in a first-order analysis (Bekker, 1983, 1985). The use of the finite element tyre model to improve the prediction of the geometry of the contact patch is discussed in Chapter 12.

It should be mentioned that the pressure p_c exerted on the terrain due to the carcass stiffness in Eqn (11.29) is difficult to determine, as it varies with the inflation pressure and normal load of the tyre. As an alternative, Bekker proposed to use the average ground pressure p_g of a tyre on hard ground to represent the sum of p_i and p_c (Bekker, 1983). The average ground pressure p_g for a given tyre with a particular load and inflation pressure can be derived from the so-called 'generalized deflection chart' normally available from the tyre manufacturer, as shown in Figure 11.13. The average ground pressure p_g is equal to the load W carried by the tyre divided by the ground contact area A shown in the figure. As an example, Figure 11.14 shows the relationships between the average ground pressure p_g and inflation pressure p_i for a 12.5/75R20 tyre at various loads. It shows that for the particular tyre under consideration, the pressure p_c exerted on the ground due to carcass stiffness is not a constant, and that its value varies with inflation pressure and load. It is interesting to note from Figure 11.14 that when the tyre load and inflation pressure are within a certain range, the average ground pressure is lower than the inflation pressure.

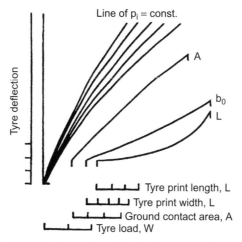

Figure 11.13: Generalized deflection chart for a pneumatic tyre

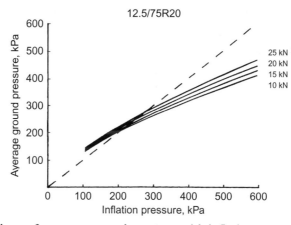

Figure 11.14: Variations of average ground pressure with inflation pressure at various loads for a 12.5/75R20 tyre (Reprinted by permission of the Department of National Defence, Canada, from Preston-Thomas and Wong, 1987)

Using the average ground pressure p_g to represent the sum of p_i and p_c, Eqn (11.29) can be rewritten as

$$z_e = \left(\frac{p_g}{k_c/b + k_\phi} \right)^{1/n} \tag{11.32}$$

B. Prediction of the Operating Mode of a Pneumatic Tyre

As mentioned previously, a pneumatic tyre with a given load may behave like a rigid wheel (referred to as the rigid mode of operation) or it may deflect significantly (referred to as the elastic mode of operation), dependent upon the rigidity of the tyre and terrain conditions. To predict the operating mode of the tyre, various methods have been developed (Bekker, 1960, 1983; Wong, 1978a).

In the 1960s, Bekker (1960) developed a method for predicting the critical inflation pressure p_{cr}, above which the tyre will behave like a rigid wheel. It was derived from the analysis of the forces and moments acting on a tyre. The critical inflation pressure p_{cr} is expressed by

$$p_{cr} = \frac{W(n+1)}{b_{ti}\left[3W/(3-n)b_{ti}(k_c/b + k_\phi)\sqrt{D}\right]^{1/(2n+1)}} \times \frac{1}{\sqrt{D - \left[3W/(3-n)b_{ti}(k_c/b + k_\phi)\sqrt{D}\right]^{2/(2n+1)}}} - p_c \quad (11.33)$$

As mentioned previously, in many cases the pressure due to carcass stiffness p_c is difficult to ascertain. If the sum of tyre inflation pressure p_i and pressure due to carcass stiffness p_c is represented by the average ground pressure p_g, as proposed by Bekker, then the critical ground pressure p_{gcr} can be expressed by

$$p_{gcr} = p_{cr} + p_c$$
$$= \frac{W(n+1)}{b_{ti}\left[3W/(3-n)b_{ti}(k_c/b + k_\phi)\sqrt{D}\right]^{1/(2n+1)}} \times \frac{1}{\sqrt{D - \left[3W/(3-n)b_{ti}(k_{c/b} + k_\phi)\sqrt{D}\right]^{2/(2n+1)}}} \quad (11.34)$$

If the average ground pressure p_g derived from a generalized deflection chart for a given tyre at a particular combination of inflation pressure and load is greater than p_{gcr} as determined by Eqn (11.34), then the tyre is assumed to be in the rigid mode of operation. On the other hand, if the value of p_g is less than p_{gcr} the tyre is assumed to be in the elastic mode of operation and the lower portion of the tyre circumference in contact with the terrain is assumed to be flattened. Under these circumstances, the ground contact pressure acting on the flat portion is assumed to be equal to p_g.

Table 11.3 shows the values of the critical ground pressure p_{gcr}, as determined by Eqn (11.34), for a 12.5/75R20 tyre with diameter of 1 m, width of 0.327 cm and load of 20 kN, over four different types of terrain: a firm soil, a medium soil, a soft soil, and a clayey soil with a high moisture content.

It should be mentioned that in deriving Eqns (11.33) and (11.34), certain assumptions and mathematical simplifications in determining the load supported by section BC of the tyre

Table 11.3: Critical ground pressures p_{gcr} for a 12.5/75R20 tyre over various types of terrain

Terrain type	Pressure–sinkage Parameters			P_{gcr} (kPa)	
	n	k_c (kN/m^{n+1})	k_ϕ (kN/m^{n+2})	As determined by Eqn (11.34)	As determined by Eqn (11.35)
Firm soil	1.2	0	122 788	1092	821
Medium soil	0.8	29.8	2083	402	292
Soft soil	0.8	16.5	911	306	214
Clayey soil	0.6	38	20.7	199	99

shown in Figure 11.6 were made (Bekker, 1960). As a result, there is a discontinuity in the transition between the rigid and elastic operating modes. To improve the accuracy in the prediction of the operating mode of a pneumatic tyre, and to provide a smooth transition from one operating mode to another, Wong (1978a) proposed the following equation for determining the critical ground pressure p_{gcr}:

$$p_{gcr} = \left[\frac{k_c}{b} + k_\phi\right]^{1/(2n+1)} \left[\frac{3W}{(3-n)b_{ti}\sqrt{D}}\right]^{2n/(2n+1)} \qquad (11.35)$$

The above equation was based on the reasoning that if the terrain is relatively soft and the sum of the inflation pressure p_i and the pressure produced by the carcass stiffness p_c is greater than the pressure exerted by the terrain at the lowest point of the tyre circumference, the tyre can be assumed to remain round like a rigid rim. On the other hand, if the ground is relatively firm and the sum of p_i and p_c is less than the critical ground pressure p_{gcr}, then the portion of the tyre in contact with the terrain will have a significant deflection and the tyre is assumed to be in an elastic mode of operation. If the tyre behaves like a rigid rim, using the Bekker pressure–sinkage equation, the normal pressure p_g at the lowest point of contact (bottom-dead-centre) is given by

$$p_g = \left[\frac{k_c}{b} + k_\phi\right] z_r^n$$

Substituting Eqn (11.22) into the above equation one obtains the expression for the critical ground pressure p_{gcr} given in Eqn (11.35).

Table 11.3 also shows the values of the critical ground pressure p_{gcr} for the 12.5/75R20 tyre with a load of 20 kN, over different terrains calculated using Eqn (11.35). It can be seen that there is a significant difference between the values of p_{gcr} determined by Eqn (11.34) and by Eqn (11.35).

Using the information shown in Figure 11.14 and that given in Table 11.3, one can predict the operating mode of the tyre. For instance, if the inflation pressure is 172 kPa, then from Figure

11.14 the average ground pressure p_g at a normal load of 20 kN will be 193 kPa. According to the method proposed by Bekker, the tyre at an inflation pressure of 172 kPa will be in the elastic mode of operation over the four types of terrain shown in Table 11.3, as the values of the critical ground pressure p_{gcr} (1092, 402, 306 and 199 kPa, respectively) are all higher than the average ground pressure p_g of 193 kPa. Following the method proposed by Wong, however, the tyre will be in the elastic operating mode only over the firm, medium and soft soils. Over the clayey soil, the tyre is predicted to behave like a rigid wheel, as the value of p_g (193 kPa) is higher than that of p_{gcr} (99 kPa).

Later, Bekker (1983) suggested another method for the prediction of the operating mode of a tyre. In this method, it is considered that if the tyre is in the elastic mode of operation, a sinkage z_e for the tyre can be determined from the average ground pressure p_g using Eqn (11.32). On the other hand, if the tyre is in the rigid mode of operation, a sinkage z_r can be determined using Eqn (11.22). A comparison is then made between the values of z_e and z_r. If z_e is found to be greater than z_r, then the tyre is considered to be in the rigid mode of operation, otherwise it is in the elastic operating mode. If z_e is equal to z_r, the tyre is at the transition between the two modes. The corresponding ground pressure is, therefore, the critical ground pressure p_{gcr}. In other words, the expression for the critical ground pressure p_{gcr} can be obtained by setting z_e equal to z_r, or by equating Eqn (11.32) to Eqn (11.22), that is

$$\left(\frac{p_{gcr}}{k_c/b + k_\phi}\right)^{1/n} = \left[\frac{3W}{b_{ti}(3-n)(k_c/b + k_\phi)\sqrt{D}}\right]^{2/(2n+1)} \tag{11.36}$$

This leads to the same expression for p_{gcr} as that given in Eqn (11.35).

In addition to the tyre compaction resistance R_c, defined by Eqn (11.23) for the rigid operating mode, or by Eqn (11.31) for the elastic operating mode, Bekker proposed that a horizontal terrain deformation resistance, commonly known as the bulldozing resistance, should be included in the calculation of the overall motion resistance of a tyre, if the tyre sinkage is significant (Bekker, 1969, 1983). He suggested that the horizontal terrain deformation resistance is analogous to the force due to the passive earth pressure acting on a retaining wall, as discussed in Section 2.2. Figure 11.15 is similar to Figure 2.20, which shows the flow patterns of soil in front of a locked wheel with a soil wedge acting as a retaining wall (Wong and Reece, 1966; Wong, 1967).

It should be pointed out, however, that in the analysis of wheel–terrain interaction and the compaction resistance described in Section 11.2.1, it is assumed that the radial (normal) pressure acting on the wheel rim is equal to that beneath a horizontal plate at the corresponding depth. The inclusion of the bulldozing resistance as part of the overall motion resistance of a wheel implies that the passive earth pressure acting on the wheel rim through a soil wedge is

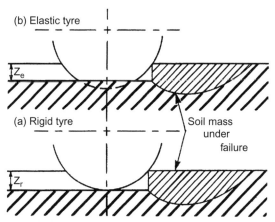

Figure 11.15: Horizontal terrain deformation in front of a rigid and an elastic tyre (From NRCC Report 22880 by M.G. Bekker, 1983)

superimposed to the radial pressure on the wheel rim based on the pressure–sinkage relationship. The superposition of the passive earth pressure over the radial stress does not seem to be consistent with the physical nature of wheel–terrain interaction. Furthermore, the soil wedge in front of a wheel shown in Figure 11.15 was only observed experimentally for a locked wheel and not for a driven or towed wheel (see Figures 2.17, 2.18 and 2.20).

C. Resistance Due to Tyre Flexing

For a tyre with notable deflection, in addition to the compaction resistance (and the bulldozing resistance as suggested by Bekker), energy is dissipated in the hysteresis of tyre material caused by the flexing of the tyre while rolling. This is reflected as a resisting force against the motion of the tyre. This resistance is referred to as the tyre flexing resistance. Based on test data, the following semi-empirical relation for predicting the tyre flexing resistance R_f was proposed by Bekker and Semonin (1975):

$$R_f = \left[3.581 \, b_{ti} \, D^2 \, p_g \, \varepsilon (0.0349\alpha - \sin 2\alpha)\right] / \alpha (D - 2\delta_t) \tag{11.37}$$

where b_{ti} is tyre width, D is tyre diameter, p_g is the average ground pressure, δ_t is tyre deflection, as shown in Figure 11.6, and ε and α are determined as follows:

$$\alpha = \cos^{-1}\left[(D - 2\delta_t)/D\right]$$

and

$$\varepsilon = 1 - \exp(-k_e \, \delta_t / h)$$

where α is the contact angle shown in Figure 11.6 in degrees, h is tyre section height, and the value of coefficient k_e is related to tyre construction. Its value is 15 for bias-ply tyres and 7 for radial-ply tyres.

■ Example 11.2

A radial tyre 12.5/75R20 is to be installed on an off-road vehicle. It is to carry a load of 25 kN at an inflation pressure of 200 kPa. The tyre has a diameter of 1 m, section height of 0.245 m and width of 0.327 m. The relationships between the inflation pressure p_i and the average ground pressure p_g for the tyre under various loads are shown in Figure 11.14. The vehicle is to operate on a sandy loam with pressure–sinkage parameters $n = 1$, $k_c = 5.7 \text{ kN/m}^2$, and $k_\phi = 2293 \text{ kN/m}^3$. Estimate the sinkage, compaction resistance, and flexing resistance, if any, of the tyre under the operating conditions described above.

Solution

(a) From Figure 11.14, at an inflation pressure of 200 kPa and normal load of 25 kN, the corresponding average ground pressure p_g is 220 kPa approximately.
The critical ground pressure p_{gcr} of the tyre for determining its operating mode, rigid or elastic, can be determined using Eqn (11.35):

$$p_{gcr} = \left[\frac{k_c}{b} + k_\phi\right]^{1/(2n+1)} \left[\frac{3W}{(3-n) b_{ti} \sqrt{D}}\right]^{2n/(2n+1)}$$

$$= (5.7/0.327 + 2293)^{1/3} \left[\frac{3 \times 25}{2 \times 0.327 \sqrt{1}}\right]^{2/3} = 312 \text{ kPa}$$

Since the average ground pressure p_g of 220 kPa is lower than the critical ground pressure p_{gcr} of 312 kPa, the tyre is in the elastic mode of operation.
In the elastic operating mode, the contact length l_t of the tyre at an average ground pressure of 220 kPa may be estimated using Eqn (11.30):

$$l_t = \frac{W}{b_{ti} p_g} = \frac{25}{0.327 \times 220} = 0.348 \text{ m}$$

Since the contact length of the tyre is greater than the width, the smaller dimension b in the term $(k_c/b + k_\phi)$ should indeed be the tyre width of 0.327 m.
The sinkage of the tyre z_e can be determined using Eqn (11.32):

$$z_e = \left[\frac{p_g}{k_c/b + k_\phi}\right]^{1/n} = \frac{220}{5.7/0.327 + 2293} = 0.095 \text{ m}$$

The compaction resistance R_c can be estimated using Eqn (11.31):

$$R_c = b_{ti}\left[\left(\frac{k_c}{b} + k_\phi\right)\frac{z_e^{n+1}}{n+1}\right] = 0.327\,(5.7/0.327 + 2293)(0.095^2/2) = 3.409\text{ kN}$$

(b) To estimate the tyre flexing resistance, the tyre deflection δ_t and the contact angle α have to be determined. From Figure 11.6, δ_t is given by

$$\delta_t = D/2 - \sqrt{(D/2)^2 - (l_t/2)^2} = 0.5 - \sqrt{0.5^2 - (0.348/2)^2} = 0.031\text{ m}$$

and

$$\alpha = \cos^{-1}[(D - 2\delta_t)/D] = 20.28°$$

For a radial tyre, $k_e = 7$ and the coefficient ε is given by

$$\varepsilon = 1 - \exp(-k_e\,\delta_t/h) = 1 - \exp(-7 \times 0.031/0.245) = 1 - \exp(-0.8857)$$
$$= 1 - \frac{1}{2.4247} = 0.5876$$

The tyre flexing resistance R_f is given by

$$R_f = \left[3.581\,b_{ti}\,D^2\,p_g\,\varepsilon(0.0349\alpha - \sin 2\alpha)\right]/\alpha(D - 2\delta_t)$$
$$= [3.581 \times 0.327 \times 1 \times 220 \times 0.5876 \times (0.0349 \times 20.28 - 0.6502)]$$
$$/20.28 \times (1 - 0.062) = 0.458\text{ kN}$$

The sum of the compaction resistance and the flexing resistance of the tyre is

$$R_c + R_f = 3.409 + 0.458 = 3.867\text{ kN}$$

D. Thrust–Slip Relationship of a Flexible Tyre

As described in Section 11.2.1, in the model for rigid wheel–terrain interaction originally developed by Bekker, the shear stress on the contact patch is not taken into account (Bekker, 1956). As a result, the model is not capable of predicting the thrust–slip relationship of a driven rigid wheel. Later, Janosi proposed a method for predicting the shear displacement beneath a wheel, based on the analysis of the cycloidal paths of various points on the wheel rim (Janosi, 1961). He assumed that the shear displacement, from which shear stress is determined, is the horizontal component of the displacement of a point on the rim along the cycloidal path in the terrain. The shape of the cycloidal path varies with wheel slip. The validity of this method was seriously questioned by Reece (1970), as the method is not compatible with the physical nature of wheel–terrain interaction. It has been shown that the shear stress along the wheel–terrain interface is related to, among other factors, the shear displacement developed along the interface and not related to the horizontal component of the cycloidal path of the wheel rim in the terrain (Onafeko and Reece, 1967; Wong and Reece, 1967a, b).

Later Bekker proposed that the thrust–slip relationship for a tyre be treated in the same way as that for a rigid track, as discussed in Section 7.2. This is illustrated in Figure 11.16 for a tyre in the elastic or rigid mode of operation (Bekker, 1983). Thus, for the shear stress–shear displacement relationship described by Eqn (5.2), the thrust–slip relationship for a tyre can be expressed by

$$F = (Ac + W \tan \phi)\left[1 - \frac{K}{il}(1 - e^{-il/K})\right] \quad (11.38)$$

where F is the thrust developed by a tyre, l is the length of the tyre contact area A measured along the direction of motion of the tyre, as shown in Figure 11.16, W is the normal load on the tyre, i is the tyre slip, defined by Eqn (6.10), and c, ϕ and K are the cohesion, angle of internal shearing resistance and shear deformation parameter of the terrain, respectively.

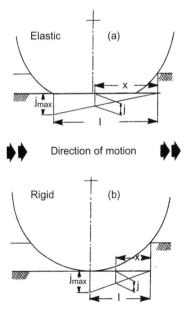

Figure 11.16: Development of shear displacement on the contact patch of a rigid and an elastic tyre (From NRCC Report 22880 by M.G. Bekker, 1983)

Equation (11.38), in essence, is the same as Eqn (7.12) for a rigid track with uniform normal pressure distribution. It should be noted, however, that on the curved portion BC of the contact patch of a tyre in the elastic operating mode shown in the right diagram of Figure 11.6, and on the contact patch of a tyre in the rigid operation mode, the normal pressure is not uniformly distributed. Furthermore, on the curved portion BC of the contact patch of a tyre in the elastic operating mode, and on the contact patch of a tyre in the rigid operating mode, the shear stress acts in the tangential direction of the wheel–terrain interface, rather than along the projected horizontal surface, as shown in Figure 11.16. These differences may cause significant variations in the prediction of tyre thrust in either the elastic or the rigid operating mode.

The wheel–terrain interaction model proposed by Bekker represents a pioneering effort in establishing a practical framework for the parametric analysis of tyre performance. However, to provide a more realistic prediction of tyre performance, improvements in the framework are required. For instance, in the original wheel–terrain interaction model, the shear stress on the interface was completely neglected, and the role of the vertical component of the shear stress in supporting tyre load was entirely omitted. The assumption that the thrust of a tyre, whether in the elastic or the rigid operating mode, can be predicted in the same way as that for a rigid track with uniform pressure distribution appears to be unrealistic. In the following chapter, improved methods for predicting the performance of tyres, as well as wheeled vehicles, are presented.

PROBLEMS

11.1 A radial tyre 12.5/75R20 is to be installed on an off-road vehicle. It is to carry a load of 25 kN. The tyre has an outside diameter of 1 m, section height of 0.245 m, and width of 0.327 m. The relationships between the inflation pressure p_i and the average ground pressure p_g for the tyre under various loads are shown in Fig. 11.14. The vehicle is to operate on a terrain with pressure-sinkage parameters $n = 0.9$, $k_c = 52.53\,\text{kN/m}^2$, and $k_\phi = 1127.97\,\text{kN/m}^3$. Estimate the sinkage, compaction resistance, and flexing resistance (if any) of the tyre at (a) inflation pressure of 172 kPa and (b) inflation pressure of 379 kPa. In the evaluation, the effect of grouser height may be neglected.

11.2 The tyre described in Problem 11.1 carrying a load of 20 kN at an inflation pressure of 379 kPa is being towed on sand. Its pressure-sinkage parameters are: $n = 1$ and $k_\phi = 1528\,\text{kN/m}^3$; and its shear strength parameters are: $c = 0$ and $\phi = 35°$. (a) Estimate the force required to tow the tyre; and (b) estimate the location of the transition point where the tangential stress changes its direction, if the skid of the towed tyre is 20%. In the evaluation, the effect of grouser height may be neglected.

CHAPTER 12

Computer-Aided Method NWVPM for Evaluating the Performance of Tyres and Wheeled Vehicles

The Bekker tyre–soil model described in the preceding chapter provides a foundation for the analysis of the mechanics of interaction between a tyre and unprepared terrain. As noted previously, however, to provide a more realistic prediction, improvements in and modifications to the Bekkar model seem necessary.

The development of a computer-aided method for predicting the performance of tyres, as well as multi-axle wheeled vehicles, from the traction perspective, was initiated in the late 1980s. Since then the method has been undergoing continual development and has become known as NWVPM. The objective of the computer-aided method NWVPM is to provide engineers in industry with a realistic and practical tool for parametric analysis of off-road wheeled vehicle performance and design. Experience in the applications of NWVPM to performance and design evaluation of off-road wheeled vehicles has demonstrated that it is a practical and useful tool for engineering practitioners.

12.1 Basic Features of NWVPM for Evaluating Tyre Performance

While the computer-aided method NWVPM for tyre performance presented in this chapter shares some of the features of Bekker's model, there are a number of significant differences between them. In NWVPM, both the normal (radial) pressure and shear stress on the tyre–terrain interface are taken into account, as shown in Figure 12.1 for a tyre in the rigid operating mode. The forces and moments exerted on the tyre by the terrain are considered to be entirely due to the normal pressure and shear stress acting on the interface. The horizontal terrain deformation resistance (commonly known as the bulldozing resistance), due to the passive earth pressure assumed in the Bekker model described in the preceding chapter, is not included. In NWVPM, the normal load on the tyre is considered to be supported by the vertical components of both the normal pressure and shear stress on the interface, whereas in the Bekker model, the normal load is assumed to be supported only by the vertical component of the radial (normal) pressure. The motion resistance acting on the tyre is due to the horizontal component of the normal pressure. The thrust developed by the tyre is due to the horizontal component of the shear stress on the

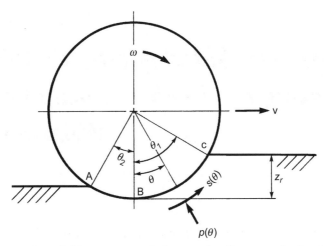

Figure 12.1: An improved model for tyre–terrain interaction for a tyre in the rigid operating mode

contact patch. The resultant vertical reaction of the terrain and the motion resistance and thrust of the tyre are obtained by integrating the appropriate components of the normal pressure or shear stress over the entire contact patch using numerical methods, thus avoiding the problem in the Bekker model using series approximation in the integration of the vertical component of radial (normal) pressure (see Section 11.2.1). Furthermore, a more realistic methodology is used in the prediction of the shear stress on the tyre–terrain interface, whereas in the extended Bekker model, the prediction of the shear stress on the contact patch of a tyre, in both rigid and elastic operating modes, is based on the methodology for predicting the shear stress on a rigid track. As the normal pressure on the contact patch of a tyre is usually not uniformly distributed, the effects of varying normal pressure on the development of shear stress, as discussed in Section 8.2, are also taken into consideration.

12.1.1 Normal and Shear Stress Distributions on the Tyre–Terrain Interface

The tyre–terrain interaction model for a tyre in the rigid operating mode is shown schematically in Figure 12.1, whereas that for a tyre in the elastic operating mode is illustrated in Figure 12.2. In the latter case, the tyre deflects and section AB of the contact patch is assumed to be flat, while sections BC and AD are assumed to maintain a circular shape. The performance of the tyre, described in terms of compaction resistance, thrust, drawbar pull, etc., is completely defined by the normal pressure $p(\theta)$ and shear stress $s(\theta)$ on the contact patch, which vary with the angular position θ.

A. Normal Pressure Distributions

As noted in Chapter 11, there are a number of methods proposed for predicting the normal pressure distribution on the tyre–terrain interface (Bekker, 1956; Wong and Reece, 1967a, b;

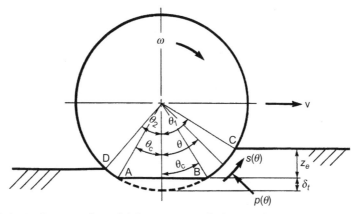

Figure 12.2: An improved model for tyre–terrain interaction for a tyre in the elastic operating mode

Harnisch et al., 2005). Each of these methods has its merits and limitations. All of them are, however, based on the assumption that the normal pressure on the tyre circumference is related to the pressure beneath a sinkage plate at the same depth. In the method developed by Wong and Reece (1967a) for a tyre in the rigid operating mode, two additional coefficients are introduced to take into account the shift of the location of the maximum normal (radial) pressure with slip, as described in Section 11.2.1. As pointed out previously, the location of the maximum normal pressure moves closer to bottom-dead-centre (i.e. the lowest point of contact on the wheel rim) with the decrease of slip. This means that if the wheel slip is not very high, the normal pressure distribution on the contact patch suggested by Bekker does not seem to be unreasonable. In the method proposed by Harnisch et al. (2005) for a tyre in the elastic operating mode, a larger substitute circle to describe the geometry of the deformed contact patch is used. The diameter of the substitute circle is calculated from the equilibrium between the vertical reaction of the terrain and the normal load applied on the tyre. Certain empirical factors are introduced in the solution process. The concept of using a larger substitute circle to represent the deformed contact patch geometry of a tyre was originally suggested by Bekker (1956). For lack of a better practical approach to predicting the normal pressure distribution on the tyre–terrain interface, the assumption adopted by Bekker that the normal pressure acting on the tyre circumference is equal to that beneath a sinkage plate at the same depth is used in the current version of the computer-aided method NWVPM. If and when a better method becomes available, it could be incorporated into the analytical framework presented in this chapter and into NWVPM.

In should be mentioned that in region BC of the contact area shown in Figures 12.1 and 12.2, the terrain is being gradually loaded, while in region AB in Figure 12.1 or in region AD in Figure 12.2, the terrain is being unloaded and the terrain is in elastic rebound. The mechanical properties in these two regions are different from that in region BC, as described in Chapter 4.

406 Chapter 12

Appropriate pressure–sinkage relations should, therefore, be used to determine the normal pressure distributions over different regions of the tyre contact patch. For instance, for mineral terrain, Eqn (4.1) or (4.2) may be used to predict the normal pressure distribution along section BC of the contact patch, while Eqn (4.23) should be employed to determine the normal pressure distribution in the region AB in Figure 12.1 or in region AD in Figure 12.2, where the terrain is in elastic rebound.

B. Shear Stress Distributions

As described in Section 5.1, the shear stress developed on the interface between the vehicle running gear and the terrain is a function of shear displacement. To predict the shear stress distribution on the tyre–terrain interface, it is necessary to examine the development of shear displacement along the contact patch.

A method for predicting the shear displacement developed along the contact patch of a tyre in the rigid operating mode was developed by Wong and Reece (1967a, b). It is based on the analysis of the slip velocity V_j on the tyre–terrain interface shown in Figure 12.3. The slip velocity V_j of a point on the tyre circumference relative to the terrain is considered to be the tangential component of the absolute velocity at the same point, as illustrated in Figure 12.3(b). The concept is analogous to the slip velocity V_j of a rigid track shown in Figure 12.3(a). The magnitude of the slip velocity V_j of a point on the tyre circumference defined by the angle θ can, therefore, be expressed by (Wong and Reece, 1967a)

$$V_j = r\omega \left[1 - (1-i)\cos\theta\right] \qquad (12.1)$$

where r is the radius of the tyre, ω is the angular speed of the tyre, and i is the slip of the tyre, defined by Eqn (6.10).

It can be seen that the slip velocity of a point on the circumference of the tyre in the rigid operating mode varies with its angular position defined by θ shown in Figure 12.3(b).

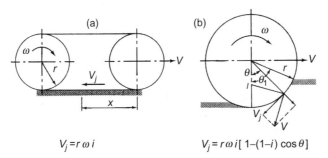

Figure 12.3: Slip velocity of a point on (a) a track and (b) the circumference of a tyre in the rigid operating mode (Reprinted by permission of ISTVS from Wong & Reece, 1967a)

The shear displacement j along the interface is expressed by

$$j = \int_0^t V_j \, dt = \int_0^{\theta_1} \left[1 - (1-i)\cos\theta\right] d\theta = r\left[(\theta_1 - \theta) - (1-i)(\sin\theta_1 - \sin\theta)\right] \quad (12.2)$$

where θ_1 is the entry angle that defines the position where a point on the tyre circumference first comes into contact with the terrain as shown in Figure 12.3b.

Figure 12.4 shows a comparison of the measured and predicted shear stress distributions along the contact patch of a rigid wheel of diameter of 1.25 m, width of 0.15 m, normal load of 8.896 kN and at a slip of 22.1% on compact sand (Wong and Reece, 1967a). The predicted shear stress distribution is obtained using the method described above, with the normal pressure distribution predicted using the method discussed in Section 11.2.1. It takes into account the shift of the location of the maximum radial (normal) pressure with wheel slip. It should be noted that in predicting the normal pressure distribution, the elastic rebound of the terrain in section AB of the contact patch (or angle θ_2) in Figure 12.1 is neglected.

When a tyre is in the elastic operating mode, the shear displacement developed along the section BC in Figure 12.5 can be determined in the same way as that described above for a tyre in the rigid operating mode. For the flat section AB, the slip velocity is considered to be a constant, similar to that under a rigid track, as shown in Figure 12.3(a). The incremental shear displacement Δj along section AB is proportional to the slip of the tyre i and the distance x between the point in question and point B in Figure 12.5. From Eqn (7.11), Δj is expressed by

$$\Delta_j = ix \quad (12.3)$$

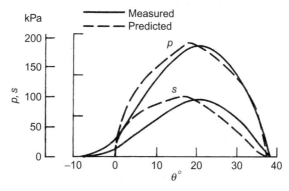

Figure 12.4: The measured and predicted normal and shear stress distribution on a rigid wheel (Reprinted by permission of ISTVS from Wong & Reece, 1967a)

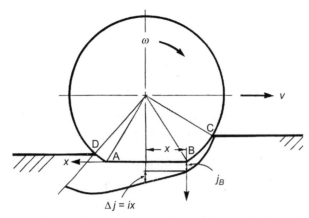

Figure 12.5: Development of shear displacement beneath a tyre in the elastic operating mode

The cumulative shear displacement j_x at a distance x from point B is then given by

$$j_x = j_B + \Delta j = j_B + ix \tag{12.4}$$

where j_B is the shear displacement at point B, which can be determined using Eqn (12.2) by treating section BC as part of a circular rim with radius r, as noted previously.

The shear displacement along section AD can again be determined in the same way as that discussed previously for a tyre in the rigid operating mode. The development of the shear displacement beneath a tyre in the elastic operating mode is schematically shown in Figure 12.5.

After the shear displacement along the tyre–terrain interface has been determined, the corresponding shear stress distribution can be defined using the shear stress–shear displacement relationships discussed in Chapter 5. It should be mentioned that the normal pressure usually varies over the contact patch. The approach to the prediction of the development shear stress under varying normal pressure described in Section 8.2 is therefore adopted in the computer-aided method NWVPM.

■ Example 12.1

A tyre with outside diameter of 1 m operates in the rigid mode on a soil, with the following shearing parameters on the tyre-soil interface: adhesion $c_a = 3.45$ kPa, angle of shearing resistance $\phi = 11°$, and shear deformation parameter $K = 0.025$ m. The tyre is being driven at 10% slip, with a contact angle $\theta_1 = 35°$ (see Figure 12.1). The grouser height is relatively small and its effect may be neglected. Estimate the shear stress on the tyre surface at the bottom-dead-centre (i.e. the point on the tyre surface below the tyre centre), where the normal pressure is 350 kPa.

Solution

The shear displacement j along the interface of a tyre operating in the rigid mode may be estimated by Eqn (12.2).

$$j = r[(\theta_1 - \theta) - (1 - i)(\sin\theta_1 - \sin\theta)]$$

With tyre outside diameter of 1 m (radius $r = 0.5$ m), contact angle $\theta_1 = 35°$ ($\theta_1 = 0.611$ radians and $\sin\theta_1 = 0.574$), and slip $i = 10\%$ (or 0.1), the shear displacement j at the bottom-dead-centre ($\theta = 0$) is calculated as follows:

$$j = 0.5[(0.611 - 0) - (1 - 0.1)(0.574 - 0)] = 0.047 \text{ m}$$

At the bottom-dead-centre, the normal pressure is 350 kPa and the shear stress s is given by

$$s = (c_a + p\tan\phi)(1 - e^{-j/k}) = (3.45 + 350 \times 0.1944)(1 - e^{-0.047/0.025})$$
$$= 60.58 \text{ kPa}$$

■

12.1.2 Tyre Performance Prediction

A. For a Tyre in the Rigid Operating Mode

If the average ground pressure p_g of a tyre is greater than the critical ground pressure p_{gcr} defined by Eqn (11.35), the tyre is considered to be in the rigid operating mode, as mentioned previously. The model for rigid wheel–terrain interaction is shown in Figure 12.1. As pointed out earlier, the performance of the wheel is completely defined by the distributions of normal pressure $p(\theta)$ and shear stress $s(\theta)$ on the interface.

The equation for the vertical force equilibrium is expressed by

$$W = \frac{b_{ti}D}{2}\left\{\int_0^{\theta_1}[p(\theta)\cos\theta + s(\theta)\sin\theta]d\theta + \int_0^{\theta_2}[p(\theta)\cos\theta - s(\theta)\sin\theta]d\theta\right\} \quad (12.5)$$

where b_{ti} is the tyre width, D is the undeflected tyre diameter, W is the normal load on the tyre, θ_1 is the front contact angle, and θ_2 is the rear contact angle. As mentioned previously, in the front contact region (section BC in Figure 12.1) the terrain is being gradually loaded, while in the rear contact region (section AB in Figure 12.1) the terrain is in elastic rebound. Appropriate pressure–sinkage relationships should, therefore, be used to determine the normal pressure distributions over different regions of the contact patch.

The sinkage z_r of a given tyre (defined by the front contact angle θ_1) and the depth of the rut resulting from the passage of the tyre (defined by the front and rear contact angles θ_1 and θ_2) can be predicted by solving Eqn (12.5). It should be noted that θ_1 and θ_2 are related and their relationship is governed by the terrain response to repetitive normal loading discussed in Chapter 4. In solving Eqn (12.5), a numerical procedure is used to perform the integration and an iterative scheme is employed to arrive at a solution.

It can be seen from Eqn (12.5) that the role of the vertical components of both the normal pressure and shear stress in supporting the normal load on the tyre is fully taken into account.

The motion resistance due to tyre–terrain interaction R_t is determined by integrating the horizontal component of the normal pressure over the contact patch, after obtaining the values of θ_1 and θ_2 from Eqn (12.5). R_t is expressed by

$$R_t = \frac{b_{ti}D}{2}\left[\int_0^{\theta_1} p(\theta)\sin\theta d\theta - \int_0^{\theta_2} p(\theta)\sin\theta d\theta\right] \quad (12.6)$$

For a tyre in the rigid operating mode, R_t is the only external motion resistance.

The thrust F of the tyre is determined by integrating the horizontal component of the shear stress over the contact area and is given by

$$F = \frac{b_{ti}D}{2}\left[\int_0^{\theta_1} s(\theta)\cos\theta d\theta + \int_0^{\theta_2} s(\theta)\cos\theta d\theta\right] \quad (12.7)$$

The drawbar pull F_d, as determined by tyre-terrain interaction, is the difference between the thrust F and the motion resistance R_t, that is

$$F_d = F - R_t \quad (12.8)$$

The torque M applied to a driven tyre is determined by integrating the shear stress over the contact patch and multiplying the resulting shear force by the moment arm $D/2$. It is expressed by

$$M = \frac{b_{ti}D^2}{4}\left[\int_0^{\theta_1} s(\theta) d\theta + \int_0^{\theta_2} s(\theta) d\theta\right] \quad (12.9)$$

B. For a Tyre in the Elastic Operating Mode

As discussed previously, if the average ground pressure p_g of a tyre is less than the critical ground pressure p_{gcr} defined by Eqn (11.35), the tyre is considered to be in the elastic mode of operation. The model for elastic tyre–terrain interaction is shown in Figure 12.2. Section AB of the tyre circumference is assumed to be flat, while sections BC and AD are assumed to maintain a circular shape.

Along the flat section AB, the normal pressure is assumed to be uniformly distributed and equal to the average ground pressure p_g of the tyre. Along section BC, the terrain is being gradually loaded, while along section AD the terrain is being unloaded (i.e. in elastic rebound). The appropriate pressure–sinkage relationships described in Chapter 4 should be used to determine the normal pressure distributions on different sections of the contact area.

The equation for the vertical force equilibrium is expressed by

$$W = \frac{b_{ti}D}{2} \int_{\theta_c}^{\theta_1} [p(\theta)\cos\theta + s(\theta)\sin\theta]d\theta + b_{ti}Dp_g \sin\theta_c \\ + \frac{b_{ti}D}{2} \int_{\theta_c}^{\theta_2} [p(\theta)\cos\theta - s(\theta)\sin\theta]d\theta \quad (12.10)$$

where θ_c represents the contact angle for half of the flat section AB of the tyre circumference shown in Figure 12.2.

The sinkage of the tyre z_e is determined by the average ground pressure p_g of the tyre and the pressure–sinkage relationship. If the pressure–sinkage relationship is described by the Bekker equation, Eqn (4.1), then z_e is given by Eqn (11.29).

The rear contact angle θ_2 of the tyre is determined by the response of the terrain to repetitive loading and the average ground pressure p_g. From the geometry shown in Figure 12.2, the following relation between θ_1 and θ_C can also be derived:

$$\cos\theta_1 = \cos\theta_c - 2z_e/D \quad (12.11)$$

Making use of these relations, the contact geometry of the tyre (as defined by the contact angles θ_1, θ_2 and θ_c) can be completely defined by solving Eqn (12.10). In the solution process, a numerical procedure is used to perform the integration and an iterative method is employed to obtain a solution.

Similar to a tyre in the rigid operating mode, the motion resistance R_t of a tyre in the elastic operating mode, due to tyre–terrain interaction, is determined by integrating the horizontal component of the normal pressure acting on sections BC and AD. It should be noted that the average ground pressure p_g acting on the flat section AB does not have a horizontal component. R_t is given by

$$R_t = \frac{b_{ti}D}{2}\left[\int_{\theta_c}^{\theta_1} p(\theta)\sin\theta d\theta - \int_{\theta_c}^{\theta_2} p(\theta)\sin\theta d\theta\right] \quad (12.12)$$

When the tyre is in the elastic mode of operation, it deflects. As a result, the total motion resistance will be the sum of the motion resistance due to tyre–terrain interaction R_t and that due to tyre carcass flexing R_f described in Section 11.2.2.

The thrust F of the tyre in the elastic mode of operation is determined by integrating the horizontal component of the shear stress over the contact patch and is expressed by

$$F = \frac{b_{ti}D}{2}\int_{\theta_c}^{\theta_1} s(\theta)\cos\theta d\theta + b_{ti}\int_0^{l_{AB}} s(\theta)dx + \frac{b_{ti}D}{2}\int_{\theta_c}^{\theta_2} s(\theta)\cos\theta d\theta \quad (12.13)$$

where l_{AB} is the contact length of the flat section AB of the tyre and is equal to $D\sin\theta_c$.

The torque M applied to a driven tyre is determined by integrating the shear stress over the contact area and multiplying the resulting shear force by the appropriate moment arm. It is given by

$$M = \frac{b_{ti}D^2}{4}\int_{\theta_c}^{\theta_1} s(\theta)\,d\theta + \frac{b_{ti}D\cos\theta_c}{2}\int_0^{l_{AB}} s(\theta)\,dx \qquad (12.14)$$
$$+ \frac{b_{ti}D^2}{4}\int_{\theta_c}^{\theta_2} s(\theta)\,d\theta$$

12.1.3 Effects of Tyre Lugs

The analysis presented in the preceding section is applicable to a 'smooth' tyre, or to a tyre with lugs operating at sinkages which are large in comparison to the lug height. In this case, the spaces between the lugs fill with terrain material and the tyre behaves like a 'smooth' one. Over soft terrain, tyre sinkage is usually considerable and hence predictions of tyre performance based on the analysis presented previously are in reasonably good agreement with experimental observations (Bekker, 1983). However, over medium to firm terrain, tyre load may be entirely supported with the tips of the lugs, or with a minimum carcass contact with the terrain. In this case, the lug effect should be taken into consideration.

A. For a Tyre in the Rigid Operating Mode

Figure 12.6 shows schematically a tyre with lugs in the rigid operating mode interacting with the terrain. In this case, it is assumed that the tyre remains cylindrical in shape with diameter D across the tips of the lugs and with diameter d across the carcass surface, where d equals $D - 2h_l$ and h_l is the lug height.

The normal pressure acting on the tips of the lugs p_t will be different from that acting on the carcass p_{ca}, because of the lug height h_l. If λ is defined as the ratio of the total lug tip area to the total tyre tread area and b_{ti} is the width of the tyre, then the pressure on the lug tips p_t acts on an effective width of λb_{ti}, whereas the pressure on the carcass p_{ca} acts on an effective width of $(1 - \lambda)b_{ti}$. The general form of the normal pressure distributions on the lug tips as well as on the carcass can be determined following an approach similar to that described in Section 12.1.1(A).

For conventional tyres, the spacing between lugs is such that the terrain between them could not fail in the same way as that in front of a cutting blade or a retaining wall (Wong, 2008). In analysing the shearing action of a tyre with lugs in the rigid mode of operation, it is, therefore, assumed that shearing between the tyre and the terrain takes place along the circular surface across the lug tips, as shown in Figure 12.6. The development of shear displacement along this shearing surface can be determined in the same way as that described in Section 12.1.1(B). However, rubber–terrain shearing characteristics should be taken into account in

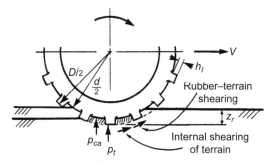

Figure 12.6: A tyre with lugs in the rigid operating mode

determining the shear stress developed on the lug tip area, whereas the internal shear strength parameters should be used in determining the shear stress on the shearing surface between lugs, as shown in Figure 12.6.

After the general forms of the normal pressure and shear stress distributions on the lug–terrain and carcass–terrain interfaces have been determined, the sinkage, motion resistance, thrust and drawbar pull of the tyre can be predicted following an approach similar to that described in Section 12.1.2(A).

B. For a Tyre in the Elastic Operating Mode

As discussed previously, in determining the operating mode of a smooth tyre, the critical ground pressure p_{gcr} defined by Eqn (11.35) is compared with the average ground pressure of the tyre p_g. The critical ground pressure p_{gcr}, in essence, represents the pressure which the terrain exerts on the lowest point of the undeflected tyre circumference, if the tyre behaves like a rigid wheel. For a tyre with lugs, if the sinkage of the lug tip is greater than the lug height, the terrain will exert pressure on both the lug tips and the tyre carcass. In other words, both the pressure on the lug tip p_t and that on the carcass p_{ca} will affect the operating mode of the tyre. Taking into account the ratio λ of the total lug tip area to the total tyre tread area, the equivalent pressure p_e exerted by the terrain that effectively determines the operating mode of a tyre with lugs can be expressed by

$$p_e = \lambda p_t + (1 - \lambda) p_{ca} \tag{12.15}$$

If the tyre with lugs behaves like a rigid wheel, the sinkage of the lug tip z_r for a given normal load on a specific terrain can be determined using the analysis described in the preceding section. When z_r is known, the pressure on the lug tip p_t and that on the carcass p_{ca}, which is related to $z_r - h_l$ (shown in Figure 12.6), can be determined. Substituting these values into Eqn (12.15), the equivalent critical ground pressure p_{ecr} can be obtained. If p_{ecr} is less than the average ground pressure p_g of the tyre derived from the generalized deflection chart, then the tyre with lugs will behave like a rigid wheel. Otherwise, it will deflect and part of its circumference will be flattened as shown in Figure 12.7.

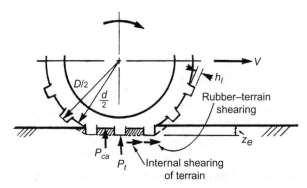

Figure 12.7: A tyre with lugs in the elastic operating mode

When a tyre is in the elastic operating mode, the lug tip sinkage (z_e) and the carcass sinkage ($z_e - h_l$) can be determined from the average ground pressure p_g by solving the following equation:

$$p_g = \lambda p_t + (1 - \lambda) p_{ca} \qquad (12.16)$$

where p_t is a function of z_e and p_{ca} is a function of ($z_e - h_l$) for a given terrain.

It should be noted that if the lug tip sinkage z_e is less than the lug height h_l, the carcass is not in contact with the terrain and p_{ca} is zero. In this case, the tyre load is entirely supported by the lug tip area and the pressure acting on the lug tip p_t is equal to p_g/λ.

With the sinkage of the lug tip z_e known, the general form of the normal pressure distributions on the lug tips and on the carcass can be determined, following an approach similar to that described in 12.1.1(A). The general form of the shear stress distributions along the shearing surfaces can also be determined in a similar way to that described in Section 12.1.1(B).

With the general forms of the normal pressure and shear stress distributions on the interfaces known, the geometry of the contact surfaces, motion resistance, thrust and drawbar pull of a tyre with lugs in the elastic operating mode can be predicted following an approach similar to that described in Section 12.1.2(B).

12.2 Basic Features of NWVPM for Evaluating Wheeled Vehicle Performance

While various methods for analysing tyre–terrain interaction and for predicting individual tyre performance have been in existence for many decades, most of them have not been integrated into an appropriate procedure for predicting the overall performance of multi-axle wheeled vehicles. Integrating the methods for predicting tyre performance outlined in Section 12.1 with vehicle dynamics, the computer-aided method NWVPM is capable of predicting the overall performance of multi-axle wheeled vehicles on unprepared terrain. The basic features

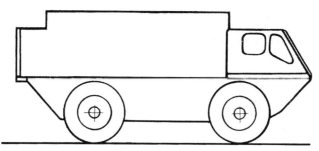

Figure 12.8: A schematic diagram of a two-axle wheeled vehicle used in the parametric analysis

of NWVPM for performance and design evaluation of multi-axle wheeled vehicles, from the traction perspective, are summarized below:

1. It can accommodate wheeled vehicles with axles up to eight, with any combination of driven or non-driven axles.

2. Each axle can have tyre pairs up to four and the parameters of each pair of tyres may be specified individually.

3. The track or tread (i.e. the transverse distance between the centres of a pair of the left and right side tyres on the same axle) of each axle may be specified individually. For instance, if the tracks of the front and rear axles of a two-axle vehicle are the same, then the tyres on the rear axle run in the ruts formed by those on the front axle in straight line motion, NWVPM will take into account the 'multi-pass effects' on the terrain (i.e. the response of the terrain to repetitive normal and shear loadings) in the prediction of the performance of the rear tyres (Wong, 1978b). NWVPM can also accommodate the situation where the tracks of the front and rear axles are different and the rear tyres run partly in the ruts formed by the front tyres and partly on the undisturbed terrain.

4. The tyre–terrain interaction models outlined in Section 12.1 are used in predicting tyre performance.

5. The dynamic load transfer between axles due to drawbar pull or gradient is taken into consideration.

6. The effects of independent suspension on the load distribution among axles are taken into account.

7. On exceeding soft terrain, when the wheel sinkage is greater than the vehicle ground clearance, the vehicle belly may be in contact with the terrain surface. NWVPM takes into account the effects of belly drag on the overall performance of the vehicle.

The terrain parameters required as input for NWVPM are the same as those for NTVPM described in Chapter 8. As an example, the vehicle parameters required as input for a two-axle vehicle shown in Figure 12.8 are given in Table 12.1.

Table 12.1: Vehicle input parameters for NWVPM

Vehicle type (or designation)
Weight, kN
Centre of gravity x-coordinate, cm*
Centre of gravity y-coordinate, cm**
Drawbar hitch, x-coordinate, cm
Drawbar hitch, y-coordinate, cm

Axle No. 1
Axle load, kN
Axle suspension stiffness, kN/m
Axle clearance (above hard ground), cm
Axle x-coordinate, cm
Driven axle (Yes/No)
Tyre type No. 1
Number of tyre pairs
Track (tread), cm

Axle No. 2
Axle load, kN
Axle suspension stiffness. kN/m
Axle clearance (above hard ground), cm
Axle x-coordinate, cm
Driven axle (Yes/No)
Tyre type No. 2
Number of tyre pairs
Track (tread), cm

Tire type No. 1
Diameter, cm
Tread width, cm
Section height, cm
Lug area/carcass area
Lug height, cm
Lug width, cm
Inflation pressure, kPa
Ground pressure, kPa
Tyre construction (radial or bias)
Parameter k_e

Tire type No. 2
Diameter, cm
Tread width, cm
Section height, cm
Lug area/carcass area
Lug height, cm
Lug width, cm
Inflation pressure, kPa
Ground pressure, kPa
Tyre construction (radial or bias)
Parameter k_e

*Origin of the x-coordinate is at the centre of the first axle, positive to the rear.
**Origin of the y-coordinate is on the ground surface, positive up.

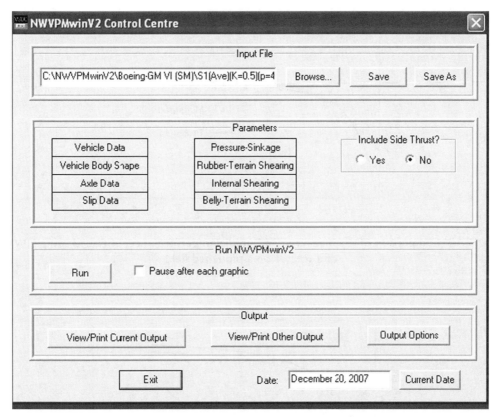

Figure 12.9: The control centre for operating the computer-aided method NWVPM, as displayed on the monitor screen

NWVPM can predict the operating mode, dynamic normal load, sinkage, motion resistance and thrust of individual tyres on the vehicle, as well as the overall tractive performance of the vehicle, including the overall motion resistance, thrust, drawbar pull and tractive efficiency as a function of slip.

NWVPMs have been undergoing continual update and development. In the latest version, the algorithms and the robustness of the numerical solution procedures have been improved. The operation of NWVPM has been made more user-friendly than previous versions. These include using the dialogue box format for inputting all vehicle, tyre and terrain data. Figure 12.9 shows the control centre, as displayed on the monitor screen, for the operation of the latest version of NWVPM.

12.3 Experimental Substantiation

The basic features of NWVPM have been substantiated by experimental data. Figures 12.10 and 12.11 show a comparison between the measured and predicted drawbar performance of

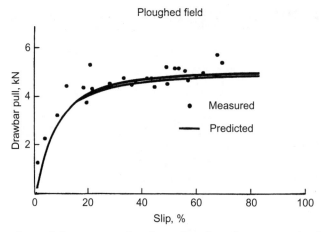

Figure 12.10: Comparison of the measured and predicted performance, obtained using NWVPM, of a tractor on a ploughed field

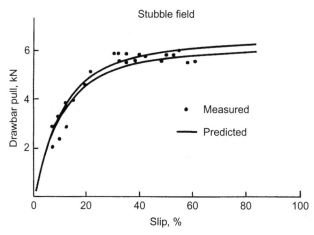

Figure 12.11: Comparison of the measured and predicted performance, obtained using NWVPM, of a tractor on a stubble field

a tractor on a ploughed field and on a stubble field, respectively. The tractor was equipped with 5.50–16 tyres in the front with a load of 2.67 kN on each tyre at an inflation pressure of 147 kPa, and with 10–28 AS tyres at the rear with a load of 6.87 kN on each tyre at an inflation pressure of 78 kPa (Bekker, 1985). The two curves shown in Figures 12.10 and 12.11 represent those predicted from the upper and lower bounds of the measured terrain values (Bekker, 1985). It can be seen from Figures 12.10 and 12.11 that there is a reasonable correlation between the measured drawbar performances and the predicted ones obtained using NWVPM.

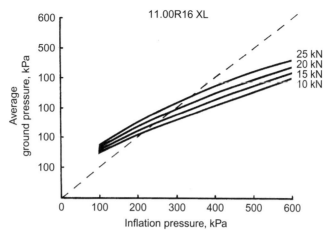

Figure 12.12: Variation of average ground pressure with inflation pressure at various normal loads for an 11.00R16 tyre (Reprinted by permission of the Department of National Defence, Canada, from Preston-Thomas and Wong, 1987)

12.4 Applications of NWVPM to Parametric Analysis of Wheeled Vehicle Performance

NWVPM is suited for the evaluation of competing off-road wheeled vehicle designs and for the examination of the impact on performance of design modifications and terrain conditions. To demonstrate its applications, NWVPM was employed to evaluate the effects of tyre type, inflation pressure and static load distribution between axles on the performance of a reference vehicle over two types of terrain.

The reference vehicle is a two-axle, all-wheel-drive vehicle similar to that shown in Figure 12.8. The static loads on the front and rear axle are 39.23 and 33.33 kN, respectively. The axle suspension stiffnesses on the front and rear axle are 2.42 and 1.57 kN/cm, respectively. The wheelbase of the vehicle is 208 cm and the centre of gravity and the drawbar hitch are 111.2 and 66.2 cm above the ground level, respectively. Two types of tyre, 11.00R16 and 12.5/75R20, with different inflation pressure combinations were examined. The relationships between average ground pressure and inflation pressure for the 11.00R16 and 12.5/75R20 tyres are shown in Figures 12.12 and 11.14, respectively.

The performance of the vehicle was simulated on two types of terrain, one referred to as the medium soil and the other as the clayey soil. For these two types of terrain, the pressure–sinkage relationships are expressed by the Bekker equation (Eqn (4.1)), the responses to repetitive normal load are described by Eqns (4.23) and (4.24), and the pertinent parameters are given in Table 12.2. The shear stress–shear displacement relationships are described by Eqn (5.2) and the shear strength parameters are given in Table 12.3.

Table 12.2: Pressure-sinkage parameters for the medium soil and clayey soil

Terrain type	Terrain parameters				
	n	k_c (kN/m^{n+1})	k_ϕ (kN/m^{n+1})	k_o (kN/m^3)	A_u (kN/m^4)
Medium soil	0.8	29.76	2083	0	192 400
Clayey soil	0.6	30.08	499.7	0	63 106

Table 12.3: Shear strength parameters for the medium soil and clayey soil

Terrain type	Shear strength parameters					
	Internal shearing			Rubber–terrain shearing		
	c (kPa)	ϕ^0	K (cm)	c (kPa)	ϕ^0	K (cm)
Medium soil	8.62	22.5	2.54	4.64	19.9	2.12
Clayey soil	7.58	14	2.54	4.08	12.4	2.12

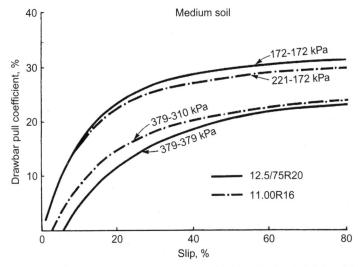

Figure 12.13: Predicted drawbar performance of a two-axle, all-wheel-drive vehicle with different tyres at various inflation pressure combinations on the medium soil

The simulated drawbar performance curves of the reference vehicle with the 11.00R16 and 12.5/75R20 tyres at different inflation pressure combinations over the medium soil and the clayey soil are shown in Figures 12.13 and 12.14, respectively. The first and second numbers in the inflation pressure combinations shown in the figures represent the inflation pressure of the tyres on the front axle and that of the tyres on the rear axle, respectively. The inflation pressure combinations selected for the simulations are consistent with the tyre manufacturer's specifications.

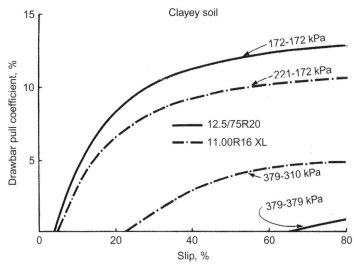

Figure 12.14: Predicted drawbar performance of a two-axle, all-wheel-drive vehicle with different tyres at various inflation pressure combinations on the clayey soil

It can be seen that changing from 11.00R16 to 12.5/75R20 tyres has an effect on the performance of the vehicle, primarily over the clayey soil. Tyre inflation pressure, however, has a more significant effect. For instance, over the medium soil, by reducing the inflation pressures of the 12.5/75R20 tyres on the two axles from 379 to 172 kPa, the drawbar pull coefficient (the ratio of the drawbar pull to vehicle weight) of the vehicle at 20% slip improves from 11.5 to 23.2%. Similarly, if the inflation pressures of the 11.00R16 tyres on the front and rear axles are reduced from 379 and 310 kPa to 221 and 172 kPa, respectively, the drawbar pull coefficient of the vehicle at 20% slip increases from 14.7 to 22.3%. Over the clayey soil, which is significantly weaker than the medium soil, the effect of tyre inflation pressure on performance is even more pronounced. For instance, when the vehicle is equipped with the 12.5/75R20 tyres at an inflation pressure of 379 kPa, it will be unable to propel itself below 65% slip (i.e. the external motion resistance is greater than the thrust that the vehicle can develop). However, if the inflation pressures of the tyres on the two axles are reduced from 379 to 172 kPa, the vehicle is able to develop a drawbar pull equivalent to 8.3% of the vehicle weight at 20% slip. A similar effect is observed for the vehicle with the 11.00R16 tyres. When the inflation pressures of the tyres on the front and rear axles are 379 and 310 kPa, respectively, the vehicle is unable to propel itself at 20% slip. However, lowering the inflation pressures of the tyres on the front and rear axles to 221 and 172 kPa, respectively, the drawbar pull coefficient of the vehicle at 20% slip achieves a value of 6.6%.

The simulated tractive efficiency curves of the reference vehicle with the 11.00R16 and 12.5/75R20 tyres at different inflation pressure combinations over the medium soil and the clayey soil are shown in Figures 12.15 and 12.16, respectively. It can be seen that the

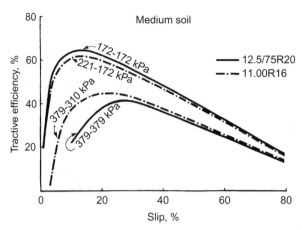

Figure 12.15: Predicted tractive efficiency of a two-axle, all-wheel-drive vehicle with different tyres at various inflation pressure combinations on the medium soil

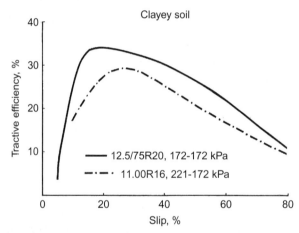

Figure 12.16: Predicted tractive efficiency of a two-axle, all-wheel-drive vehicle with different tyres at various inflation pressure combinations on the clayey soil

inflation pressures of the tyres have a significant effect on the tractive efficiency of the vehicle. The tractive efficiency is defined as the ratio of the power available at the drawbar to that input to the tyres. For instance, over the medium soil, by lowering the inflation pressures of the 12.5/75R20 tyres on the two axles from 379 to 172 kPa, the maximum tractive efficiency increases from 41 to 64%. Similarly, by lowering the inflation pressures of the 11.00R16 tyres on the front and rear axles from 379 and 310 kPa to 221 and 172 kPa, respectively, the maximum tractive efficiency of the vehicle increases from 45 to 62%. Figure 12.16 shows that over the clayey soil, the vehicle equipped with the 12.5/75R20 tyres at an inflation pressure of 172 kPa has somewhat higher tractive efficiency than that with the 11.00R16 tyres on the front and rear axles at inflation pressures of 221 and 172 kPa, respectively.

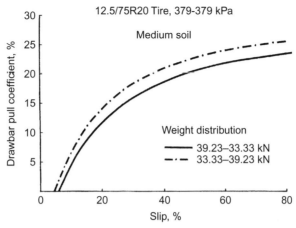

Figure 12.17: Effect of static load distribution on the drawbar performance of a two-axle, all-wheel-drive vehicle with 12.5/75R20 tyres at inflation pressure of 379 kPa on the medium soil

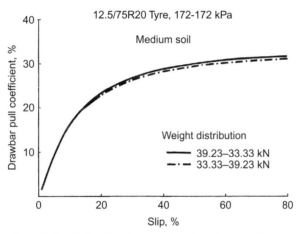

Figure 12.18: Effect of static load distribution on the drawbar performance of a two-axle, all-wheel-drive vehicle with 12.5/75R20 tyres at inflation pressure of 172 kPa on the medium soil

Figures 12.17 and 12.18 show the effect of static load distribution between the two axles on the performance of the vehicle over the medium soil. It can be seen that by changing the static load distribution between the front and rear axle from 39.23 and 33.33 kN to 33.33 and 39.23 kN, respectively, the tractive performance improves somewhat when the tyre inflation pressure is 379 kPa. However, when the tyre inflation pressure is 172 kPa, the effect of static load distribution between the two axles has a negligible effect on performance.

NWVPM has been successfully used in the evaluation of the design and performance of off-road wheeled vehicles for vehicle manufacturers and governmental agencies in North America and elsewhere. For instance, NWVPM has been employed in the assessment of the effects of different types of tyre on the cross-country performance of a family of 6×6 and 8×8 armoured wheeled vehicles for a Canadian governmental agency. It has also been used in the evaluation of the mobility of a variety of logistic military wheeled vehicles for a US governmental agency. In collaboration with the Surface Mobility Technology Team, Glenn Research Center, National Aeronautics and Space Administration (NASA), USA, a study was carried out to evaluate the potential applications of NWVPM to the prediction of the performance of lunar vehicle wheels, which is discussed in the following section (Wong & Asnani, 2008).

12.5 Applications of NWVPM to the Evaluation of Lunar Vehicle Wheels

Since the announcement in 2004 that the USA will undertake extended human missions to the Moon in preparation for future exploration of Mars, and the announcement by the European Space Agency (ESA) that it will deploy an autonomous robotic rover, ExoMars, to the Martian surface in the coming decade, efforts in the development of a new generation of extraterrestrial vehicles have been intensified. Other countries, such as Russia, China, Japan and India, have also announced their plans for exploration of the Moon, Mars and beyond.

To develop suitable wheels (or other forms of running gear) for a new generation of extraterrestrial vehicles for future lunar or planetary exploration in an expeditious and cost-effective manner, a realistic analytical tool that can be used to quantitatively evaluate their performances and designs is highly desirable. This will shorten the process of evaluating design concepts prior to prototyping and testing; hence, it will considerably reduce development costs of extraterrestrial rovers.

To evaluate the potential applications of NWVPM to the prediction of the performance of the wheels for lunar vehicles, a study was conducted in collaboration with the Surface Mobility Technology Team, Glenn Research Center, National Aeronautics and Space Administration (NASA), USA (Wong and Asnani, 2008). It is to examine the correlation between the predicted performances of wheel candidates for the lunar roving vehicle for the NASA Apollo programme in the 1970s using NWVPM and the corresponding test data obtained under earth gravity in the soil bin and documented in the US Army Engineer Waterways Experiment Station (WES) Technical Report M-70-2 (Freitag et al., 1970). While the results of tests of lunar vehicle wheels conducted in earth environment may not necessarily be representative of those on the lunar surface, because of the differences in gravity and in environmental conditions – for instance, the lunar soil is subject to gravity equal to approximately 1/6 of that on earth and the atmospheric pressure on the lunar surface is much lower than that on earth – it is still a valid approach to use test data obtained in earth environment to evaluate the predictive capability of NWVPM, and its potential applications to predicting the performances of wheels for future generation of wheeled rovers on extraterrestrial bodies.

12.5.1 Lunar Vehicle Wheels

Two-wheel candidates for the lunar roving vehicle for the NASA Apollo programme, namely the wire-mesh wheel developed by the Boeing-General Motors team and the hoop-spring wheel by Bendix, were used in this correlation study. The wire-mesh wheel was eventually chosen for the personnel-carrying lunar roving vehicle that was successfully operated by astronauts on the lunar surface during the Apollo 15, 16 and 17 missions.

The baseline version of the wire-mesh wheel tested at WES is shown in Figure 12.19(a). This wire-mesh wheel was similar in design, but slightly larger in diameter, than the wheel installed on the lunar roving vehicle used in the Apollo 15, 16 and 17 missions. It consists of a hub formed by a spun aluminium disc attached to the electric traction drive, an aluminium rim to which a woven wire wheel is attached, and an inner titanium bump-stop ring (Cowart, 1971). The hub is riveted to the rim with the wheel constructed of woven zinc-coated steel wire. For flotation, the external surface of the wheel is covered by a herringbone pattern of titanium tread strips. Since some of the data on the baseline version of the wire-mesh wheel required as input to NWVPM, such as the shearing characteristics between the zinc-coated steel wire and terrain and that between the titanium tread strips and terrain, are not available in the US Army Engineer Waterways Experiment Station Technical Report M-70-2, two modified versions of this wire-mesh wheel, designated as the Boeing-GM IV and VI wheels, are used in the correlation study. These two modified versions have the same basic structure as the baseline version described above. However, their surfaces are covered with fabric (Gray tape) coated with sand. As a result, the traction of these two wheels is derived from the internal shearing of the sand, the characteristics of which are available in the WES report. This makes it possible to use NWVPM to predict their tractive performances.

The major difference between the Boeing-GM IV and VI wheels is that in the latter 50% of the wire structure is removed; hence it is more flexible and has greater contact area and lower contact (ground) pressure than the former under the same normal load. As shown in Table 12.4, the

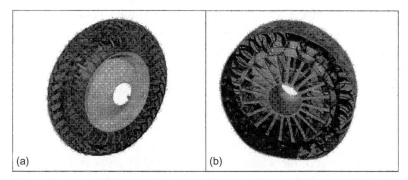

Figure 12.19: Two wheel candidates, wire-mesh wheel and loop-spring wheel, for the lunar roving vehicle for the NASA Apollo programme

Table 12.4: Basic parameters of the lunar vehicle wheels used in the study

Wheel type	Load (N)	Carcass diameter (cm)	Loaded section height (cm)	Loaded section width (cm)	Contact area, (cm^2)	Contact (ground) pressure (kPa)
Boeing-GM IV	311	102.29	18.80	28.36	234.10	13.31
Boeing-GM VI	311	99.83	14.48	32.41	710.19	4.42
Boeing-GM VI	311	100.28	14.00	32.45	778.77	4.04
Bendix I	67	101.60	15.88	25.40	243.54	2.76
Bendix I	133	101.60	14.86	25.40	519.35	2.58
Bendix I	311	101.60	12.32	25.40	722.58	3.93

Source: Freitag et al. (1970).

Boeing-GM IV wheel has a contact (ground) pressure on a hard surface of 13.31 kPa at a normal load of 311 N, while the Boeing-GM VI wheel has a contact (ground) pressure of 4.42 or 4.04 kPa under the same normal load, dependent upon the set of wheel data selected. The difference in contact (ground) pressure between these two wheels leads to a significant difference in tractive performance which is discussed later. It should be noted that the normal loads and the contact (ground) pressures of the wheels during tests were corresponding to those that are expected on the lunar surface with gravity equal to approximately 1/6 of that on earth.

The hoop-spring wheel, designated as the Bendix I wheel, consists of a titanium outer band, hoop-spring elements, and a rigid aluminium hub with spokes, as shown in Figure 12.19(b). The surface of this wheel is coated with sand and no treads are installed. The traction is derived from the internal shearing of the sand, similar to that of the Boeing-GM IV and VI wheels described above. The relevant parameters of the Bendix I wheel are given in Table 12.4. As can be seen from the table, under a normal load of 311 N, it has a contact (ground) pressure of 3.93 kPa.

12.5.2 Procedures and Soil Conditions for Testing Lunar Vehicle Wheels

Wheel performance tests were performed under earth gravity in soil bins in the laboratories of the Mobility Research Branch of the US Army Engineer Waterways Experiment Station. The soil used in the tests was air-dry sand from the desert near Yuma, Arizona, USA. The upper bound (maximum), the lower bound (minimum) and the average Bekker soil values of the sand, designated as sand S_1, are given in Table 12.5. It is noted that there are significant differences in the maximum and minimum soil values, which indicate the variability (or inconsistency) of soil conditions in the tests. For instance, the maximum value of the angle of internal shearing resistance ϕ_{max} is 30° (tan 30° = 0.5774) and the minimum value ϕ_{min} is 20.5° (tan 20.5° = 0.3739). The difference between tan ϕ_{max} and tan ϕ_{min} normalized with respect to tan ϕ_{min} is 54.4%. As the thrust of the wheel is closely related to tan ϕ, this indicates that such a wide variability in soil conditions in the tests would have significant effects on the variation of the measured data, as well as the predictions, hence the correlation between them.

Table 12.5: Bekker soil values for loose, air-dry sand S_1 used in the study

Category	Moisture content %	Dry density (g/cm³)	Pressure–sinkage parameters			Shear strength parameters	
			n	k_c (kN/m^{n+1})	k_ϕ (kN/m^{n+2})	$\phi°$	c (kPa)
Maximum	0.6	1.527	0.96	6.65	1935.52	30	0
Minimum	0.4	1.446	0.84	−3.83	212.51	20.5	0
Average	0.5	1.484	0.91	−0.66	754.13	27.4	0

Source: Freitag et al. (1970).

The Bekker soil values, including the pressure–sinkage parameters, k_c, k_ϕ and n, and shear strength parameters, cohesion c and angle of internal shearing resistance ϕ, are required as terrain inputs by NWVPM. To predict the thrust–slip (or drawbar pull–slip) relationship of the wheel, the shear deformation parameter K of the soil is also required. This parameter for sand S_1 is, however, not given in the WES report. To obtain a reasonable value of K as input to NWVPM, an attempt was made at NASA Glenn Research Center to replicate the conditions under which the original shear tests were performed at the US Army Engineer Waterways Experiment Station. Using a sand with similar particle size distribution, density and moisture content as those of sand S_1, and employing shearing rings with the same dimensions as the original ones, a series of shear tests was conducted at normal pressures corresponding to those under the wheels examined in this study. Based on the test results, it is estimated that the value of K for sand S_1 is approximately 0.5 cm. Accordingly, this estimated value of K was used as input to NWVPM in the simulations, the results of which are used in this correlation study. The value of K generally affects the shape of the initial part of the shear stress and shear displacement curve of the exponential form, hence the predicted thrust–slip (or drawbar pull–slip) relationship in the low slip range. The value of K has little or no effect on the prediction of the thrust (or drawbar pull) in the high slip range or of its maximum thrust (or maximum drawbar pull) (Wong, 2008).

The wheel tests were conducted on a dynamometer, with continuous recording of wheel load, drawbar pull, torque, sinkage, slip and speed. The accuracy of drawbar pull and torque measurements was estimated to be ±3%. The wheel speed was no higher than 0.5 m/s (Freitag et al., 1970).

As noted previously, normal loads applied to the wheels during tests were based on what would be expected on the lunar surface, which has acceleration due to gravity equal to approximately 1/6 of that on the earth surface, while the soil used in the tests was subject to earth gravity. This would indicate that the tractive performances of the wheels obtained from the tests are not necessarily representative of those on the lunar surface, because of the differences in the acceleration due to gravity and in environmental conditions. Predicting the performances of wheels on the lunar surface based on test results obtained on earth with soil simulants subject to earth gravity is an issue that requires further investigation.

As mentioned previously, a comparison of the test data obtained under earth gravity and predictions made by NWVPM using appropriate input is still a valid procedure for evaluating the predictive capability of NWVPM and its potential applications to predicting the performances of wheels for future generation of wheeled rovers on extraterrestrial bodies.

12.5.3 Correlations between the Measured Performance of Lunar Vehicle Wheels and Predicted Performance by NWVPM

The ratio of the drawbar pull to normal load (usually referred to as the drawbar pull coefficient) obtained on level ground is an indication of the ability of a traction device or a vehicle to pull (or push) external load, to overcome grade resistance, or to accelerate. At 20% slip, the traction device usually is in the desirable range of operating efficiency. Consequently, the ratio of the drawbar pull to normal load at 20% on level ground is a widely used performance indicator of a traction device or of a vehicle, as noted previously. In the WES report, the ratio of the drawbar pull to normal load at 20% slip (designated as P_{20}/W) was used as a basic parameter to represent the measured performances of wheels for lunar vehicles. Accordingly, in this study, P_{20}/W is also chosen as the basic parameter for evaluating the correlation between the measured performances and the predicted ones obtained using NWVPM. The measured data on the 'first pass' performance of the wheels (i.e. the wheel performance on undisturbed soil) given in the WES report are used in the following analysis of the correlation between the predicted and measured wheel performances.

A. Boeing-GM IV and VI Wheels

The values of the measured and predicted P_{20}/W for the Boeing-GM IV wheel and those for the Boeing-GM VI wheel, with two different sets of wheel input data given in Table 12.4, are presented in Table 12.6 (Wong and Asnani, 2008). The predicted values are based on the average Bekker soil values given in Table 12.5 and on the estimated value of $K = 0.5$ cm, as noted previously.

Table 12.6: Comparison of the predicted drawbar pull to normal load ratio at 20% slip (P_{20}/W) with the measured for the Boeing-GM IV and VI wheels on dry sand S_1

Wheel type	Normal load (N)	Contact (ground) pressure (kPa)	Measured* P_{20}/W (%)	Predicted** P_{20}/W (%)	Predicted/ measured (%)
Boeing-GM IV	311	13.31	28	29.8	106.4
Boeing-GM VI	311	4.04	38.4	41.5†	108.1
Boeing-GM VI	311	4.42	38.4	40.9†	106.5

*Source: Freitag et al. (1970).
**Based on average Bekker soil values and estimated $K = 0.5$ cm.
†Taking into account the estimated coefficient of motion resistance due to wheel flexing of 4.3%.

As described in Section 11.2.2, in the prediction of the drawbar pull, both the motion resistance R_c due to soil compaction and the motion resistance R_f due to wheel flexing should be taken into account, if the wheel is in the elastic operating mode. The values of the parameter k_e in Eqn (11.37) characterizing the internal losses caused by flexing of the lunar vehicle wheels examined in this study are, however, not known.

For the Boeing-GM IV wheel, an examination of the simulation results reveals that it has relatively little deflection (less than 1 mm) on sand S_1. Consequently, it is not unreasonable to neglect the motion resistance R_f due to wheel flexing in determining the predicted value of P_{20}/W for the Boeing-GM IV wheel.

The Boeing-GM VI wheel is more flexible than the Boeing-GM IV wheel, as mentioned previously. Simulation results indicate that its deflection is an order of magnitude higher than that of the Boeing-GM IV wheel (approximately 1.2 cm). Consequently, its motion resistance R_f due to wheel flexing should be taken into account in determining the predicted value of P_{20}/W. It was reported by Bekker and Semonin (1975) that the measured coefficient of motion resistance due to wheel flexing (i.e. the ratio of the motion resistance R_f to normal load) for the Boeing-GM wheel used in the lunar roving vehicle for the Apollo 15, 16 and 17 missions is between 4.17 and 4.49%, for an average of 4.3%, approximately. While it is recognized that the Boeing-GM VI wheel has a slightly different geometric configuration than the wire-mesh wheel used in the lunar roving vehicle and that it is covered with fabric, the measured average value of 4.3% cited above is taken as the motion resistance coefficient due to flexing for the Boeing-GM VI wheel, for lack of a better estimate. This value is used in determining the predicted values of P_{20}/W for the Boeing-GM VI wheel presented in Table 12.6.

As can be seen from Table 12.6, the measured value of P_{20}/W for the Boeing-GM IV wheel at a contact (ground) pressure of 13.31 kPa is 28%, whereas the predicted one, with the coefficient of motion resistance due to wheel flexing being neglected, is 29.8%. This represents a difference of +6.4%, as indicated by the ratio of the predicted value to the measured one shown in the table.

With the coefficient of motion resistance due to flexing of 4.3% being taken into account, the predicted values of P_{20}/W for the Boeing-GM VI wheel with contact (ground) pressures of 4.04 and 4.42 kPa are 41.5 and 40.9%, respectively. The average of the two predicted values of P_{20}/W is 41.2%, whereas the measured value is 38.4%, which represents a difference of +7.3%.

As mentioned previously, the major difference between the Boeing-GM IV and VI wheels is that in the latter 50% of the wire structure is removed; hence, the Boeing-GM VI wheel is more flexible and has greater contact area and lower contact (ground) pressure than the Boeing-GM IV wheel under the same load. As a result, the predicted motion resistance due to soil compaction of the Boeing-GM VI wheel is much lower than that of the Boeing-GM IV wheel, whereas its thrust is higher. For instance, the average predicted coefficient of motion resistance due to soil compaction (i.e. the ratio of the motion resistance due to soil compaction to normal load) for the Boeing-GM VI wheel is 0.8% at a load of 311 N and at 20% slip, whereas that for the Boeing-GM IV wheel is 7.6%. The average predicted thrust coefficient (i.e. the ratio of the thrust to normal load) for the Boeing-GM VI wheel is 46.3% at 20% slip, whereas that for the Boeing-GM IV wheel is 37.4%. Consequently, the value P_{20}/W of the Boeing-GM VI wheel is much higher than that of the Boeing-GM IV wheel, as shown in Table 12.6, even though in calculating the value of P_{20}/W of the Boeing-GM VI wheel the estimated motion resistance coefficient due to wheel flexing of 4.3% is included.

In summary, it appears that the correlation between the measured and predicted tractive performances of the Boeing-GM IV and VI wheels may be considered reasonable, in view of the variability of soil conditions and of the accuracy of measurements in the tests. It is also shown that NWVPM can effectively differentiate the tractive performances of the Boeing-GM IV and VI wheels, as indicated by the data shown in Table 12.6.

B. Bendix 1 Wheel

The tractive performances on sand S_1 of the Bendix 1 wheel under normal loads of 67, 133 and 311 N and with basic parameters given in Table 12.4 were predicted using NWVPM. The average Bekker soil values given in Table 12.5 with the estimated value of $K = 0.5$ cm were used in the predictions. As noted previously, because of the value of parameter k_e for this wheel is not known, its motion resistance R_f due to wheel flexing cannot be predicted by NWVPM. However, considering its design features described earlier, it is estimated that the coefficient of motion resistance due to wheel flexing of the Bendix 1 wheel is much less than 1% and is therefore neglected in computing the predicted values of P_{20}/W.

The predicted values of P_{20}/W and the corresponding measured ones under various normal loads are presented in Table 12.7. It can be seen that

- under a normal load of 67 N, the measured value of P_{20}/W is 42.4%, whereas the predicted one is 37.2%. This represents a difference of -12.3%, as indicated in Table 12.7;
- under a normal load of 133 N, the measured value of P_{20}/W is 45.8%, whereas the predicted one is 45.2%, which represents a difference of merely -1.3%;
- under a normal load of 311 N, the measured value of P_{20}/W is 46.3%, whereas the predicted one is 47.3%, which represents a difference of only $+2.2\%$.

The average measured value of P_{20}/W for all three normal loads under consideration is 44.8%, whereas the average predicted value of P_{20}/W for all three normal loads is 43.2%, which represents a difference of only -3.6%.

Table 12.7: Comparison of the predicted drawbar pull to normal load ratio at 20% slip (P_{20}/W) with the measured for the Bendix I wheel on dry sand S_1

Normal load (N)	Contact (ground) pressure (kPa)	Measured* P_{20}/W (%)	Predicted** P_{20}/W (%)	Predicted/ measured (%)
67	2.76	42.4	37.2	87.7
133	2.58	45.8	45.2	98.7
311	3.93	46.3	47.3	102.2
		Average 44.8	Average 43.2	Average 96.4

*Source: Freitag et al. (1970).
**Based on average Bekker soil values and estimated $K = 0.5$ cm.

In summary, it appears that the correlation between the measured and predicted tractive performances of the Bendix 1 wheel under three normal loads may be considered reasonable, in view of the variability of soil conditions and of the accuracy of measurements in the tests.

C. Concluding Remarks

1. The tractive performances on air-dry sand of two wheel candidates for the lunar roving vehicle for the Apollo programme predicted by the computer-aided method NWVPM correlate reasonably well with the measured data presented in the WES report (Freitag et al., 1970), considering the variability of soil conditions and of the accuracy of measurements in the tests.

2. Results of this correlation study indicate that NWVPM has the potential as an engineering tool for the evaluation of tractive performances of wheel candidates for a future generation of extraterrestrial vehicles, provided that appropriate input data are available.

3. While it is shown in this study that the performances of the lunar vehicle wheels predicted by NWVPM correlate reasonably well with available test data, further evaluation of NWVPM under a wider range of operating conditions is recommended. Furthermore, obtaining a complete and reliable set of relevant terrain data, including the characteristics of shearing between wheel surface material and terrain, shear deformation parameter K, and responses of the terrain to repetitive normal and shear loadings, is of importance to predictions made by NWVPM.

4. There has been a hypothesis that methods developed for predicting the performance of heavily loaded off-road vehicles of large size may not be applicable to lightly loaded off-road vehicles of small size. This may be so for prediction methods based on empirical relationships. As is well known, empirical relations, in general, should not be extrapolated beyond the conditions upon which they were derived. Consequently, it is indeed uncertain that empirical methods based on test data obtained with heavily loaded off-road vehicles with large size could be applied to predicting the performances of lightly loaded off-road vehicles of small size.

For the computer-aided method NWVPM, it is based on the understanding of the physical nature of vehicle–terrain interaction and on the principles of terramechanics. Consequently, it should be applicable to both heavily loaded off-road vehicles, as well as lightly loaded ones. The results of this study provide evidence for supporting this conclusion, as there is a reasonable correlation between the performances of lightly loaded wheels predicted by NWVPM and test data.

5. Test results obtained on earth, with loads applied to the wheels corresponding to those on extraterrestrial bodies while soil simulants are subject to earth gravity, are not necessarily representative of the performances of the wheels on extraterrestrial bodies. The development of an appropriate methodology for predicting the performances of wheels on extraterrestrial bodies, based on test results obtained on earth with soil simulants subject to earth gravity, is required.

It should be noted, however, that it is still a valid approach to use test data obtained under earth gravity to evaluate the predictive capability of NWVPM and its potential applications to predicting the performances of wheels for a future generation of wheeled rovers on extraterrestrial bodies.

6. Because of the variability of soil conditions, even in laboratory soil bins, and of the complexity of soil responses to vehicular loadings, one should have a proper perspective for the role of simulation models in the prediction of performances of terrestrial or extraterrestrial vehicles. It is too ambitious to expect at the present time that simulation models can replicate, in every respect, the actual vehicle–terrain interaction or can precisely predict vehicle performances in the field (or even under laboratory conditions). It is suggested that the primary role of simulation models is to provide engineers with reliable tools (with their basic features experimentally validated) for comparing the performances or designs of terrestrial or extraterrestrial vehicles on a relative basis and in a consistent manner.

Because of the capabilities of NWVPM demonstrated above, the Surface Mobility Technology team of NASA Glenn Research Center has adopted NWVPM in evaluating various design configurations for the next generation of lunar roving vehicles, which will be used in future NASA extended human missions to the Moon.

12.6 Applications of the Finite Element Technique to Tyre Modelling in the Analysis of Tyre–Terrain Interaction

In the tyre–terrain interaction model proposed by Bekker, as well as in the computer simulation model described in Section 12.2, it is assumed that when the tyre is in the elastic operating mode, the lower portion of the tyre circumference will be flattened, whereas the remainder maintains a circular shape. It has been shown that this assumption can lead to reasonable predictions of tyre performance in many cases (Bekker, 1983). However, to further improve the accuracy of predictions, it is necessary to more accurately define the geometry of the contact patch.

A pneumatic tyre is a complex structure similar in shape to a toroid and filled with pressurized air. Its carcass is made up of a number of layers of cords and belts of high modulus of elasticity, encased in a matrix of low modulus rubber. In the tyre industry, the finite element method has been widely used in the design of tyres and sophisticated computer models have been developed. However, these models generally require detailed information on the mechanical properties of the rubber compounds, cords, belts, etc. This kind of information is usually proprietary and is not accessible to the tyre user.

For the purpose of improving the prediction of the contact geometry in the study of tyre–terrain interaction, a new approach to tyre modelling using the finite element technique has been explored (Chen and Wong, 1988; Nakashima and Wong, 1993). The basic objective is: on the one hand, the tyre model should provide a realistic representation of the geometry of the tyre contact patch on deformable terrain, and on the other hand, it should be convenient for the engineering practitioner to use.

A method based on the finite element technique for two-dimensional modelling of the pneumatic tyre for tyre–terrain interaction study was developed (Chen and Wong, 1988). In this method, a pneumatic tyre is represented by a system of finite elements, as shown in Figure 12.20. The elastic property of the tyre model, such as its equivalent Young's modulus, is derived from the load–deflection relationship given in the generalized deflection chart of the pneumatic tyre to be modelled. As noted previously, this chart is available from the tyre manufacturer, as shown in Figure 11.13. This ensures that the tyre model exhibits a similar load–deflection relationship to that of the actual pneumatic tyre to be studied.

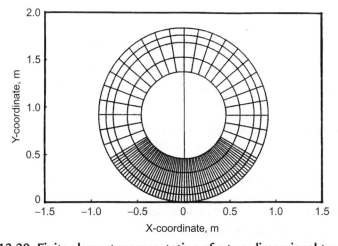

Figure 12.20: Finite element representation of a two-dimensional tyre model

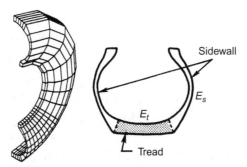

Figure 12.21: A three-dimensional finite element tyre model with different elastic properties for the tyre tread and sidewall (Reprinted by permission of ISTVS from Nakashima and Wong, 1993)

An improved three-dimensional tyre model was later developed (Nakashima and Wong, 1993). The tyre is modelled as a toroid inflated by pressurized air and with the same dimensions and cross-section as the pneumatic tyre to be studied. In this model, the tyre tread and the sidewall are assumed to be made of two different materials with different Young's moduli, as shown in Figure 12.21. The values of these two moduli, one for the tyre tread and the other for the tyre sidewall, are derived from the load–deflection and deflection–contact area relationships given in the generalized deflection chart for the pneumatic tyre to be modelled. This ensures that the tyre model exhibits similar load–deflection and deflection–contact area relationships to that of the actual pneumatic tyre.

In the analysis of tyre–terrain interaction, the pressure–sinkage and shear stress–shear displacement relationships of the terrain obtained by the bevameter are used as input. In addition, the responses of the terrain to repetitive normal and shear loadings are taken into account. Using an iterative, numerical technique and considering the equilibrium conditions of a moving tyre under steady-state conditions, the deflected shape of the tyre, sinkage and normal and shear stress distributions on the tyre–terrain interface can be determined. From these, the motion resistance, thrust, drawbar pull and tractive efficiency as functions of slip for a given pneumatic tyre can be more realistically predicted.

To demonstrate the application of this methodology to the study of tyre–terrain interaction, the two-dimensional tyre model described above is used as an example.

12.6.1 Modelling Tyre Behavior using the Finite Element Technique

The procedures for developing an improved two-dimensional model for the pneumatic tyre, based on a finite element representation, are summarized as follows:

(a) The tyre model is assumed to be an homogeneous, isotropic, elastic, hollow cylinder, with inside and outside diameters and width equal to those of the pneumatic tyre to be analysed. The model is discretized with a suitable finite element mesh as shown in Figure 12.20.

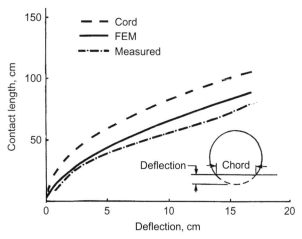

Figure 12.22: Comparison of the measured and predicted contact length–deflection relationships of a 29.5-25 tyre

(b) The tyre model is assumed to have a Poisson's ratio of 0.5, as the volume change of the material of an actual tyre under load is usually not significant.

(c) A finite element code is used and an incremental approach is followed to predict the relationship between deflection and contact length of the tyre model on a flat rigid surface. Figure 12.22 shows a comparison between the predicted deflection–contact length relationship using the finite element technique and the measured one for a 29.5-25 tyre on hard ground. The measured data are taken from the generalized deflection chart published by the tyre manufacturer.

It can be seen from the figure that the deflection–contact length relationship predicted using the finite element technique is much closer to the measured one than that using the chord to represent the contact length.

(d) Since the tyre is modelled as an elastic, hollow cylinder, its contact area with a flat, rigid surface will be rectangular. In reality, the contact area of a pneumatic tyre will usually have an elliptical shape. To model the characteristics of the tyre in a realistic manner, the rectangular contact area of the tyre model is set equal to the measured elliptical area of the pneumatic tyre given in the generalized deflection chart published by the tyre manufacturer. Based on the measured contact area and the contact length predicted by the finite element technique, an equivalent contact width of the tyre model for a given deflection can be derived. The equivalent contact width–deflection relationship is subsequently used in the analysis of tyre–terrain interaction.

(e) With the relationship between contact geometry and deflection of the tyre model known, the finite element technique and an iterative procedure are used to derive the value of

Young's modulus for the tyre model that yields a load–deflection relationship similar to that for the pneumatic tyre at a given inflation pressure. The load–deflection curves for a given pneumatic tyre at various inflation pressures can be obtained from the generalized deflection chart.

With Young's modulus, Poisson's ratio and the geometry known, the elastic behaviour of the tyre model is completely defined. This model is then used in the analysis of tyre–terrain interaction.

It should be pointed out that although the physical structure of the tyre model is quite different from that of an actual pneumatic tyre, its basic elastic behaviour and contact geometry are similar to those of the given pneumatic tyre under the same operating conditions. Consequently, using this tyre model in the analysis of tyre–terrain interaction should yield an improved prediction of its performance.

12.6.2 Analysis of Tyre–Terrain Interaction using the Tyre Model

For a given pneumatic tyre, an appropriate tyre model can be formulated following the procedures outlined in the preceding section. The tyre model can then be used in the analysis of tyre–terrain interaction. The pressure–sinkage and shear stress–shear displacement relationships, and the responses to repetitive normal and shear loadings of the terrain described in Chapters 4 and 5 are used to characterize the behaviour of the terrain. A finite element code and an iterative procedure have been developed for predicting the deformed shape and sinkage of the tyre, and the normal pressure and shear stress distributions on the tyre–terrain interface. Figure 12.23 shows the deformed shape and the sinkage of the front tyre (29.5-25) of a two-axle vehicle operating over a soft soil, with a tyre load of 119.5 kN and an inflation pressure of 206.8 kPa. The pressure–sinkage and shear strength parameters of the soft soil are given in Table 12.8.

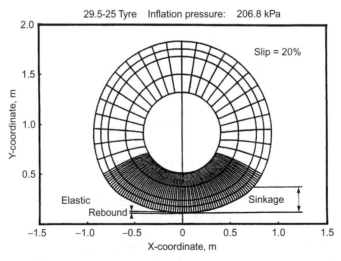

Figure 12.23: Deformed shape and sinkage of the front tyre of a two-axle vehicle on soft soil

Figures 12.24 and 12.25 show the normal pressure and shear stress distributions on the front tyre–terrain interface at 1 and 20% slip, respectively. It is interesting to note that at 1% slip, the shear stress changes sign in the rear contact region, whereas at 20% slip the shear stress acts in one direction throughout the contact patch.

Figure 12.26 shows the deflected shape and sinkage of the rear tyres (29.5-25) of the two-axle vehicle, running in the ruts formed by the front tyres. Each rear tyre carries a load of 118.03 kN at an inflation pressure of 206.8 kPa. It can be seen that in this case the sinkage of the rear tyres is significantly less than that of the front tyres shown in Figure 12.23. Figure 12.27 shows the normal pressure and shear stress distributions on the rear tyre–terrain interface at 20% slip. Because of the smaller sinkage, the contact length of the rear tyres is shorter than that of the front tyres.

Figure 12.28 shows the predicted tractive performance of the two-axle vehicle over the soft soil.

It can be seen that the application of the finite element technique for modelling pneumatic tyres to the prediction of tyre performance described above has the following features:

Table 12.8: Pressure–sinkage and shear strength parameters for soft soil

Terrain type	Pressure–sinkage parameters					Shear strength parameters		
	n	k_c (kN/m^{n+1})	k_ϕ (kN/m^{n+2})	k_o (kN/m^3)	A_u (kN/m^4)	c (kPa)	ϕ (degrees)	K (cm)
Soft soil	0.8	16.5	911	0	86,000	3.71	25.6	2.1

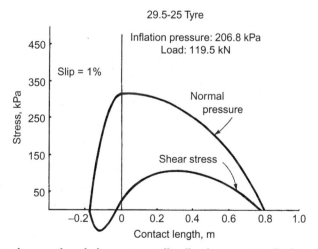

Figure 12.24: Predicted normal and shear stress distributions at 1% slip beneath the front tyre of a two-axle vehicle on soft soil

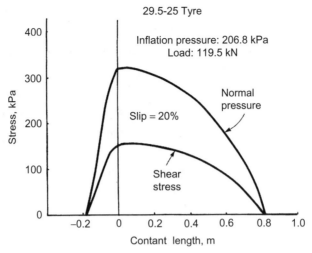

Figure 12.25: Predicted normal and shear stress distributions at 20% slip beneath the front tyre of a two-axle vehicle on soft soil

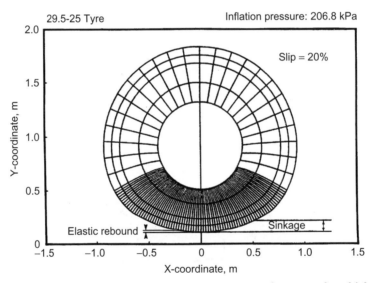

Figure 12.26: Deformed shape and sinkage of the rear tyre of a two-axle vehicle on soft soil

(a) The elastic property of the finite element tyre model is derived from the measured data provided by the tyre manufacturer. Thus, this method should yield a more realistic prediction of the geometry of the tyre contact patch.

(b) The application of the finite element technique for modelling tyres described above to the analysis of tyre–terrain interaction would lead to a more realistic prediction of tyre performance, because of the improvements in predicting the geometry of the tyre contact patch.

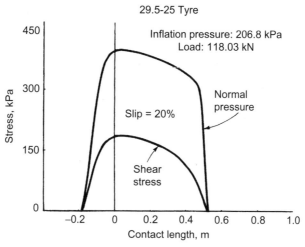

Figure 12.27: Predicted normal and shear stress distributions at 20% slip beneath the rear tyre of a two-axle vehicle on soft soil

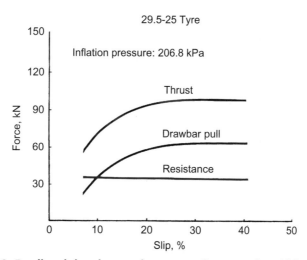

Figure 12.28: Predicted drawbar performance of a two-axle vehicle on soft soil

12.7 Wheeled Vehicles vs Tracked Vehicles from the Traction Perspective

The issue of wheeled vehicles vs tracked vehicles for off-road operations has been a subject of debate for a long period of time. While a number of experimental studies comparing the performances of specific wheeled vehicles with those of comparable tracked vehicles under selected operating environments have been performed, it appears that relatively little fundamental analysis on this subject has been published in the open literature. As adequate mobility over unprepared terrain is a fundamental requirement of off-road vehicles, it is of importance to analyse the basic factors that determine the tractive capability of the wheeled and tracked vehicles and the differences in the mobility between them (Wong & Hung, 2006c).

12.7.1 General Analysis of Wheeled Vehicles vs Tracked Vehicles

The thrust that propels an off-road vehicle is an indication of its tractive capability on a given terrain. To evaluate wheeled vehicles vs tracked vehicles from the traction perspective, one may begin with the examination of the thrusts that the wheeled and tracked vehicles develop on a range of terrains of interest. The approach to comparing the tractive capability of a wheeled vehicle with that of a tracked vehicle is illustrated through an example given below. It examines the thrusts developed by an 8×8 off-road wheeled vehicle and a comparable tracked vehicle.

Figure 12.29 shows a schematic of the contact patches of the four tyres on one side of an 8×8 wheeled vehicle and the contact patch of a track on one side of a comparable tracked vehicle. The wheelbase B of the wheeled vehicle is assumed to be the same as the contact length L_{tr} of the tracked vehicle. To highlight the difference in the development of thrust between a wheeled and a tracked vehicle, it is further assumed that shear stress–shear displacement relationship of the terrain is described by Eqn (5.2), and that the contact patches of the tyres and the track are of rectangular shape with uniform contact pressure.

Based on these assumptions, the thrust developed by a tyre and that by a track may be expressed in the form of Eqn (7.12), and the ratio of the thrust of a wheeled vehicle to that of a tracked vehicle, hereinafter called the thrust ratio, is given by:

$$\frac{F_{ti}}{F_{tr}} = \frac{n_{ti}\left[cb_{ti}L_{ti} + (W/n_{ti})\tan\phi\right]\left[1 - \frac{k}{iL_{ti}}(1 - \exp(-iL_{ti}/K))\right]}{n_{tr}\left[cb_{tr}L_{tr} + (W/n_{tr})\tan\phi\right]\left[1 - \frac{k}{iL_{tr}}(1 - \exp(-iL_{tr}/K))\right]} \quad (12.17)$$

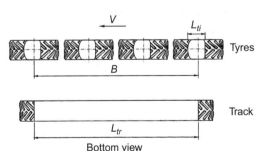

Figure 12.29: Schematic diagram of the contact patches of four tyres on an 8×8 wheeled vehicle and the contact patch of a comparable tracked vehicle (Reprinted by permission of ISTVS from Wong & Huang, 2006c)

where F_{ti} and F_{tr} are the thrust developed by the wheeled vehicle and the tracked vehicle, respectively, n_{ti} and n_{tr} are number of tyres on the wheeled vehicle and the number of tracks on the tracked vehicle, respectively, b_{ti} and b_{tr} are the tyre width and track width, respectively, L_{ti} and L_{tr} are the contact length of a tyre and that of a track, respectively, W is the total load on the vehicle and is assumed to be the same for both the wheeled and the tracked vehicle, i is the slip and is assumed to be the same for the tyre and the track, and c, ϕ and K are the cohesion, angle of shearing resistance and shear deformation parameter of the terrain, respectively. For the example shown in Figure 12.29, it is assumed that $b_{ti} = b_{tr} = 0.38$ m, $B = L_{tr} = 3.3$ m, and the tyre outside diameter of 0.984 m (corresponding to that of a tyre 325/85R16), and that vehicle weight W of 110.57 kN and slip i of 20% are the same for both the wheeled and tracked vehicles. Under these circumstances, the variations of the thrust ratio F_{ti}/F_{tr} with the contact length ratio L_{ti}/L_{tr} and the shear deformation parameter of the terrain K on four types of soil, namely, sand, clay with medium moisture content (MMC), clay with high moisture content (HMC) and loam, are shown in Figures 12.30, 12.31, 12.32 and 12.33, respectively. The cohesion and angle of shearing resistance for each of the four types of soil are shown in the respective figure. Different tyre inflation pressures produce various tyre contact lengths. Thus different contact length ratios indicate different inflation pressures of the tyres on the wheeled vehicle.

It can be seen from Figures 12.30, 12.31, 12.32 and 12.33 that on the four types of soil studied, the thrust ratio F_{ti}/F_{tr} is always lower than one, even at a relatively high contact length ratio L_{ti}/L_{tr} of 0.11, which corresponds to a tyre inflation pressure lower than 100 kPa for a 325/85R16 tyre. For instance, on the sand shown in Figure 12.30, with the value of shear deformation parameter $K = 0.015$ m and contact length ratio $L_{ti}/L_{tr} = 0.11$, the thrust ratio

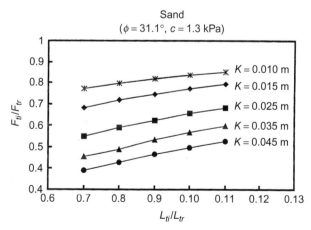

Figure 12.30: Variations of the ratio of wheeled vehicle thrust to tracked vehicle thrust at 20% slip with contact length ratio on sand (Reprinted by permission of ISTVS from Wong & Huang, 2006c)

F_{ti}/F_{tr} is 0.792, which indicates that the thrust developed by the 8×8 wheeled vehicle is only 79.2% of that developed by the tracked vehicle at 20% slip. On the clay with medium moisture content shown in Figure 12.31, with $K = 0.025$ m and $L_{ti}/L_{tr} = 0.11$, F_{ti}/F_{tr} is 0.541, which indicates that the thrust developed by the 8×8 wheeled vehicle is 54.1% of that developed by the tracked vehicle. On the clay with high moisture content shown in Figure 12.32,

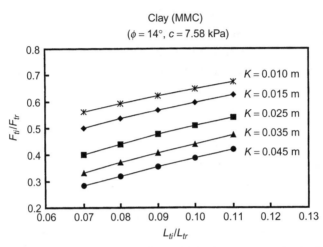

Figure 12.31: Variations of the ratio of wheeled vehicle thrust to tracked vehicle thrust at 20% slip with contact length ratio on clay with medium moisture content (MMC) (Reprinted by permission of ISTVS from Wong & Huang, 2006c)

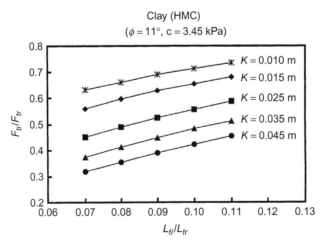

Figure 12.32: Variations of the ratio of wheeled vehicle thrust to tracked vehicle thrust at 20% slip with contact length ratio on clay with high moisture content (HMC) (Reprinted by permission of ISTVS from Wong & Huang, 2006c)

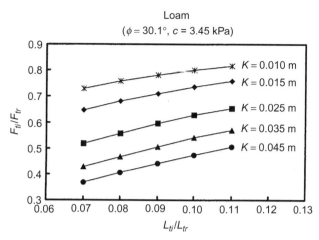

Figure 12.33: Variations of the ratio of wheeled vehicle thrust to tracked vehicle thrust at 20% slip with contact length ratio on loam (Reprinted by permission of ISTVS from Wong & Huang, 2006c)

with $K = 0.025$ m and $L_{ti}/L_{tr} = 0.11$, F_{ti}/F_{tr} is 0.588, which indicates that the thrust developed by the 8×8 wheeled vehicle is 58.8% of that developed by the tracked vehicle. On the loam shown in Figure 12.33, with $K = 0.045$ m and $L_{ti}/L_{tr} = 0.11$, F_{ti}/F_{tr} is 0.504, which indicates that the thrust developed by the 8×8 wheeled vehicle is 50.4% of that developed by the tracked vehicle. The lower thrust of the 8×8 wheeled vehicle, in comparison with the comparable tracked vehicle, is primarily due to the fact that with a much shorter contact length, the shear displacement at the rear of the contact patch of a tyre will be much lower than that of a track at the same slip. This will limit the full development of shear stress on the tyre contact patch in many cases, particularly when the value of K is high. Since the thrust developed by a vehicle running gear is the integration of the shear stress over the contact area, with lower shear stress and lower contact area, the thrust developed by a wheeled vehicle will generally be lower than that developed by a comparable tracked vehicle. This can be illustrated by Figure 12.34, in which the development of shear stress s under the tyres of a wheeled vehicle is compared with that under a rigid track. It should be noted, however, that with the same weight (or normal load) but having a lower total contact area, the average normal pressure under the tyres of a wheeled vehicle will be higher than that under a comparable track. This would affect the magnitude of the shear stress on the tyre contact patch, in comparison with that under a comparable track. This factor has been taken into account in the results shown in Figures 12.30, 12.31, 12.32 and 12.33.

On a given terrain with a particular value of K, the thrust ratio F_{ti}/F_{tr} increases with the increase of contact length ratio L_{ti}/L_{tr}, as can be seen from Figures 12.30, 12.31, 12.32 and 12.33. For instance, on the sand shown in Figure 12.30, with the value of shear deformation parameter $K = 0.015$ m, increasing the contact length ratio L_{ti}/L_{tr} from 0.07 to 0.11, the thrust

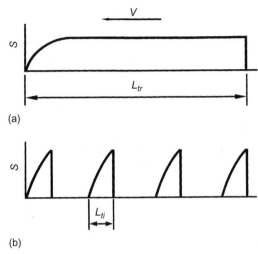

Figure 12.34: Comparison of the idealized shear stress distributions under (a) a rigid track and (b) under four tyres with uniform contact pressure (Reprinted by permission of ISTVS from Wong & Huang, 2006c)

ratio F_{ti}/F_{tr} increases from 0.683 to 0.792. On the clay with medium moisture content shown in Figure 12.31, with $K = 0.025$ m, increasing L_{ti}/L_{tr} from 0.07 to 0.11, F_{ti}/F_{tr} increases from 0.399 to 0.541. On the clay with high moisture content shown in Figure 12.32, with $K = 0.025$ m, increasing L_{ti}/L_{tr} from 0.07 to 0.11, F_{ti}/F_{tr} increases from 0.449 to 0.588. On the loam shown in Figure 12.33, with $K = 0.045$ m, increasing L_{ti}/L_{tr} from 0.07 to 0.11, F_{ti}/F_{tr} increases from 0.368 to 0.505. For a given tyre, the contact length is related to inflation pressure. Lowering tyre inflation pressure generally increases tyre contact length, which allows the shear stress on the contact patch to be more fully developed at a given slip, as explained earlier. It should also be noted that the increase in tyre contact length leads to the increase in tyre contact area. This in turn will increase the thrust component caused by the cohesion of the terrain, hence contributing to the increase in the thrust ratio on a terrain with cohesion.

For the same contact length ratio L_{ti}/L_{tr}, the thrust ratio F_{ti}/F_{tr} increases significantly with the decrease in the value of the shear deformation parameter K, as can be seen from Figures 12.30, 12.31, 12.32 and 12.33. For instance, on the sand shown in Figure 12.30, with the contact length ratio $L_{ti}/L_{tr} = 0.11$, the thrust ratio F_{ti}/F_{tr} increases significantly from 0.526 to 0.792 when the value of the shear deformation parameter K decreases from 0.045 to 0.015 m. On the clay with medium moisture content shown in Figure 12.31, with $L_{ti}/L_{tr} = 0.11$, F_{ti}/F_{tr} increases from 0.417 to 0.628 when the value of K decreases from 0.045 to 0.015 m. On the clay with high moisture content shown in Figure 12.32, with $L_{ti}/L_{tr} = 0.11$, F_{ti}/F_{tr} increases from 0.454 to 0.683 when the value of K decreases from 0.045 to 0.015 m. On the loam shown in Figure 12.33, with $L_{ti}/L_{tr} = 0.11$, F_{ti}/F_{tr} increases from 0.505 to 0.759 when the value of K decreases from 0.045 to 0.015 m. As explained earlier, the lower value of K allows the shear stress on the tyre contact patch to be more fully developed than that at a higher value of K. Therefore, the thrust ratio increases with the decrease in the value of K.

On a soil with significant cohesion, the thrust ratio F_{ti}/F_{tr} is more sensitive to the contact length ratio L_{ti}/L_{tr} than on a frictional soil. For instance, as shown in Figure 12.31, on a clay with $c = 7.58$ kPa, $\phi = 14°$ and $K = 0.025$ m, increasing the contact length ratio from 0.07 to 0.11, the thrust ratio F_{ti}/F_{tr} increases from 0.399 to 0.541, representing an increase of 35.6%, whereas on a frictional soil, such as the sand shown in Figure 12.30, with $c = 1.3$ kPa, $\phi = 31.1°$ and $K = 0.025$ m, increasing the contact length ratio from 0.07 to 0.11, the thrust ratio increases from 0.547 to 0.683, representing an increase of only 24.9%. This is primarily due to the fact that on a soil with significant cohesion, a notable portion of the thrust is derived from cohesion of the soil and is dependent on tyre contact area, hence tyre contact length. On the other hand, on a soil with a significant angle of internal shearing resistance, a major portion of the thrust is derived from friction and is independent of tyre contact area. Consequently, the increase in tyre contact length will have less effect on thrust on a frictional soil, such as sand, than on a cohesive soil, such as clay.

In summary, based on the examples given above, it can be said that the thrust (or propelling force) that can be developed by a wheeled vehicle is generally lower than that of a comparable tracked vehicle. This is primarily due to the fact that the tyre contact length is much shorter than the track contact length of a comparable tracked vehicle. As a result, the shear stress on the tyre contact patch is usually less fully developed than that under a comparable track, other conditions being equal. Among the vehicle design parameters, tyre contact length, which is related to tyre inflation pressure, has a noticeable effect on the thrust of a wheeled vehicle. Lowering tyre inflation pressure, hence increasing tyre contact length, would lead to improved traction of a wheeled vehicle, particularly on soil with significant cohesion. Among the terrain parameters, the shear deformation parameter K has a significant influence on the development of thrust of a wheeled vehicle, because of its relatively short contact length. Other conditions being equal, on a firm sandy (or frictional) soil, the traction of a wheeled vehicle is less dependent on tyre contact area. Therefore, the traction of a wheeled vehicle is closer to that of a comparable tracked vehicle on a firm sandy (or frictional) soil than on a clayey (or cohesive) soil.

The general analysis given above is useful in providing a physical insight into the traction capabilities of wheeled and tracked vehicles. It elucidates the differences between a tyre and a track in generating thrust to propel the vehicle. It also identifies the vehicle and terrain parameters that would significantly affect the tractive capabilities of both types of vehicle. It should be pointed out, however, that in the above analysis, the issue of wheels vs tracks has been examined solely from the thrust (or propelling force) standpoint. The mobility of a vehicle on unprepared terrain is, however, determined by not only its thrust, but also its motion resistance. Furthermore, in the above general evaluation, a number of simplified assumptions have been made. These include the vehicle weight being uniformly distributed among the tyres or the tracks, the contact area of the tyre or the track being flat, the contact length being the same for all tyres on a wheeled vehicle, etc. These simplifying assumptions in many cases may not necessarily be realistic. Furthermore, for a tyre, the thrust is usually developed in part by rubber–terrain shearing between the lug surface and the terrain, and in part by the internal shearing of the terrain between lugs (grousers). All of these point to the need for computer-aided methods, so that all these factors can be taken into account and the mobility of both the wheeled and the tracked vehicles on unprepared terrain can be realistically evaluated.

12.7.2 Using Computer-Aided Methods for the Analysis of Wheeled Vehicles vs Tracked Vehicles

To illustrate the approach to adequately addressing the issue of wheeled vehicles vs tracked vehicles from the traction perspective, an example is presented below. The performance of an 8×8 wheeled vehicle, similar to a widely used light armoured vehicle (LAV), was evaluated using the computer-aided method NWVPM described previously, and was compared with that of a tracked vehicle with link tracks (segmented metal tracks) having relatively short track pitch, similar to a widely used armoured personnel carrier (APC), predicted using the computer-aided method NTVPM described in Chapter 8. The schematic diagrams of the wheeled vehicle and that of the tracked vehicle are shown in Figures 12.35 and 12.36, respectively. The basic parameters of these two vehicles are given in Table 12.9. The tractive performance of the wheeled vehicle and that of the tracked vehicle were predicted using NWVPM and NTVPM, respectively, on four types of soil. The basic pressure–sinkage and shear strength parameters of these four types of soil are shown in Table 12.10.

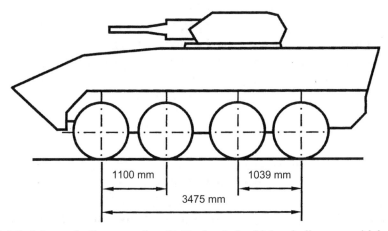

Figure 12.35: Schematic diagram of an 8×8 wheeled vehicle, similar to a widely used light armoured vehicle (LAV), used in the study (Reprinted by permission of ISTVS from Wong & Huang, 2006c)

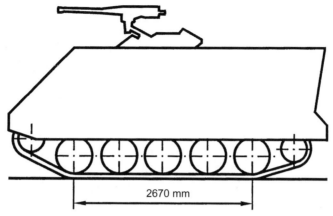

Figure 12.36: Schematic diagram of a tracked vehicle, similar to a widely used armoured personnel carrier (APC), used in the study (Reprinted by permission of ISTVS from Wong & Huang, 2006c)

Table 12.9: Basic parameters of the wheeled and tracked vehicle used in the analysis

Vehicle parameters	Wheeled vehicle (8×8)*	Tracked vehicle**
Total weight, kN	127.48	110.57
Wheelbase of the wheeled vehicle or track nominal contact length, m	3.475	2.67
Number of tyres or roadwheels on one side	4	5
Tyre or roadwheel outside diameter, m	0.984	0.61
Tyre or track width, m	0.393	0.38

*Similar to a widely used light armoured vehicle.
**Similar to a widely used armoured personnel carrier.

Table 12.10: Basic soil parameters

Soil type	Pressure–sinkage parameters			Shear strength parameters		
	n	k_c (kN/m^{n+1})	k_ϕ (kN/m^{n+2})	c (kPa)	$\phi°$	K (cm)
Sand	0.79	102	5301	1.30	31.1	0.012
Clay (MMC)*	0.85	43.08	499.7	7.58	14	0.025
Clay (HMC)**	1.0	20.68	814.3	3.45	11	0.025
Loam	0.66	8.7	816	3.45	30.1	0.042

*Medium moisture content.
**High moisture content.

As the two vehicles have different weights and the wheelbase of the wheeled vehicle is longer than the track contact length of the tracked vehicle, to provide a common basis for comparison, two non-dimensional performance parameters, thrust coefficient (the ratio of the thrust to vehicle weight), and drawbar pull coefficient (the ratio of the drawbar pull to vehicle weight) at 20% slip are used.

The predicted values of the thrust coefficient of the wheeled vehicle $(F/W)_{ti}$ at tyre inflation pressures of 103, 207 and 310 kPa and those of the tracked vehicle $(F/W)_{tr}$ at 20% slip on the four types of soil are shown in Table 12.11. To compare their traction capabilities, the ratios of the thrust coefficient of the wheeled vehicle $(F/W)_{ti}$ to that of the tracked vehicle $(F/W)_{tr}$ are also shown in the table. It can be seen that on sand at a tyre inflation pressure of 103 kPa, the ratio of $(F/W)_{ti}$ to $(F/W)_{tr}$ is the highest at 0.972, and that on clay with medium moisture content (MMC) at a tyre inflation pressure of 310 kPa, the ratio of $(F/W)_{ti}$ to $(F/W)_{tr}$ is the lowest at 0.545. As noted previously, the thrust coefficient only indicates the capability of the vehicle in developing propelling force and does not take into account the motion resistance of the vehicle. To characterize the mobility of the vehicle, the drawbar pull coefficient is used, as it represents the difference between the thrust coefficient and motion resistance coefficient, and its ability to develop net force to pull (or push) implements or working machinery, to overcome grade resistance, or to accelerate.

The values of the drawbar pull coefficient of the wheeled vehicle $(F_d/W)_{ti}$ at tyre inflation pressures of 103, 207 and 310 kPa and those of the tracked vehicle $(F_d/W)_{tr}$ at 20% slip on the four types of soil are shown in Table 12.12. To compare their performance, the ratios of the drawbar pull coefficient of the wheeled vehicle $(F_d/W)_{ti}$ to that of the tracked vehicle $(F_d/W)_{tr}$ are also shown in the table. Figure 12.37 shows the variations of the ratio of $(F_d/W)_{ti}$ to $(F_d/W)_{tr}$ at 20% slip with tyre inflation pressure on the four types of soil. As shown in the figure and the table, the mobility of the 8×8 wheeled vehicle in general cannot match that of the tracked vehicle on the four types of soil studied. It is interesting to note that the 8×8

Table 12.11: Comparison of the thrust coefficient at 20% slip of the tracked and wheeled vehicle

Soil type	Tracked vehicle $(F/W)_{tr}$	Wheeled vehicle $(F/W)_{ti}$			$(F/W)_{ti}/(F/W)_{tr}$		
		Tyre inflation pressure (kPa)			Tyre inflation pressure (kPa)		
		103	207	310	103	207	310
Sand	0.465	0.452	0.413	0.387	0.972	0.888	0.832
Clay (MMC)*	0.290	0.199	0.170	0.158	0.686	0.586	0.545
Clay (HMC)**	0.176	0.143	0.125	0.117	0.813	0.710	0.665
Loam	0.472	0.356	0.303	0.278	0.754	0.642	0.589

*Medium moisture content.
**High moisture content.

wheeled vehicle will be immobile on the clay with high moisture content (HMC), if the tyre inflation pressure is higher than approximately 180 kPa. This is because with higher inflation pressure, tyre deflection and tyre contact area decrease. This leads to the increase in sinkage and motion resistance and to the reduction in the component of the thrust derived from cohesion of the terrain, as explained previously.

In comparison with the simplified evaluation described in Section 12.7.1, the computer-aided methods NWVPM and NTVPM can provide a realistic and quantitative assessment of the mobility of wheeled vehicles and tracked vehicles over unprepared terrain. Thus it demonstrates that NWVPM and NTVPM are useful tools for adequately addressing the issue of wheeled vehicles vs tracked vehicles from the traction perspective.

Table 12.12: Comparison of the drawbar pull coefficient at 20% slip of the tracked and wheeled vehicle

Soil type	Tracked vehicle $(F_d/W)_{tr}$	Wheeled vehicle $(F_d/W)_{ti}$			$(F_d/W)_{ti}/(F^d/W)_{tr}$		
		Tyre inflation pressure (kPa)			Tyre inflation pressure (kPa)		
		103	207	310	103	207	310
Sand	0.463	0.401	0.343	0.300	0.866	0.741	0.648
Clay (MMC)*	0.231	0.116	0.054	0.008	0.502	0.234	0.035
Clay (HMC)**	0.094	0.048	−0.018	−0.065	0.511	−0.191	−0.691
Loam	0.450	0.281	0.197	0.133	0.624	0.438	0.296

*Medium moisture content.
**High moisture content.

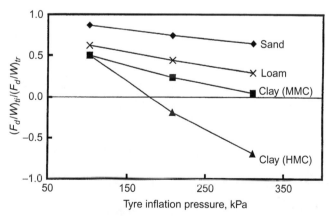

Figure 12.37: Comparison of the drawbar pull to weight ratio of an 8×8 wheeled vehicle with different tyre inflation pressures to that of a tracked vehicle at 20% slip on four types of soil (Reprinted by permission of ISTVS from Wong & Huang, 2006c)

PROBLEMS

12.1 A tyre has an outside diameter of 0.873 m and a width of 0.279 m. It operates in the rigid mode on a soil with the following shearing parameters on the tyre-soil interface: adhesion $c_a = 9.65$ kPa, angle of shearing resistance $\phi = 20°$, and shear deformation parameter $K = 0.025$ m. The tyre is being driven at 20% slip, with a contact angle $\theta_1 = 30°$ (see Fig. 12.1). The grouser height is relatively small and its effect may be neglected. Estimate the shear stress on the tyre surface at the angular position $\theta = 10°$, where the normal pressure is 300 kPa.

12.2 The tyre with an outside diameter of 0.85 m operates in the elastic mode on the soil described in the above problem. The tyre contact angle θ_1 is 30° and θ_c is 20° (see Fig. 12.2). The grouser height is relatively small and its effect may be neglected. The tyre is driven at 20% slip. Estimate the shear stress on the tyre surface at the bottom-dead-centre (i.e. the point on the tyre surface below the tyre centre), where the normal pressure is 200 kPa.

References

Anderson, G., Pidgeon, J. D., & Spencer, H. B. (1980). A new hand-held recording penetrometer for soil study. *Journal of Soil Science, 31*(2), 279–296.

Asaf, Z., Rubinstein, D., & Shmulevich, I. (2006). Evaluation of link-track performances using DEM. *Journal of Terramechanics, 43*(2), 141–161.

Asnani, V., Delap, D., & Creager, C. (2009). The development of wheels for the Lunar Roving Vehicle. *Journal of Terramechanics, 46*(3), 89–103.

Bekker, M. G. (1956). *Theory of land locomotion*. Ann Arbor, Michigan: The University of Michigan Press.

Bekker, M. G. (1960). *Off-the-road locomotion*. Ann Arbor, Michigan: The University of Michigan Press.

Bekker, M. G. (1964). Mechanics of locomotion and lunar surface vehicle concepts. *SAE Transactions, 72*, 549–569.

Bekker, M. G. (1967). Evolution of approach to off-road locomotion. *Journal of Terramechanics, 4*(1), 49–58.

Bekker, M. G. (1969). *Introduction to terrain–vehicle systems*. Ann Arbor, Michigan: The University of Michigan Press.

Bekker, M. G. (1981). A search for the mechanics of off-road locomotion in system analyses. *Bulletin of the Association of Polish Engineers in Canada, 35*, 29–56.

Bekker, M. G. (1983). Prediction of design and performance parameters in agro-forestry vehicles. NRCC Report No. 22880, National Research Council of Canada.

Bekker, M. G. (1985). The effect of tire tread in parametric analyses of tire–soil systems. NRCC Report No. 24146, National Research Council of Canada.

Bekker, M. G., & Semonin, E. V. (1975). Motion resistance of pneumatic tires. *Journal of Automotive Engineering*, April issue.

Besselink, B. C. (2003). Tractive efficiency of four-wheel-drive vehicles: An analysis for non-uniform traction conditions. *Proceedings of the Institution Mechanical Engineers, 217*(D5), 363–374.

Bodin, A. (2002). Improving the mobility performance of tracked vehicles in deep snow. Doctoral thesis (ISSN:1402-1544), Sweden: Department of Applied Physics and Mechanical Engineering, Lulea University of Technology.

Chen, S. T., & Wong, J. Y. (1988). An approach to improving the prediction of pneumatic tire performance over unprepared terrain. *Proceedings of the 2nd Asia-Pacific Conference of the International Society for Terrain–Vehicle Systems*, 293–302.

Cowart, E. G. (1971). Lunar roving vehicle: Spacecraft on wheels. *Proceedings of the Institution Mechanical Engineers*, *187*(45/73), 463–491.

Cundall, P. A., & Strack, O. D. (1979). Discrete numerical model for granular assemblies. *Geotechnique*, *29*(1), 47–65.

Dagan, G., & Tulin, M. P. (1969). A study of the steady flow of a rigid-plastic clay beneath a driven wheel. *Journal of Terramechanics*, *6*(2), 9–27.

Engeler, A. (1994). Parallel tests with a 6 and 7 roadwheel tracked vehicle. *Proceedings of the 6th European Conference of the International Society for Terrain-Vehicle Systems and the 4th OVK Symposium on Off-Road Vehicles in Theory and Practice*, (2), 603–617.

Fervers, C. W. (2004). Improved FEM simulation model for tire–soil interaction. *Journal of Terramechanics*, *41*(2 & 3), 87–100.

Freitag, D. R. (1965). A dimensional analysis of the performance of pneumatic tires on soft soils. Technical Report No. 3-688, Vicksburg, Mississippi, USA: US Army Engineer Waterways Experiment Station.

Freitag, D. R., Green, A. J., & Melzer, K. J. (1970). Performance evaluation of wheels for lunar vehicles. Technical Report M-70-2, prepared for George C. Marshall Space Flight Center, NASA, by US Army Engineer Waterways Experiment Station, Vicksburg, Mississippi, USA.

Gao, Y., & Wong, J. Y. (1994). The development and validation of a computer-aided method for design evaluation of tracked vehicles with rigid links. *Proceedings of the Institution Mechanical Engineers*, *208*(D3), 207–215.

Garber, M., & Wong, J. Y. (1981a). Prediction of ground pressure distribution under tracked vehicles – Part I. An analytical method for predicting ground pressure distribution. *Journal of Terramechanics*, *18*(1), 1–23.

Garber, M., & Wong, J. Y. (1981b). Prediction of ground pressure distribution under tracked vehicles – Part II. Effects of design parameters of track–suspension system on ground pressure distribution. *Journal of Terramechanics*, *18*(2), 71–79.

Gee-Clough, D. (1978). A comparison of the mobility number and Bekker approaches to traction mechanics and recent advances in both methods at N.I.A.E. *Proceedings of the 6th International Conference of the International Society for Terrain–Vehicle Systems*, *II*, 735–755.

Golob, T. B. (1981). Development of a terrain strength measuring system. *Journal of Terramechanics*, *18*(2), 109–118.

Hallonborg, U. (1996). Super ellipse as tyre–ground contact area. *Journal of Terramechanics*, *33*(3), 125–132.

Harnisch, C., Lach, B., Jakobs, R., Troulis, M., & Nehls, O. (2005). A new tyre–soil interaction model for vehicle simulation on deformable ground. *Vehicle System Dynamics*, *43*(Suppl.), 384–394.

Harrison, W. L. (1975). Vehicle performance over snow – math-model validation study. Technical Report 268, US Army Corps of Engineers. Hanover, New Hampshire, USA: Cold Regions Research and Engineering Laboratory

Hettiaratchi, D. R. P., & Reece, A. R. (1974). The calculation of passive soil resistance. *Geotechnique, 24*(3), 289–310.

Hettiaratchi, D. R. P., & Liang, Y. (1987). Monograms for the estimation of soil strength from indentation test. *Journal of Terramechanics, 24*(3), 187–198.

Horner, D. A., Peters, J. F., & Carrillo, A. (2001). Large scale discrete element modeling of vehicle–soil interaction. *Journal of Engineering Mechanics, Proceedings of American Society Civil Engineers, 127*(10), 1027–1032.

Janosi, Z. (1961). An analysis of pneumatic tyre performance on deformable soil. *Proceedings of the 1st International Conference on the Mechanics of Soil-Vehicle Systems*. Edizioni Minerva Tecnica, Troino, Italy.

Janosi, Z., & Hanamoto, B. (1961). The analytical determination of drawbar pull as a function of slip for tracked vehicles in deformable soils. *Proceedings of the 1st International Conference on the Mechanics of Soil–Vehicle Systems,* Torino, Italy: Edizioni Minerva Tecnica.

Jurkat, M. P., Nuttall, C. J., & Haley, P. W. (1975). The U.S. Army mobility model (AMM-75). *Proceedings of the 5th International Conference of the International Society for Terrain–Vehicle Systems, IV,* 1-1-1-48.

Kacigin, V. V., & Guskov, V. V. (1968). The basis of tractor performance theory. *Journal of Terramechanics, 5*(3), 43–66.

Karafiath, L. L. (1971). Plasticity theory and the stress distribution beneath wheels. *Journal of Terramechanics, 8*(2), 49–60.

Karafiath, L. L. (1984). Finite element analysis of ground deformation beneath moving track loads. *Proceedings of the 8th International Conference of the International Society for Terrain–Vehicle Systems, I,* 277–290.

Karafiath, L. L., & Nowatzki, E. A. (1978). *Soil Mechanics for Off-road Vehicle Engineering.* Clausthal, Germany: Trans Tech Publication.

Keira, H. M. S. (1979). Effects of vibration on the shearing characteristics of soil engaging machinery. Unpublished Ph.D. Thesis, Ottawa, Canada: Carleton University

Krick, G. (1969). Radial and shear stress distribution under rigid wheels and pneumatic tires operating on yielding soils with consideration of tire deformation. *Journal of Terramechanics, 6*(3), 73–98.

Kurtay, T., & Reece, A. R. (1970). Plasticity theory and critical state soil mechanics. *Journal of Terramechanics, 7*(3 & 4), 23–56.

Liu, C. H., & Wong, J. Y. (1996). Numerical simulations of tire–soil interaction based on critical state soil mechanics. *Journal of Terramechanics, 33*(5), 209–221.

Liu, C. H., Wong, J. Y., & Mang, H. A. (2000). Large strain finite element analysis of sand: Model, algorithm and application to numerical simulation of tire–soil interaction. *Computers and Structures, 74,* 253–265.

MacFarlane, I. E. (Ed.). (1969). *Muskeg engineering handbook*. Toronto, Canada: The University of Toronto Press.

Meyerhof, G. G. (1960). Bearing capacity of floating ice sheets. *Journal of Engineering Mechanics, Proceeding of American Society Civil Engineers*, *86*(EM5), 90–115.

Micklethwaite, E. W. E. (1944). *Soil mechanics in relation to fighting vehicles*. Royal Military College of Science. Chertsey, UK.

Mulqueen, J., Stafford, J. V., & Tanner, D. W. (1977). Evaluation of penetrometers for measuring soil strength. *Journal of Terramechanics*, *14*(3), 137–151.

Nakashima, H., & Wong, J. Y. (1993). A three-dimensional tire model by the finite element method. *Journal of Terramechanics*, *30*(1), 21–34.

Nakashima, H., Fujii, H., Oida, A., Momozu, M., Kawase, Y., Kanamori, H., Aoki, S., & Yokoyama, T. (2007). Parametric analysis of lugged wheel performance for a lunar microrover by means of DEM. *Journal of Terramechanics*, *44*(2), 153–162.

Notwazki, E. A., & Karafiath, L. L. (1974). General yield conditions in a plastic analysis of soil–wheel interaction. *Journal of Terramechanics*, *11*(1), 29–44.

Nuttall, C. J., Jr. (1964) Some notes on the steering of tracked vehicles by articulation. *Journal of Terramechanics*, *1*(1), 38–74.

Nuttall, C. J., Jr., Rula, A. A., & Dugoff, H. J. (1974) Computer model for comprehensive evaluation of cross-country mobility Paper 740426. *SAE Transactions*.

Oida, A. (1979). Study on equation of shear stress–displacement curves. Report No. 5, Japan: Farm Power and Machinery Laboratory, Kyoto University

Olsen, H. J. (1987). Electronic cone penetrometer for field test. *Proceedings of the 9th International Conference of the International Society for Terrain–Vehicle Systems*, *I*, 20–27.

Onafeko, O., & Reece, A. R. (1967). Soil stresses and displacements beneath rigid wheels. *Journal of Terramechanics*, *4*(1), 59–80.

Ogorkiewicz, R. M. (1991). *Technology of tanks*. Surrey, UK: Jane's Information Group.

Osman, M. S. (1964). The mechanics of soil cutting blades. *Journal of Agricultural Engineering Research*, *9*(4).

Perumpral, J. V., Liljedahl, J. B., & Perloff, W. H. (1971). A numerical method for predicting the stress distribution and soil deformation under a tractor wheel. *Journal of Terramechanics*, *8*(1), 9–22.

Preston-Thomas, J., & Wong, J. Y. (1987). *Computer simulation study of the mobility of the Armoured Vehicle General Purpose (AVGP) with different tires* Unpublished report of Vehicle Systems Development Corporation, prepared for the Directorate of Combat Mobility Engineering and Maintenance, Department of National Defence, Canada.

Reece, A. R. (1964). The effect of grousers on off-the-road vehicle performance. *Journal of Agricultural Engineering Research*, *9*(4), 360–371.

Reece, A. R. (1965). Principles of soil–vehicle mechanics. *Proceedings of the Institution Mechanical Engineers*, *180*(Part 2A 2), 45–67.

Reece, A. R. (1970). Review of *introduction to terrain–vehicle systems*. *Journal of Terramechanics*, *7*(1), 75–77.

Reece, A. R., & Peca, J. O. (1981). An assessment of the value of the cone penetrometer in mobility prediction. *Proceedings of the 7th International Conference of the International Society for Terrain–Vehicle Systems*, *III*, A1–A33.

Rohani, B., & Baladi, G. Y. (1981). Correlation of mobility cone index with experimental engineering properties of soil. *Proceedings of the 7th International Conference of the International Society for Terrain–Vehicle Systems*, *III*, 959–990.

Rohrbach, S. E., & Jackson, G. J. (1982). Tracked vehicle tractive performance prediction – a case study in understanding the soil/tool interface Technical Paper No. 820655. *SAE*.

Roscoe, K. H., Schofield, A. N., & Wroth, C. P. (1958). On yielding of soils. *Geotechnique*, *8*(1), 22–53.

Rowland, D. (1972). Tracked vehicle ground pressure and its effect on soft ground performance. *Proceedings of the 4th International Conference of the International Society for Terrain–Vehicle Systems*, *I*, 353–384.

Rowland, D. (1975). A review of vehicle design for soft ground operation. *Proceedings of the 5th International Conference of the International Society for Terrain–Vehicle Systems*, *I*, 179–219.

Rula, A. A., & Nuttall, C. J. (1971). An analysis of ground mobility models (ANAMOB). Technical report M-71-4, Vicksburg, Mississippi, USA: US Army Corps of Engineers Waterways Experiment Station

Schofield, A. N., & Wroth, C. P. (1968). *Critical State Soil Mechanics*. London: McGraw-Hill.

Schreiber, M., & Kutzbach, H. D. (2007). Comparison of different zero-slip definitions and a proposal to standardize tire traction performance. *Journal of Terramechanics*, *44*(1), 75–79.

Sela, A. D. (1964). The shear stress–deformation relationship of soils. *Journal of Terramechanics*, *1*(1), 31–37.

Seta, E., Kamegawa, T., & Nakajima, Y. (2003). Prediction of snow/tire interaction using explicit FEM and FVM. *Tire Science and Technology, TSTCA*, *31*(3), 173–188.

Society of Automotive Engineers (SAE). (1967). *Off-Road Mobility Evaluation – SAE J939*.

Sohne, W. (1958). Fundamentals of pressure distribution and soil compaction under tractor tires and 290. *Agricultural Engineering, May*, 276–281.

Sohne, W. (1968). Four-wheel drive or rear-wheel drive for high power farm tractors. *Journal of Terramechanics*, *5*(3), 9–28.

Sohne, W. (1976). Terramechanics and its influence on the concepts of tractors, tractor power development, and energy consumption. *Journal of Terramechanics*, *13*(1), 27–43.

Tanaka, H., Momozu, M., Oida, A., & Yamazaki, M. (2000). Simulation of soil deformation and resistance at bar penetration by the distinct element method. *Journal of Terramechanics*, *37*(1), 41–56.

Turnage, G. W. (1972). Performance of soils under tire loads; application of test results to tire selection for off-road vehicles. Technical Report No. 3-666, Vicksburg, Mississippi, USA: US Army Engineer Waterways Experiment Station.

Turnage, G. W. (1978). A synopsis of tire design and operational considerations aimed at increasing in-soil tire drawbar performance. *Proceedings of the 6th International Conference of the International Society for Terrain–Vehicle Systems, II*, 757–810.

Turnage, G. W. (1984). Prediction of in-sand tire and wheeled vehicle drawbar performance. *Proceedings of the 8th International Conference of the International Society for Terrain–Vehicle Systems, I*, 121–150.

Wills, B. M. D. (1963). The measurement of soil shear strength and deformation moduli and a comparison of the actual and theoretical performance of a family of rigid tracks. *Journal of Agricultural Engineering Research, 8*(2), 115–131.

Wismer, R. D., & Luth, H. J. (1972). Off-road traction prediction for wheeled vehicles. Paper No. 72-619, *American Society Agricultural Engineers*, Michigan, USA: St Joseph.

Wong, J. Y. (1967). Behaviour of soil beneath rigid wheels. *Journal of Agricultural Engineering Research, 12*(4), 257–269.

Wong, J. Y. (1970). Optimization of the tractive performance of four-wheel-drive off-road vehicles. *SAE Transactions, 79*, 2238–2246.

Wong, J. Y. (1972a). Performance of the air-cushion-surface-contacting hybrid vehicle for overland operation. *Proceedings of the Institution Mechanical Engineers, 186*(50/72), 613–623.

Wong, J. Y. (1972b). Discussion on 'Stress field under slipping rigid wheels'. *Journal of Soil Mechanical and Foundation Division, Proceedings of American Society of Civil Engineers, 98*(SM9), 977–981.

Wong, J. Y. (1975). System energy in high speed ground transportation. *High Speed Ground Transportation Journal, 9*(1), 307–320.

Wong, J. Y. (1977). Discussion on 'Prediction of wheel–soil interaction and performance using the finite element method'. *Journal of Terramechanics, 14*(4), 240–250.

Wong, J. Y. (1978a). *Theory of ground vehicles* (1st ed.). New York, NY: John Wiley (Russian translation published by Machinostroenie Publishing House, Moscow, USSR, 1982; Chinese translation published by Machinery Industry Publishing House, Beijing, China, 1985).

Wong, J. Y. (1978b). Predictions of multiple-pass performance of tires – a review. *Proceedings of the 6th International Conference of the International Society for Terrain-Vehicle Systems, II*, 541–554.

Wong, J. Y. (1979). Review of *Soil mechanics for off-road vehicle engineering* also *Journal of Terramechanics*, 16(4): 191–194. *Canadian Geotechnical Journal, 16*(3), 624–626.

Wong, J. Y. (1980). Data processing methodology in the characterization of the mechanical properties of terrain. *Journal of Terramechanics, 17*(1), 13–41.

Wong, J. Y. (1983). Evaluation of soil strength measurements. NRCC Report No. 22881, National Research Council of Canada.

Wong, J. Y. (1984a). An introduction to terramechanics. *Journal of Terramechanics, 21*(1), 5–17.

Wong, J. Y. (1984b). An improved method for predicting tracked vehicle performance. *Journal of Terramechanics*, *21*(1), 35–43.

Wong, J. Y. (1984c). On the study of wheel–soil interaction. *Journal of Terramechanics*, *21*(2), 117–131.

Wong, J. Y. (1986a). Computer-aided analysis of the effects of design parameters on the performance of tracked vehicles. *Journal of Terramechanics*, *23*(2), 95–124.

Wong, J. Y. (1986b). A comprehensive computer simulation model for the off-road performance of tracked vehicles. *Proceedings of the Symposium on Simulation and Control of Ground Vehicles and Transportation Systems*, L. Segel, J. Y. Wong, E. H. Law & D. Hrovat (Eds.), *American society mechanical engineers,* AMD (80) DSC No. 2: 63–79.

Wong, J. Y. (1990). Some recent developments in the computer-aided methods for design evaluation of off-road vehicles – a summary. *Proceedings of the Symposium on Transportation Systems*, J. Y. Wong, J. J. Moskwa & S. A. Velinsky (Eds.), *Am. Soc. Mech. Engrs*, AMD (108): 183–189.

Wong, J. Y. (1991). Some recent developments in vehicle–terrain interaction studies. *Journal of Terramechanics*, *28*(4), 269–288.

Wong, J. Y. (1992a). Optimization of the tractive performance of articulated tracked vehicles using an advanced computer simulation model. *Proceedings of the Institution Mechanical Engineers*, *206*(D1), 29–45.

Wong, J. Y. (1992b). Expansion of the terrain input base for Nepean Tracked Vehicle Performance Model, NTVPM, to accept Swiss rammsonde data from deep snow. *Journal of Terramechanics*, *29*(3), 341–357.

Wong, J. Y. (1992c). Computer-aided methods for the optimization of the mobility of single-unit and two-unit articulated tracked vehicles. *Journal of Terramechanics*, *29*(4/5), 395–421.

Wong, J. Y. (1994a). On the role of mean maximum pressure as an indicator of cross-country mobility for tracked vehicles. *Journal of Terramechanics*, *31*(3), 197–213.

Wong, J. Y. (1994b). Computer-aided methods for design evaluation of track systems. *SAE Transactions*, Section 2, *Journal of Commercial Vehicles*, Paper No. 941675, 72–80.

Wong, J. Y. (1994c). Terramechanics – its present and future. *Proceedings of the 6th European Conference of the International Society for Terrain–Vehicle Systems* and *the 4th OVK Symposium on Off-Road Vehicles in Theory and Practice*, Vienna, Austria (1): 1–21.

Wong, J. Y. (1994d). Computer simulation models for evaluating the performance and design of tracked and wheeled vehicle. *Proceedings of the 1st North American Workshop on Modeling the Mechanics of Off-Road Mobility,* sponsored by the US Army Research Office, and held at the US Army Corps of Engineers Waterways Experiment Station, Vicksburg, Mississippi, USA: A3–A24.

Wong, J. Y. (1995). Application of the computer simulation model NTVPM-86 to the development of a new version of the infantry fighting vehicle ASCOD. *Journal of Terramechanics*, *32*(1), 53–61.

Wong, J. Y. (1997). Dynamics of tracked vehicles. *Vehicle System Dynamics*, *28*(2 & 3), 197–219.

Wong, J. Y. (1998). Optimization of design parameters of rigid-link track systems using an advanced computer-aided method. *Proceedings of the Institution Mechanical Engineers*, *212*(D3), 153–167.

Wong, J. Y. (1999). Computer-aided methods for design evaluation of tracked vehicles and their applications to product development. *International Journal of Vehicle Design*, *22*(1/2), 73–92.

Wong, J. Y. (2007). Development of high-mobility tracked vehicles for over snow operations. Keynote address. *Proceeding International Society for Terrain–Vehicle Systems Joint North American, Asia-Pacific Conference and Annual Meeting of Japanese Society for Terramechanics*, Fairbanks, Alaska, June 23–26. Also in *Journal of Terramechanics*, *46*(4), 141–155, 2009.

Wong, J. Y. (2008). *Theory of ground vehicles* (4th ed). New York, NY: John Wiley.

Wong, J. Y., & Asnani, V. M. (2008). Study of the correlation between the performances of lunar vehicles predicted by the Nepean wheeled vehicle performance model and test data. *Proceedings of the Institution Mechanical Engineers*, *222*(D), 1939–1954.

Wong, J. Y., & Chiang, C. F. (2001). A general theory for skid steering of tracked vehicles on firm ground. *Proceedings of the Institution Mechanical Engineers*, *215*(D), 343–355.

Wong, J. Y., & Gao, Y. (1994). Applications of a computer-aided method to parametric study of tracked vehicles with rigid links. *Proceedings of the Institution Mechanical Engineers*, *208*(D), 251–257.

Wong, J. Y., & Huang, W. (2004). Model behaviour. *Industrial Vehicle Technology International*, April/May, 28–33.

Wong, J. Y., & Huang, W. (2005). Evaluation of the effects of design features on tracked vehicle mobility using an advanced computer simulation model. *International Journal Heavy Vehicle Systems*, *12*(4), 344–365.

Wong, J. Y., & Huang, W. (2006a). An investigation into the effects of initial track tension on soft ground mobility of tracked vehicles using an advanced computer simulation model. *Proceedings of the Institution Mechanical Engineers*, *220*(D), 695–711.

Wong, J. Y., & Huang, W. (2006b). Study of detracking risks of track systems. *Proceedings of the Institution Mechanical Engineers*, *220*(D), 1235–1253.

Wong, J. Y., & Huang, W. (2006c). Wheel vs. tracks – A fundamental evaluation from the traction perspective. *Journal of Terramechanics*, *43*(1), 27–42.

Wong, J. Y., & Huang, W. (2008). Approaches to improving the mobility of military tracked vehicles on soft terrain. *International Journal Heavy Vehicle Systems*, *15*(2/3/4), 127–151.

Wong, J. Y., & Preston-Thomas, J. (1983a). On the characterization of the shear stress–displacement relationship of terrain. *Journal of Terramechanics*, *19*(4), 107–127.

Wong, J. Y., & Preston-Thomas, J. (1983b). On the characterization of the pressure–sinkage relationship of snow covers containing an ice layer. *Journal of Terramechanics*, *20*(1), 1–12.

Wong, J. Y., & Preston-Thomas, J. (1986). Parametric analysis of tracked vehicle performance using an advanced computer simulation model. *Proceedings of the Institution Mechanical Engineers*, *200*(D2), 101–114.

Wong, J. Y., & Preston-Thomas, J. (1988). Investigation into the effects of suspension characteristics and design parameters on the performance of tracked vehicles using an advanced computer simulation model. *Proceedings of the Institution Mechanical Engineers, 202*(D3), 143–161.

Wong, J. Y., & Reece, A. R. (1966). Soil failure beneath rigid wheels. *Proceedings of the 2nd International Conference of the International Society for Terrain–Vehicle Systems*, 425–445.

Wong, J. Y., & Reece, A. R. (1967a). Prediction of rigid wheel performance based on the analysis of soil–wheel stresses – Part I. Performance of driven rigid wheels. *Journal of Terramechanics, 4*(1), 81–98.

Wong, J. Y., & Reece, A. R. (1967b). Prediction of rigid wheel performance based on the analysis of soil–wheel stresses – Part II. Performance of towed rigid wheels. *Journal of Terramechanics, 4*(2), 7–25.

Wong, J. Y., Garber, M., & Preston-Thomas, J. (1984). Theoretical prediction and experimental substantiation of ground pressure distribution and tractive performance of tracked vehicles. *Proceedings of the Institution Mechanical Engineers, 198*(D15), 265–285.

Wong, J. Y., Harris, P. S., & Preston-Thomas. (1981). Development of a portable automatic data processing system for terrain evaluation. *Proceedings of the 7th International Conference of the International Society for Terrain–Vehicle Systems, III*, 1067–1091.

Wong, J. Y., Radforth, J. R., & Preston-Thomas, J. (1982). Some further studies of the mechanical properties of muskeg. *Journal of Terramechanics, 19*(2), 217–223.

Wong, J. Y., Garber, M., Radforth, J. R., & Dowell, J. T. (1979). Characterization of the mechanical properties of muskeg with special reference to vehicle mobility. *Journal of Terramechanics, 16*(4), 163–180.

Wong, J. Y., McLaughlin, N. B., Knezevic, Z., & Burtt, S. (1998). Optimization of the tractive performance of four-wheel-drive tractors: theoretical analysis and experimental substantiation. *Proceedings of the Institution Mechanical Engineers, 212*(D4), 285–297.

Wong, J. Y., McLaughlin, N. B., Zhao, Zhiwen., Li, Jianqiao., & Burtt, S. (1999). Optimization of the tractive performance of four-wheel-drive tractors – theory and practice. *Proceedings of the 13th International Conference of the International Society for Terrain–Vehicle Systems, II*, 628–631.

Wong, J. Y., Zhao, Zhiwen., Li, Jianqiao., McLaughlin, N. B., & Burtt, S. (2000). Optimization of the tractive performance of four-wheel-drive tractors – correlation between analytical predictions and experimental data. *SAE Transactions,* Section 2, *Journal of Commercial Vehicles,* Paper 2000-01-2596; also in *SAE Journal of Off-Highway Engineering,* February 2001: 46–50.

Wu, S. X., Hu, J. H., & Wong, J. Y. (1984). Behaviour of soil under a lugged wheel. *Proceedings of the 8th International Conference of the International Society for Terrain–Vehicle Systems, II*, 545–559.

Yong, R. N., & Fattah, E. A. (1976). Prediction of wheel–soil interaction and performance using the finite element method. *Journal of Terramechanics, 13*(4), 227–240.

Yong, R. N., & Muro, T. (1981). Plate loading and vane–cone measurements for fresh and sintered snow. *Proceedings of the 7th International Conference of the International Society for Terrain–Vehicle Systems, III*, 1093–1118.

Yong, R. N., Youssef, A. F., & Fattah, E. A. (1975). Vane–cone measurements for assessment of tractive performance in wheel–soil interaction. *Proceedings of the 5th International Conference of the International Society for Terrain–Vehicle Systems, III*, 769–788.

Zhang, R., & Li, Jiangqiao (2006). Simulation on mechanical behaviour of cohesive soil by distinct element method. *Journal of Terramechanics, 43*(3), 303–316.

Zhang, T., Lee, J. H., Liu, Q. (2005). Finite element simulation of tire–snow interaction under combined longitudinal and lateral slip condition. *Proceedings of the 15th International Conference of the International Society for Terrain–Vehicle Systems*, Paper 3A05.

Index

A
Adhesion, 38
Aerodynamic drag (resistance), 129–31
Angle of interface (soil-metal) friction, 38–9, 43–4
Angle of shearing resistance, 31, 39
Articulated tracked vehicles, 295–305

B
Bekker's pressure-sinkage equation, 77
Belly
  drag, 193, 223–4
  load, 222–3
  trim angle, 217, 242, 248, 250
Belly-terrain interaction, 193, 216
Bevameter technique, 68–72
Boussinesq's equation, 22
Bulldozing
  effect, 217
  force, 36

C
Centre of gravity location, effect of, 247–57
Coefficient
  belly drag, 223–4
  belly load, 222–3
  drawbar pull, 135, 226–7, 234–6, 243, 246, 251, 255–6, 261–2, 264, 266, 272–3, 277–8, 285–6, 290
  thrust, 226, 234–5
  total motion resistance, 225
  track motion resistance, 224–5, 232–4
Cohesion, 31
Compaction resistance
  for rigid track, 164
  for tyre
    in the elastic operating mode, 392
    in the rigid operating mode, 380, 382
Computer-aided methods
  NTVPM for vehicles with flexible tracks, 177–209
    applications, 211–331
    basic approach, 178–95
    experimental validation, 195–209
  NWVPM for off-road wheeled vehicles
    applications
      to evaluation of lunar vehicle wheels, 424–432
      to parametric analysis, 419–24
    basic features, 414–17
    experimental validation, 417–18
  RTVPM for vehicles with long-pitch link tracks
    applications, 353–368
    basic approach, 333–49
    experimental validation, 349–53
Concentration factor, 25
Cone index, 65–6
  gradient, 374, 376–7
  rating, 66
Cone penetrometer technique, 65–8
Critical ground pressure, 394–5
Critical sinkage of muskeg, 91
Critical State Soil Mechanics, 45–8

D
Detracking risk, 305–319
Discrete (distinct) element method (DEM), 55–62
Drawbar
  efficiency, *See* Efficiency, tractive
  power, 135
  pull, 129–30
    coefficient, 135, 226–7
    to weight ratio, 135

E
Earth pressure
  passive, 35
  theory of, 38
Efficiency
  drawbar, *See* Efficiency, tractive
  motion, 136
  propulsive, 148
  slip, 136
  structural, 148
  tractive, 135–8, 228, 236, 243, 246–7, 252, 256, 262, 266–7, 273–4, 278–9, 286–7, 291
  transmission, 133
  transport, 148
Elasticity, theory of, 22
Empirical methods
  for tracked vehicle performance, 155–63
  for tyre performance, 374–8
  for wheeled vehicle performance, 372–4

F

Failure criterion, Mohr-Coulomb, 31
Finite element method (FEM), 48–55
   applications to tyre modelling, 432–6
Flow patterns of soil
   under grouser, 13
   under wheel, 41–3
      driven, 13, 41
      locked, 42
      spinning (100% slip), 42
      towed, 42
Flow value, 35
Four-wheel-drive off-road vehicle, 139–46

G

Ground clearance, 193, 217
Grouser (lug) effect, 35–7

H

Hydraulic diameter, 96
Hysteresis
   muskeg, 101
   tyre materials, 397

I

Initial track tension, 216
   effect of, 214–39, 364–6
   regulating system (device), 238
Instantaneous centre of wheel rotation, 389

L

Lift to drag ratio, 148
Load
   collapse, for an ice layer, 106–7
   ultimate, for an ice layer, 108–10

M

Mean ground pressure (MGP), 171, 174
Mean maximum pressure (MMP), 159, 230–2, 244, 253, 263, 274–5, 287–8
   empirical methods for predicting, 159
Mineral terrain
   pressure-sinkage relationship, 76–85
   response to repetitive normal loading, 85–8
   shear strength, 124–5
Mobility
   index
      for tracked vehicle, 156–7
      for wheeled vehicle, 372–3
   map, 150–1
   profile, 150–1
Modulus, Young's, 436
Mohr circle, 31
Mohr-Coulomb failure criterion, *See* Failure criterion, Mohr-Coulomb
Multi-pass effect, 415
Muskeg
   pressure-sinkage relationship, 88–100
   response to repetitive normal loading, 100–2
   shear strength, 126

N

Normal pressure distribution
   under flexible track, 178–88, 191–92
   in steady-state operations, 203–4, 218–20, 229–31
   in static conditions, 171–4
   under long-pitch link track, 350, 352, 356–7, 361–2
   under rigid track, 166–8
   under rigid wheel, 385, 387
   under tyre in elastic operating mode, 13–4, 437–9

P

Parametric analysis
   tracked vehicle performance, 163–9
   wheeled vehicle performance, 419–23
Passive earth pressure theory, 38
Passive failure, 34
Peat
   shear strength, 126
   stiffness, 94
Plastic equilibrium, 31, 33–7
   Theory of, 7, 43–5
Plastic flow, 31
Pressure
   average tyre ground, 392–5
   critical ground, *See* Critical ground pressure
   on tyre carcass, 412–14
   on tyre lug, 412–14
Pressure-sinkage relationship
   mineral terrain, 76–85
   muskeg, 90–100
   snow-covered terrain, 110–2
Principal stress
   major, 35
   minor, 35

R

Rankine
   active state, 34
   passive failure zone, 38
   passive state, 34
Resistance
   bulldozing, *See* Bulldozing force
   compaction, *See* Compaction resistance
   horizontal terrain deformation, 396–7
   obstacle, 131
   total motion, 133, 225
   track internal, 131–2
   tyre flexing, 397–8

S

Shear deformation parameter, 116
Shear displacement
   under flexible track, 189
   under rigid track, 166
   under tyre in elastic operating mode, 407–8
   under tyre in rigid operating mode, 407
Shear strength, 31
   under repetitive shear loading, 123–8
Shear stress distribution
   under flexible track, 188–91
   under long-pitch link track, 349–50, 352–3
   under rigid track, 165–6
   under rigid wheel, 385, 387, 406–7
   under tyre in elastic operating mode, 13–14, 437–9

Shear stress-shear displacement relationship, characterization of, 115–23
  using exponential equation, 120–1
  using "hump" equation, 121–3
  using Oida's equation, 117–8
  using Wong's equation, 118–20
Sinkage
  under flexible track, 187–8, 218–20, 229–31
  under long-pitch link track, 358, 363
  under rigid track, 163
  under tyre in elastic operating mode, 390, 393
  under tyre in rigid operating mode, 381–2
Skid, 389
Slip, 133
  efficiency, *See* Efficiency, slip
  lines, 35, 43–4
  sinkage, 192, 385
  velocity (speed), 165, 188–9, 406–7
Snow-covered terrain
  pressure-sinkage relationship, 102–12
  response to repetitive normal loading, 112–3
  shear strength, 126

Speed
  actual, 134
  theoretical, 134
Sprocket location, effect of, 281–93
Suspension
  Setting, 239–40
    effect of, 239–47
    regulating system, 247
  tracked vehicle, 239–40
  wheeled vehicle, 416
Surcharge, 35, 37

T

Terrain, modelling of
  as a plastic medium, 31–45
  as an elastic medium, 22–30
  based on the critical state soil mechanics, 45–8
  using the discrete (distinct) element method, 55–62
  using the finite element method, 48–55
Thrust (tractive effort), 32, 129, 132–3
Track
  band (rubber belt), 177
  flexible, 177
  initial tension of, *See* Initial track tension
  internal resistance, 131–2
  long-pitch link, 333
  short-pitch link, 177

  tension-elongation relationship of, 178–9, 282
Track system configurations, 214–6
Track width, effect of, 268–81
Tracked vehicle, high-mobility, 293–4, 366–8
Tractive efficiency, *See* Efficiency, tractive
Transport
  efficiency, *See* Efficiency, transport
  productivity, 147–8
Tyre
  in elastic operating mode, 391–401, 404–5, 407–8, 410–12, 413–14
  in rigid operating mode, 379–91, 397, 400, 403–4, 406–7, 409–10, 412–13

V

Vane-cone, 68–9
Vehicle cone index
  for tracked vehicles, 157–8
  for wheeled vehicles, 373–4

W

Weight, effects of, 257–268
Wheeled vehicles vs. tracked vehicles, 439–49